T0257129

SPECIAL PAPERS IN PALAEONTOLOGY NO. 81

PATTERNS AND PROCESSES IN EARLY VERTEBRATE EVOLUTION

EDITED BY

MARCELLO RUTA, JENNIFER A. CLACK
and ANGELA C. MILNER

with 74 text-figures and 4 tables

THE PALAEONTOLOGICAL ASSOCIATION
LONDON

June 2009

CONTENTS

[Special Papers in Palaeontology 81, 2009, pp. 5–13]

FOREWORD

by MARCELLO RUTA*, JENNIFER A. CLACK† *and* ANGELA C. MILNER‡

*Department of Earth Sciences, University of Bristol, Wills Memorial Building, Queen's Road, Bristol BS8 1RJ, UK; e-mail: m.ruta@bristol.ac.uk
†University Museum of Zoology, University of Cambridge, Downing Street, Cambridge CB2 3EJ, UK; e-mail: j.a.clack@zoo.cam.ac.uk
‡Department of Palaeontology, The Natural History Museum, Cromwell Road, London SW7 5BD, UK; e-mail: a.milner@nhm.ac.uk

THE present volume is a collection of nine papers from 14 international specialists (from Australia, England, France, Germany, India, Italy, Japan, Morocco and Slovak Republic) working in various fields of vertebrate palaeontology. Topics range from comparative morphology to phylogeny and from palaeobiogeography to macroevolutionary analysis. A variety of tetrapod groups are covered, including chroniosuchians (Clack and Klembara), nectrideans (Milner and Ruta), seymouriamorphs (Klembara), squamates (Evans and Manabe) and temnospondyls (Ruta; Schoch and Witzmann; Sequeira; Steyer and Jalil; Warren, Damiani and Sengupta). The contributions bring together new anatomical data, revised patterns of evolutionary relationships and novel applications in analytical palaeobiology. In addition, two new genera and three new species of temnospondyls are established.

This volume celebrates Dr Andrew Milner (formerly at Birkbeck College, University of London and currently based at The Natural History Museum, London) and honours both his scientific achievements and his enlightening roles as a colleague, friend and mentor. The articles embrace a tiny fraction of Andrew's broad range of interests, but we hope they will appeal to him and initiate further stimulating discussions and additional research.

We thank Professor Michael Bassett (former President of the Palaeontological Association) and the Association Council for their enthusiastic support and for granting permission to produce this *Special Papers* issue. Professors David Batten and Svend Stouge (respectively, former and current Chief Editor of the Association publications) provided generous advice and invaluable assistance on numerous editorial topics. We are grateful to them for their immense patience and for answering countless queries promptly and effectively. Dr Louise Robb, Senior Production Editor at Wiley-Blackwell, offered considerable help through all stages of completion of this volume.

Last but not least, we are indebted to all authors for having made this volume possible and to the referees of the various papers for having acted with competence, erudition and professionalism.

© The Palaeontological Association

doi: 10.1111/j.1475-4983.2009.00882.x

DR ANDREW R. MILNER – AN APPRECIATION

Andrew began his studies for a BSc degree in the Department of Zoology at the University of Newcastle upon Tyne with the intention of becoming a marine biologist. However, a temporary summer job at the government Fisheries Laboratory (then part of the Ministry of Agriculture, Fisheries and Food) in Lowestoft, Suffolk, involving study of herring population dynamics, combined with the negative effects of motor-boating in the North Sea and the positive influence of Dr Alec Panchen's undergraduate courses in vertebrate palaeontology, convinced him that aquatic life 300 million years ago was a more comfortable and stimulating alternative. Andrew remained at Newcastle and began his research career in 1968 as a student supervised by Alec Panchen. He obtained his PhD in 1974, writing up his dissertation while holding a demonstratorship in the Department of Pure and Applied Zoology at the University of Leeds, from 1971 to 1974, working under Professor R. McNeill Alexander. Whilst at Leeds, Andrew was responsible for setting one of his undergraduate students, David Norman (now Director of the Sedgwick Museum, Cambridge), on the path to a career in vertebrate palaeontology.

In 1974, Andrew obtained a lectureship in the Department of Biology, Birkbeck College, University of London, and remained there for the rest of his teaching career, progressing via a senior lectureship to Reader in Vertebrate Palaeontology in 1995. Birkbeck College caters exclusively for students with daytime jobs who pursue first degrees through evening classes. Hence, Andrew's teaching took place in the night-owl slot from 6pm to 9pm in the evenings. While evening lectures might not be ideal for some, Andrew thoroughly enjoyed it, not being an early bird by nature or inclination. His classes included, especially in the first half of his Birkbeck career, many school teachers and numerous Natural History Museum staff who had begun their careers as nongraduate junior staff. He found it immensely rewarding to teach mature and enthusiastic students, not to mention the opportunities for socializing over a pint in the college bar.

Andrew was a popular and highly respected lecturer and taught a wide variety of zoological topics in addition to his areas of primary research expertise, and supervised or co-supervised an equally wide range of PhD students. These were: Heather Brand [*The reproductive behaviour and endocrinology of the cotton-topped tamarin* Saguinus oedipus oedipus (*Callitrichidae: Primates*) (1984)]; Christopher Sanford [*Relationships and phylogeny of the Salmonidae* (1987)]; John Cookson [*Applications of computer-based three-dimensional reconstruction to biological and palaeontological structures* (1994)]; Willian Blows [*The exoskeleton of ankylosaurian dinosaurs* (1996)]; Marcello Ruta [*Interrelationships of the anomalocystitid mitrates* (1998)] and Samuel Davis [*Phylogeny of acanthodian fishes* (2002)].

During his early years at Birkbeck, Andrew contributed lectures on primate evolution to Professor John Napier's University of London Intercollegiate Course entitled 'Primate Biology and Anthropogenesis'. His other teaching duties have encompassed courses at introductory and advanced levels on Animal Diversity, Invertebrates, Vertebrates, Mammal Biology, Ecology and Evolution of Terrestrial Vertebrates, Evolutionary Biology, Historical Evolution, Vertebrate Ecology and Conservation, Biology and Ethics, and Vertebrate Palaeontology. Andrew also lent his considerable entomological knowledge and expertise to undergraduate courses in Insect Taxonomy, Insect Evolution and Master's degrees in Entomology and Biology, and taught the entomological half of a 1 week course on Field Biology at Slapton Ley, Devon, for which he was course organizer for 11 years.

Andrew took voluntary early retirement from the night life in October 2005, prompted by the discovery that he had been volunteered to teach applied molecular entomology on a Master's course. However, he did return to Birkbeck on a part-time basis for two more years to teach vertebrate courses. Since retiring, Andrew has continued his active research as a Scientific Associate in the Department of Palaeontology at the Natural History Museum, London. He has, however, allowed himself some time off to pursue his long standing interests in wildlife photography and military aviation.

Andrew's PhD thesis encompassed a thorough review of branchiosaur temnospondyls and the role of neoteny and metamorphosis in shaping the evolution of the group (Milner 1974). He was well into his thesis research when he discovered that Jürgen Boy at the University of Mainz was researching the same topic for his dissertation – and had a 2 year start. Andrew visited Jürgen in Mainz in 1970 and both were delighted to discover that they had independently reached the same conclusions about the nature of branchiosaurs as neotenous animals, overturning the long-held views of D. M. S. Watson and A. S. Romer. Jürgen's work was already being written up for publication in 1970. As there was little point in duplication, Andrews's thesis results were not published and he turned his attention to other temnospondyl groups. Work on temnospondyls has been a continuing primary focus of his research on a broad front, with publications on their morphology, evolution, biogeography, palaeobiology and the circumstances of their death and preservation, all of which led to his reputation as an internationally pre-eminent temnospondyl specialist. In 1980, he published a groundbreaking and influential study of the palaeoecology of the tetrapod assemblage from the late Westphalian Nýřany locality in the Czech Republic, in which he reviewed the fauna and identified associations of open-water, shallow swamp-lake and terrestrial/marginal assemblages (Milner 1980b). One anomalous member of the Nýřany fauna, the terrestrially-adapted nectridean *Scinosaurus crassus*, specimens of which were shown to represent a breeding population of adults, is reviewed in the present volume (Milner and Ruta, pp. 71–89). Andrew has worked his way systematically through the temnospondyls with a long list of single authored papers and collaborative projects jointly with Sandy Sequeira, variously a technician and his grant-funded research assistant at Birkbeck College, Professor Jennifer Clack (University of Cambridge), Dr Jozef Klembara (Comenius University, Bratislava) and Dr Rainer Schoch (Stuttgart Museum für Naturkunde). In 1982, he produced a description of small temnospondyls from Mazon Creek in Illinois, which further supported the hypothesis that branchiosaurids were neotenous dissorophoid temnospondyls (Milner 1982b), in line with the results of his own and Boy's work on branchiosaurs from German Permian localities.

When the renowned collector Stan Wood discovered the Viséan terrestrial fossil assemblage at East Kirkton, near Bathgate, Scotland, in the mid-1980s, Andrew's interest in the early evolution of temnospondyls led to his involvement in the 'East Kirkton Project' under the auspices of the National Museums of Scotland. He was one of the founder members of this project and a co-organizer of the resulting symposium on 'Volcanism and Early Terrestrial Biotas' sponsored by the Royal Society of Edinburgh in 1992, in which he published on the earliest temnospondyl *Balanerpeton woodi* (Milner and Sequeira 1994a). This work provided a base line for temnospondyl morphology and evolution as well as an analysis of the unique hot spring preservation and palaeoecology at East Kirkton (Clarkson, Milner and Coates 1994). An invitation to co-author a *Nature* review article on the origin and diversification of tetrapods (Ahlberg and Milner 1994) followed shortly thereafter. Other particularly significant publications through the 1990s included a widely cited review of the temnospondyl radiation (Milner 1990e) carried out in conjunction with a complete revision of the Stereospondyli in collaboration with Dr Rainer Schoch. That revision was published as a volume in the Handbuch der Paläoherpetologie (Schoch and Milner 2000) and stands as the definitive reference work on the group. Rainer and Andrew revisited branchiosaurs more recently and published the first definitive cladistic analysis of all the best known larval or paedomorphic species from Permo–Carboniferous lakes across Europe. They went on to propose an evolutionary scenario suggesting that the origin and radiation of branchiosaurids were underpinned by a key innovation, namely the acquisition of specialized filter-feeding pharyngeal denticles housed in gill clefts (Schoch and Milner 2008b).

With the mission of furthering research in the area of early amphibian studies, Andrew has organized several symposia and has edited or co-edited their resulting published volumes. In 1992, he organized a symposium on Permo-Carboniferous continental faunas held at the National Museums of Scotland, published later as an edited volume (Milner 1996a). He was also co-editor of a Festschrift volume for Dr Alec Panchen (Norman *et al.* 1998) which brought together a further body of recent work on early amphibians including a benchmark description of the cochleosaurid temnospondyl *Adamanterpeton ohioensis* from Linton, Ohio (Milner and Sequeira 1998). He convened with Dr Jason Anderson (University of Calgary) a special symposium on 'The Dissorophoidea – focus on an early amphibian radiation' at the 67th Annual Meeting of the Society of Vertebrate Paleontology held in Austin, Texas, in October 2007, that drew together the latest research on the group closest to the origin of modern amphibians.

A second related strand of Andrew's research is the origin and early diversification of the modern amphibian groups, a controversial topic on which he has made several much cited theoretical and synthetic contributions, advocating lissamphibian monophyly and their origin from among dissorophoids (e.g Milner 1988*b*; Schoch and Milner 2004). Several recent large-scale cladistic analyses have reached the same conclusions (e.g. Ruta *et al.* 2003; Ruta and Coates 2007). A highly productive collaboration with Professor Susan Evans (University College London) through the 1980s to mid-1990s resulted in a series of papers on early fossil frogs and salamanders from the Jurassic of Oxfordshire and led to invitations to work on other similar material from Dorset and Wyoming, on salamanders from the Cretaceous of Spain and on salamanders and caecilians from the mid-Cretaceous of the Sudan (e.g. Evans *et al.* 1988*c*, 1996*c*; Evans and Milner 1993*b*, 1996*d*). These studies led to the characterization of the Jurassic/Cretaceous amphibian faunas of the northern hemisphere and Africa in an important and well-received major review (Milner 2000*d*), the first since the Gymnophiona and Caudata volume of the Handbuch der Paläeoherpetologie (Estes 1981). Andrew's interest in biogeography led to a major review of Mesozoic and early Caenozoic salamander biogeography employing for the first time a cladistic-vicariance model to explain their distribution (Milner 1983).

In addition to his mainstream research areas, Andrew has published collaborative papers on various reptile groups. He co-authored the first cladistic-vicariance model for ornithopod dinosaur biogeography (Milner and Norman 1984), co-authored work on pterosaurs (Howse and Milner 1995*b*) and undertook the first modern review of the systematics of Purbeck turtles (Milner 2004*e*). The latter research followed his organizing a symposium on 'The Fauna and Flora of the Purbeck Limestone' together with Paul Ensom in 1999 and co-editing the subsequent volume of conference papers (Milner and Batten 2002*e*). Furthermore, Andrew has served as an editor or associate editor for *Zoological Journal of the Linnean Society*, *Palaeontology*, *Herpetological Journal* and *Journal of Vertebrate Paleontology*.

ADDITIONAL REFERENCES

ESTES, R. 1981. Teil 2. Gymnophiona, Caudata. *In* WELLNHOFER, P. (ed.). *Handbuch der Paläoherpetologie*. Pfeil, Munich, 115 pp.

MILNER, A. R. 1974. *A revision of the 'branchiosaurs' and some associated Palaeozoic Temnospondyl Amphibia*. PhD Thesis, University of Newcastle upon Tyne, 488 pp.

RUTA, M. and COATES, M. I. 2007. Dates, nodes and character conflict: addressing the lissamphibian problem. *Journal of Systematic Palaeontology*, **5**, 69–122.

———— and QUICKE, D. L. J. 2003. Early tetrapod relationships revisited. *Biological Reviews of the Cambridge Philosophical Society*, **78**, 251–345.

ANDREW R. MILNER'S BIBLIOGRAPHY

1. 1973 MILNER, A. R. and PANCHEN, A. L. Chapter 3.9. Geographical variation in the tetrapod faunas of the Upper Carboniferous and Lower Permian. 353–368. *In* TARLING, D. H. and RUNCORN, S. K. (eds). *Implications of continental drift to the earth sciences*. Academic Press, London, 1184 pp.

2. 1977*a* MILNER, A. R. Triassic extinction or Jurassic vacuum? (News and Views). *Nature*, **265**, 402.

3. 1977*b* MILNER, A. R. Early assemblage of Carboniferous amphibians (News and Views). *Nature*, **265**, 495–496.

4. 1977*c* MILNER, A. R. Review of: Andrews, S. M., Miles, R. S. and Walker A. D. (eds) 1977. *Problems in vertebrate evolution*. Academic Press, London, 411 pp. *Palaeontological Association Circular*, **89**, 16–17.

5. 1978 MILNER, A. R. A reappraisal of the Early Permian amphibians *Memonomenos dyscriton* and *Cricotillus brachydens*. *Palaeontology*, **21**, 667–686.

6. 1979 MILNER, A. R. Review of: Pearson, R. 1978. *Climate and evolution*. Academic Press, London, 274 pp. *Journal of Arid Environments*, **2**, 292–293.

7. 1980*a* MILNER, A. R. The temnospondyl amphibian *Dendrerpeton* from the Upper Carboniferous of Ireland. *Palaeontology*, **23**, 125–141.

8. 1980*b* MILNER, A. R. Chapter 17. The tetrapod assemblage from Nýřany, Czechoslovakia. 439–496. *In* PANCHEN, A. L. (ed.). *The terrestrial environment and the origin of land vertebrates*. The Systematics Association, Special Volume 15. Academic Press, London, 633 pp.

9. 1981*a* MILNER, A. R. On the identity of 'Ptyonius bendai' (Amphibia) from the Lower Permian of Košt'álov, Czechoslovakia. *Neues Jahrbuch für Geologie und Paläontologie, Monatshefte*, **1981** (6), 346–352.

10. 1981*b* MILNER, A. R. Review of: Lillegraven, J. A., Kielan–Jaworowska, Z. and Clemens W. A. (eds). 1980. *Mesozoic mammals: the first two-thirds of mammalian history*. University of California Press, 321 pp. *Mammal Reviews*, **11**, 89–90.

11. 1981*c* MILNER, A. R. Flamingos, stilts and whales. (News and Views). *Nature*, **289**, 347.

12. 1982*a* MILNER, A. R. Review of: Grande, L. 1980. *Paleontology of the Green River formation, with a review of the fish fauna. Geological Survey of Wyoming Bulletin* **63**, 333 pp. University of Wyoming. *Palaeontological Association Circular*, **108**, 9.

13. 1982*b* MILNER, A. R. Small temnospondyl amphibians from the Middle Pennsylvanian of Illinois. *Palaeontology*, **25**, 635–664.

14. 1982*c* MILNER, A. R. A small temnospondyl amphibian from the Lower Pennsylvanian of Nova Scotia. *Journal of Paleontology*, **56**, 1302–1305.

15. 1982*d* MILNER, A. C., MILNER, A. R. and ESTES, R. Amphibians and squamates from the Upper Eocene of Hordle Cliff, Hampshire – a preliminary report. *Tertiary Research*, **4**, 149–154.

16. 1983 MILNER, A. R. Chapter 15. The biogeography of salamanders in the Mesozoic and early Caenozoic, a cladistic-vicariance model. 431–468. *In* SIMS, R., PRICE, J. and WHALLEY, P. (eds). *Evolution, time and space: the emergence of the biosphere*. The Systematics Association, Special Volume 23. Academic Press, London, 492 pp.

17. 1984 MILNER, A. R. and NORMAN, D. B. The biogeography of advanced ornithopod dinosaurs (Archosauria: Ornithischia) – a cladistic-vicariance model. 145–150. *In* REIF, W.-E. and WESTPHAL, F. (eds). *Third symposium on Mesozoic terrestrial ecosystems. Short papers*. Attempto Verlag, Tübingen, 259 pp.

18. 1985*a* MILNER, A. R. Scottish window on terrestrial life in the Lower Carboniferous (News and Views). *Nature*, **314**, 320–321.

19. 1985*b* MILNER, A. R. *Cosesaurus* – the last proavian? (News and Views). *Nature*, **315**, 544.

20. 1985*c* MILNER, A. R. Discussion. 320. *In* CHALONER, W. G. and LAWSON, J. D. (eds). *Evolution and environment in the Late Silurian and Early Devonian*. Royal Society, London, 342 pp.

21. 1985*d* MILNER, A. R. On the identity of *Trematopsis seltini* (Amphibia: Temnospondyli) from the Lower Permian of Texas. *Neues Jahrbuch für Geologie und Paläontologie, Monatshefte*, **1985**(6), 357–367.

22. 1985*e* MILNER, A. R. On the identity of the amphibian *Hesperoherpeton garnettense* from the Upper Pennsylvanian of Kansas. *Palaeontology*, **28**, 767–776.

23. 1985*f* MILNER, A. R. Review of: Martin, P. S. and Klein, R. G. (eds). 1984. *Quaternary extinctions. A prehistoric revolution*. University of Arizona Press, 892 pp. *Endeavour*, **9** (2), 109–110.

24. 1986*a* MILNER, A. R., SMITHSON, T. R., MILNER, A. C., COATES, M. I. and ROLFE, W. D. I. The search for early tetrapods. *Modern Geology*, **10**, 1–28.

25. 1986*b* MILNER, A. R. Dissorophoid amphibians from Nýřany, Czechoslovakia. 671–674. *In* ROČEK, Z. (ed.). *Studies in Herpetology. Proceedings of the Third European Herpetological Meeting*. Charles University, Prague, 754 pp.

26. 1986*c* MILNER, A. R. Vertebrate form and function. Reviews of: Hildebrand, M, Bramble, D. M., Liem, K. F. and Wake, D. B. (eds) 1985. *Functional vertebrate morphology*. Harvard University Press, 430 pp.; and Rogers, E. 1986. *Looking at vertebrates: a practical guide to vertebrate adaptations*. Longman, 195 pp. *Trends in Ecology and Evolution*, **1**, 170–171.

27. 1987 MILNER, A. R. The Westphalian tetrapod fauna; some aspects of its geography and ecology. *Journal of the Geological Society of London*, **144**, 495–506.

28. 1988*a* MILNER, A. R. Review of: Humphries, C. J. and Parenti, L. R. (eds) 1986. *Cladistic biogeography*. Clarendon Press, Oxford, 98 pp. *Palaeontological Association Newsletter*, **1**, 25–26.

29. 1988*b* MILNER, A. R. Chapter 3. The relationships and origin of living amphibians. 59–102. *In* BENTON, M. J. (ed.). *The phylogeny and classification of the tetrapods. Vol. 1: amphibians, reptiles, birds*. The Systematics Association, Special Volume 35A. Clarendon Press, Oxford, 392 pp.

30. 1988*c* EVANS, S. E., MILNER, A. R. and MUSSETT, F. The earliest known salamanders (Amphibia: Caudata): a record from the Middle Jurassic of England. *Geobios*, **21**, 539–552.

31. 1989*a* EVANS, S. E. and MILNER, A. R. *Fulengia*, a supposed early lizard reinterpreted as a prosauropod dinosaur. *Palaeontology*, **32**, 223–230.

32. 1989*b* MILNER, A. R. Late extinctions of amphibians (News and Views). *Nature*, **338**, 117.

33. 1989*c* MILNER, A. R. The East Kirkton temnospondyl and the relationships of early temnospondyls. *First World Congress of Herpetology, University of Kent at Canterbury. Abstract Volume*, 201.

34. 1990*a* EVANS, S.E., MILNER, A. R. and MUSSETT, F. A discoglossid frog from the Middle Jurassic of England. *Palaeontology*, **33**, 299–311.

35. 1990*b* MILNER, A. R. The relationships of the eryopoid-grade temnospondyl amphibians from the Permian of Europe. *Acta Musei Reginae–hradecensis, Series A*, **22** (for 1989), 131–137.

36. 1990*c* MILNER, A. R., GARDINER, B. G., FRASER, N. C. and TAYLOR, M. A. Vertebrates from the Middle Triassic Otter Sandstone Formation of Devon. *Palaeontology*, **33**, 873–892.

37. 1990*d* ZAJÍC, J., MILNER, A. R. and KLEMBARA, J. The first partially articulated amphibian (Temnospondyli: Dissorophoidea) from the Liné Formation (Stephanian C, central Bohemia). *Věstník Ústředního ústavu geologického*, **65**, 329–337.

38. 1990e MILNER, A. R. Chapter 15. The radiations of temnospondyl amphibians. 321–349. *In* TAYLOR, P. D. and LARWOOD, G. P. (eds). *Major evolutionary radiations*. The Systematics Association, Special Volume 42. Clarendon Press, Oxford, 437 pp.

39. 1991a ENSOM, P. C., EVANS, S. E. and MILNER, A. R. Amphibians and reptiles from the Purbeck Limestone Formation (Upper Jurassic) of Dorset. *5th Symposium on Mesozoic Terrestrial Ecosystems and Biota, Extended Abstracts. Contributions of the Paleontological Museum, University of Oslo*, **364**, 19–20.

40. 1991b EVANS, S. E. and MILNER, A. R. Middle Jurassic microvertebrate faunas from the British Isles. *5th Symposium on Mesozoic Terrestrial Ecosystems and Biota, Extended Abstracts. Contributions of the Paleontological Museum, University of Oslo*, **364**, 21–22.

41. 1991c MILNER, A. R. Lydekkerinid temnospondyls – relationships and 'extinction'. *5th Symposium on Mesozoic Terrestrial Ecosystems and Biota, Extended Abstracts. Contributions of the Paleontological Museum, University of Oslo*, **364**, 49–50.

42. 1991d MILNER, A. R. and EVANS, S. E. The Upper Jurassic diapsid *Lisboasaurus estesi* – a maniraptoran theropod. *Palaeontology*, **34**, 503–514.

43. 1992 MILNER, A. R. and SEQUEIRA, S. E. K. Temnospondyl amphibians from the Viséan of East Kirkton, Scotland. *Journal of Vertebrate Paleontology*, **12**, supplement to no. 3 (Abstracts of papers), 44A.

44. 1993a MILNER, A. R. Chapter 13: Biogeography of Palaeozoic tetrapods. 324–353. *In* LONG, J. A. (ed.). *Palaeozoic vertebrate biostratigraphy and biogeography*. Belhaven Press, London, 369 pp.

45. 1993b EVANS, S. E. and MILNER, A. R. Frogs and salamanders from the Upper Jurassic Morrison Formation (Quarry Nine, Como Bluff) of North America. *Journal of Vertebrate Paleontology*, **13**, 24–30.

46. 1993c MILNER, A. R. The Paleozoic relatives of lissamphibians. *In* CANNATELLA, D. and HILLIS, D. (eds). *Amphibian relationships. Phylogenetic analysis of morphology and molecules. Herpetological Monograph*, **7**, 8–27.

47. 1993d HOWSE, S. C. B. and MILNER, A. R. *Ornithodesmus* – a maniraptoran theropod dinosaur from the Lower cretaceous of the Isle of Wight, England. *Palaeontology*, **36**, 425–437.

48. 1993e SEQUEIRA, S. E. K. and MILNER, A. R. The temnospondyl amphibian *Capetus* from the Upper Carboniferous of Nýřany, Czech Republic. *Palaeontology*, **36**, 657–680.

49. 1993f MILNER, A. R. Chapter 38. Amphibian–grade Tetrapoda. 663–677. *In* BENTON, M. J. (ed.). *The fossil record 2*. Palaeontological Association/Chapman and Hall, London, 845 pp.

50. 1993g EVANS, S. E. and MILNER, A. R. Mesozoic salamander assemblages in Europe. *Second World Congress of Herpetology, Adelaide, South Australia. Abstracts*. 84.

51. 1994a MILNER, A. R. and SEQUEIRA, S. E. K. The temnospondyl amphibians from the Viséan of East Kirkton, West Lothian, Scotland. *Transactions of the Royal Society of Edinburgh, Earth Sciences*, **84** (for 1993), 331–361.

52. 1994b CLARKSON, E. N. K., MILNER, A. R. and COATES, M. I. Palaeoecology of the Viséan of East Kirkton, West Lothian, Scotland. *Transactions of the Royal Society of Edinburgh, Earth Sciences*, **84** (1993), 417–425.

53. 1994c AHLBERG, P. E. and MILNER, A. R. The origin and early diversification of tetrapods. *Nature*, **368**, 507–514.

54. 1994d CLACK, J. A. and MILNER, A. R. *Platyrhinops* from the Upper Carboniferous of Linton and Nýřany and the family Amphibamidae (Amphibia: Temnospondyli). *In* HEIDTKE, U. (ed.). *New research on Permo–Carboniferous faunas. Pollichia–buch*, **29** (for 1993), 185–191.

55. 1994e MILNER, A. R. and SEQUEIRA, S. E. K. *Capetus* and the problems of primitive temnospondyl relationships. *In* HEIDTKE, U. (ed.). *New Research on Permo–Carboniferous Faunas. Pollichia–buch*, **29** (for 1993), 193–199.

56. 1994f ENSOM, P. C., EVANS, S. E., FRANCIS, J. E., KIELAN-JAWOROWSKA, Z. and MILNER, A. R. The fauna and flora of the Sunnydown Farm footprint site and associated sites: Purbeck Limestone Formation, Dorset. *Proceedings of the Dorset Natural History and Archaeological Society*, **115**, 181–182.

57. 1994g MILNER, A. R. Chapter 1. Late Triassic and Jurassic amphibians: fossil record and phylogeny. 5–22. *In* FRASER, N. C. and SUES, H.-D. (eds). *In the shadow of the dinosaurs: early Mesozoic tetrapods*. Cambridge University Press, Cambridge, 435 pp.

58. 1994h EVANS, S. E. and MILNER, A. R. Chapter 18. Middle Jurassic microvertebrate assemblages from the British Isles. 303–321. *In* FRASER, N. C. and SUES, H.-D. (eds). *In the shadow of the dinosaurs: early Mesozoic tetrapods*. Cambridge University Press, Cambridge, 435 pp.

59. 1995a HOWSE, S. C. B. and MILNER, A. R. The pterodactyloids from the Purbeck Limestone Formation of dorset. *Bulletin of the British Museum (Natural History), Geology Series*, **51** (1), 73–88.

60. 1995b EVANS, S. E. and MILNER, A. R. Early Cretaceous salamanders (Amphibia: Caudata) from Las Hoyas, Spain. 63–65. In: *2nd International Symposium on Lithographic Limestones; Lleida-Cuenca (Spain)*. Extended Abstracts. Ediciones de la Universidad Autónoma de Madrid, Madrid, 89 pp.

61. 1995c EVANS, S. E., McGOWAN, G., MILNER, A. R. and SANCHÍZ, B. IV.4 Amphibians. 51–53. *In* MELÉNDEZ M. N. (ed.). *2nd International Symposium on Lithographic Limestones; Las Hoyas. Field Trip Guide Book*. Ediciones de la Universidad Autónoma de Madrid, Madrid, 89 pp.

62. 1996*a* MILNER, A. R. (ed.) *Studies on Carboniferous and Permian vertebrates (Proceedings of the Fourth International Symposium on Permo–Carboniferous continental faunas). Special Papers in Palaeontology*, **52**, 1–148.

63. 1996*b* MILNER, A. R. A revision of the temnospondyl amphibians from the Upper Carboniferous of Joggins, Nova Scotia. 81–103. *In* MILNER, A. R. (ed.). *Studies on Carboniferous and Permian vertebrates. Special Papers in Palaeontology*, **52**, 1–148.

64. 1996*c* EVANS, S. E., MILNER, A. R. and WERNER, C. Sirenid salamanders and a gymnophionan amphibian from the Cretaceous of the Sudan. *Palaeontology*, **39**, 77–95.

65. 1996*d* EVANS. S. E. and MILNER, A. R. A metamorphosed salamander from the Lower Cretaceous of Spain. *Philosophical Transactions of the Royal Society of London, Series B*, **351**, 627–646.

66. 1996*e* MILNER, A. R. Early amphibian globetrotters? (News and Views). *Nature*, **381**, 741–742.

67. 1996*f* MILNER, A. R. Systematics of the genus *Eryops* (Amphibia: Temnospondyli) and its possible biostratigraphical value. *Journal of Vertebrate Paleontology*, **16**, suppt. to no. 3 (Abstracts of papers), 53A.

68. 1996*g* SEQUEIRA, S. E. K. and MILNER, A. R. A revision of the Family Saurerpetontidae (Amphibia: Temnospondyli) from the Pennsylvanian and Permian of North America. *Journal of Vertebrate Paleontology*, **16**, supplement to no. 3, (Abstracts of papers), 64A.

69. 1996*h* MILNER, A. R., DUFFIN, C. and DELSATE, D. Plagiosaurid and capitosaurid amphibian material from the Late Triassic of Medernach, Grand–Duchy of Luxembourg: preliminary note. *Bulletin de la Société Belge de Géologie*, **104** (1–2), 43–53.

70. 1996*i* MILNER, A. R. Review of: Benton, M. J. and Spencer, P. S. 1995. *Fossil reptiles of Great Britain*. Geological Conservation Review Series **10**. Chapman Hall, London, 386 pp. *Herpetological Journal*, **6**, 150.

71. 1997*a* MACLEOD, N., RAWSON, P. F., FOREY, P. L., BANNER, F. T., BOUDAGHER–FADEL, M. K., BOWN, P. R., BURNETT, J. A., CHAMBERS, P., CULVER, S., EVANS, S. E., JEFFERY, C., KAMINSKI, M. A., LORD, A. R., MILNER, A. R., MILNER, A. C., MORRIS, N., OWEN, E., ROSEN, B. R., SMITH, A. B., TAYLOR, P. D., URQUHART, E. and YOUNG, J. R. The Cretaceous–Tertiary biotic transition. *Journal of the Geological Society of London*, **154**, 265–292.

72. 1997*b* MILNER, A. R. and SEQUEIRA, S. E. K. Slender-snouted temnospondyls – three radiations or one long lineage? Abstract for ICVM97. *Journal of Morphology*, **232**, 295.

73. 1997*c* MILNER, A. R. The origin of the modern amphibian groups. 143. *In* ROČEK, Z. and HART, S. (eds). *Abstracts of the Third World Congress of Herpetology*. Prague.

74. 1997*d* MILNER, A. C. and MILNER, A. R. A new specimen of *Baphetes* from Nýřany, Czech Republic and the intrinsic relationships of the Baphetidae. *Journal of Vertebrate Paleontology*, **17**, supplement to no. 3 (Abstracts of papers), 65A.

75. 1997*e* MILNER, A. R. and SEQUEIRA, S. E. K. The systematic position of *Perryella olsoni* (Amphibia: Temnospondyli) from the Lower Permian of Oklahoma. *Journal of Vertebrate Paleontology*, **17**, supplement to no. 3 (Abstracts of papers), 65A.

76. 1998*a* NORMAN, D. B., MILNER, A. R. and MILNER, A. C. (eds). *A study of fossil vertebrates. Essays in honour of Alec Panchen. Zoological Journal of the Linnean Society*, **122**, 1–384.

77. 1998*b* MILNER, A. R. and MILNER, A.C. Dr A. L. Panchen FRSE: an appreciation. 1–7. *In* NORMAN, D. B., MILNER, A. R. and MILNER, A. C. (eds). *A study of fossil vertebrates. Essays in honour of Alec Panchen. Zoological Journal of the Linnean Society*, **122**, 384 pp.

78. 1998*c* MILNER, A. R. and SEQUEIRA, S. E. K. A cochleosaurid temnospondyl amphibian from the Middle Pennsylvanian of Linton, Ohio, USA. 261–290. *In* NORMAN, D. B., MILNER, A. R. and MILNER, A. C. (eds). *A study of fossil vertebrates. Essays in honour of Alec Panchen. Zoological Journal of the Linnean Society*, **122**, 384 pp.

79. 1999. MILNER, A. R. Review of : Unwin, D. (ed.) 1999. *Cretaceous fossil vertebrates. Special Papers in Palaeontology*, **60**, 219 pp. *Cretaceous Research*, **20**, 659–660.

80. 2000*a* AHLBERG, P. E. and MILNER, A. R. The origin and early diversification of tetrapods. 267–287. *In* GEE, H. (ed.). *Shaking the tree, readings from nature in the history of life*. University of Chicago Press. Chicago and London. 411 pp. (Reprinting of Ahlberg and Milner 1994).

81. 2000*b* MILNER, A. C., MILNER, A. R. and EVANS, S. E. 21. Amphibians, reptiles and birds: a biogeographical review. 316–332. *In* CULVER, S. J. and RAWSON, P. F. (eds). *Global change and the biosphere*. Cambridge University Press. Cambridge, 501 pp.

82. 2000*c* COATES, M. I., RUTA, M. and MILNER, A. R. Early tetrapod evolution. (Commentary on Laurin, Girondot and de Ricqlès). *Trends in Ecology and Evolution*, **15** (8), 327–328.

83. 2000*d* MILNER, A. R. Chapter 18. Mesozoic and Tertiary Caudata and Albanerpetontidae. 1412–1444. *In* HEATWOLE, H. and CARROLL, R. L. (eds). *Biology of Amphibia. Volume 4. Palaeontology*. Surrey Beatty, Chipping Norton, NSW, 523 pp.

84. 2000*e* SCHOCH, R. R. and MILNER A. R. Teil 3B Stereospondyli. *In* WELLNHOFER, P. (ed.). *Handbuch der Paläoherpetologie*. Pfeil, Munich, 202 pp.

85.   2000f   RUTA, M., MILNER, A. R. and COATES, M. I. The Scottish Carboniferous tetrapod *Caerorhachis bairdi*. *Journal of Vertebrate Paleontology*, **20**, supplement to no. 3 (Abstracts of papers), 66A.

86.   2001a   HOWSE, S. C. B., MILNER, A. R. and MARTILL, D. M. Chapter 11. Pterosaurs. 324–335. *In* MARTILL, D. and NAISH, D. (eds). *Palaeontological Association field guides to fossils: number 10. Dinosaurs of the Isle of Wight.* Palaeontological Association, 464 pp.

87.   2001b   MILNER, A. R. Article 1538. Caudata (Salamanders). In: *Encyclopaedia of Life Sciences.* Macmillan Reference Ltd, London [electronic publication at http://www.els.net/].

88.   2001c   MILNER, A. R. Article 1539. Gymnophiona (Caecilians). In: *Encyclopaedia of Life Sciences.* Macmillan Reference Ltd, London [electronic publication at http://www.els.net/].

89.   2002a   RUTA, M., MILNER, A. R. and COATES, M. I. The tetrapod *Caerorhachis bairdi* Holmes and Carroll from the Lower Carboniferous of Scotland. *Transactions of the Royal Society of Edinburgh, Earth Sciences*, **92** (for 2001), 229–261.

90.   2002b   SCHOCH, R. R., MILNER, A. R. and HELLRUNG, H. The last trematosaurid amphibian *Hyperokynodon keuperinus* revisited. *Stuttgarter Beiträge zur Naturkunde Serie B (Geologie und Paläontologie)*, **321**, 1–9.

91.   2002c   MILNER, A. R., SEQUEIRA, S. E. K. and SCHOCH, R. R. Dvinosaurian temnospondyls in the Permian. *Journal of Vertebrate Paleontology*, **22**, supplement to no. 3 (Abstracts of papers), 88A.

92.   2002d   SCHOCH, R. R. and MILNER, A. R. The origin of lissamphibians. *Journal of Vertebrate Paleontology*, **22**, supplement to no. 3 (Abstracts of papers), 104A.

93.   2002e   MILNER, A. R. and BATTEN, D. J. (eds). *Life and environment in Purbeck times. Special Papers in Palaeontology*, **68**, 1–268.

94.   2003a   MILNER, A. R. and SEQUEIRA, S. E. K. Revision of the amphibian genus *Limnerpeton* (Temnospondyli) from the Upper Carboniferous of the Czech Republic. *Acta Paleontologica Polonica*, **48**, 31–50.

95.   2003b   MILNER, A. R. and SEQUEIRA, S. E. K. On a small *Cochleosaurus* described as a large *Limnogyrinus* (Amphibia, Temnospondyli) from the Upper Carboniferous of the Czech Republic. *Acta Paleontologica Polonica*, **48**, 51–55.

96.   2003c   MILNER, A. R. and SEQUEIRA, S. E. K. Branchiosaurs, larvae and metamorphosis. *Journal of Vertebrate Paleontology*, **23**, supplement to no. 3 (Abstracts of papers), 79A.

97.   2003d   MILNER, A. R. Before the Rotliegendes – the Nýřany tetrapod fauna. *Endogene und Exogene Hintergrunde der Biodiversität. Terra Nostra*, **2003**/5 (Abstracts of papers), 112–113.

98.   2003e   MILNER, A. R. *Longiscitula houghae* DeMar 1966 (Amphibia: Temnospondyli), a junior synonym of *Dissorophus multicinctus* Cope 1895. *Journal of Vertebrate Paleontology*, **23**, 941–944.

99.   2004a   MILNER, A. R. and SEQUEIRA, S. E. K. *Slaugenhopia texensis* (Amphibia: Temnospondyli) from the Permian of Texas is a primitive tupilakosaurid. *Journal of Vertebrate Paleontology*, **24**, 320–325.

100.   2004b   MILNER, A. R. and SCHOCH, R. R. The latest metoposaurid amphibians from Europe. *Neues Jahrbuch für Geologie und Paläontologie, Abhandlungen*, **232** (2–3), 231–252.

101.   2004c   SCHOCH, R. R. and MILNER, A. R. Intrarelationships, monophyly and evolution of the Branchiosauridae. *Journal of Vertebrate Paleontology*, **24**, supplement to no. 3 (Abstracts of papers), 110A.

102.   2004d   SCHOCH, R. R. and MILNER, A. R. Structure and implications of theories on the origins of lissamphibians. 345–377. *In* WILSON, M., CLOUTIER, R. and ARRATIA, G. (eds). *Recent advances in studies of lower vertebrates.* Pfeil, Munich, 703 pp.

103.   2004e   MILNER, A. R. The turtles of the Purbeck Limestone Group of Dorset, Southern England. *Palaeontology*, **47**, 1441–1467.

104.   2004f   MILNER, A. R. Mesozoic amphibians and other non–amniote tetrapods. 516–522. *In* SELLEY, R. C., COCKS, L. R. M. and PLIMER, I. R. (eds). *Encyclopedia of Geology Volume 2.* Elsevier, Amsterdam, xxxvii + 545 pp.

105.   2004g   MILNER, A. R. Cenozoic amphibians. 523–526. *In* SELLEY, R. C., COCKS, L. R. M. and PLIMER, I. R. (eds). *Encyclopedia of Geology Volume 2.* Elsevier, Amsterdam, xxxvii + 545 pp.

106.   2006   MILNER, A. R. and SCHOCH, R. R. *Stegops*, a problematic spiky–headed temnospondyl. *Journal of Vertebrate Paleontology*, **26**, supplement to no. 3 (Abstracts of papers), 101A.

107.   2007a   MILNER, A. C., MILNER, A. R. and WALSH, S. A. A new specimen of *Baphetes* from Nýřany, Czech Republic and the intrinsic relationships of the Baphetidae. *Ichthyolith Issues Special Publication*, **10**, 69 (Abstract).

108.   2007b   SEQUEIRA, S. E. K. and MILNER, A. R. The postcranial skeleton of *Cochleosaurus bohemicus*, a basal temnospondyl from Nýřany, Czech Republic. *Ichthyolith Issues Special Publication*, **10**, 81 (Abstract).

109.   2007c   MILNER, A. R. Temnospondyli. In: *McGraw–Hill Encyclopedia of Science and Technology.* 10th edn. **18**, 263–264. McGraw–Hill, New York.

110.   2007d   MILNER, A. R., KLEMBARA, J. and DÓSTAL, O. A zatrachydid temnospondyl from the Lower Permian of the Boskovice Furrow in Moravia (Czech Republic). *Journal of Vertebrate Paleontology*, **27**, 711–715.

111. 2007*e* MILNER, A. R. *Mordex* and the base of the Trematopidae. *Journal of Vertebrate Paleontology*, **27**, supplement to no. 3 (Abstracts of papers), 118A.

112. 2007*f* CLACK, J. A. and MILNER, A. R. The amphibamid *Platyrhinops*, morphology and metamorphosis. *Journal of Vertebrate Paleontology*, **27**, supplement to no. 3 (Abstracts of papers), 59A.

113. 2007*g* WITZMANN, F., SCHOCH, R. R. and MILNER, A. R. The origin of Dissorophoidea – an alternative perspective. *Journal of Vertebrate Paleontology*, **27**, supplement to no. 3 (Abstracts of papers), 167A.

114. 2008*a* MILNER, A. R. The tail of *Microbrachis* (Tetrapoda; Microsauria). *Lethaia*, **41**, 257–261.

115. 2008*b* PARDO, J. D., HUTTENLOCKER, A. K., SMALL, B. J. and MILNER, A. R. Biotic responses to climate change in the Permo-Carboniferous transition, Part I: vertebrate faunal distributions and regional provincialism. *Journal of Vertebrate Paleontology*, **28**, supplement to no. 3 (Abstracts of papers), 125A.

116. 2008*c* HUTTENLOCKER, A. K., PARDO, J. D., SMALL, B. J. and MILNER, A. R. Biotic responses to climate change in the Permo-Carboniferous transition, Part II: beta diversity, regional evolutionary responses, and Vaughn's faunal cline. *Journal of Vertebrate Paleontology*, **28**, supplement to no. 3 (Abstracts of papers), 94A.

117. 2008*d* SCHOCH, R. R. and MILNER, A. R. The intrarelationships and evolutionary history of the temnospondyl family Branchiosauridae. *Journal of Systematic Palaeontology*, **6** (4), 409–431.

118. 2008*e* BARRETT, P. M., BUTLER, R. J., EDWARDS, N. and MILNER, A. R. Pterosaur distribution in time and space - an atlas. 61–107. *In* HONE, D.W.E. and BUFFETAUT, E. (eds). *Flugsaurier: pterosaur papers in honour of Peter Wellnhofer. Zitteliana Series B*, **28**, 255 pp.

119. 2009*a* MILNER, A. R. Carboniferous amphibians and reptiles – experimentation and diversification. 71. *In* GODEFROIT, P. and LAMBERT, O. (eds). *Tribute to Charles Darwin and Bernissart Iguanodons: new perspectives on vertebrate evolution and Early Cretaceous ecosystems.* (Abstracts of papers and field guide). Brussels 2009. 145 pp.

120. 2009*b* RAYFIELD, J., BARRETT, P. M. and MILNER, A. R. Utility and validity of Middle and Late Triassic 'Land Vertebrate Faunachrons'. *Journal of Vertebrate Paleontology*, **29**, 80–87.

121. 2009*c* MILNER, A. R. Comments on the proposed conservation of *Buettneria* Case, 1922 (Amphibia). *Bulletin of Zoological Nomenclature*, **66** (1), 76–77.

122. 2009*d* ENSOM, P. C., CLEMENTS, R. G., FEIST-BURKHARDT, S., MILNER, A. R., CHITOLIE, J., JEFFERY, P. A. and JONES, C. The age and identity of an ichthyosaur reputedly from the Purbeck Limestone Group, Lower Cretaceous, Dorset, southern England. *Cretaceous Research*, **30**, 699–709.

123. 2009*e* MILNER, A. C., MILNER, A. R. and WALSH, S. A. A new baphetid specimen from the Upper Carboniferous of the Czech Republic and the intrinsic relationships of the Baphetidae. *Acta Zoologica*, **90** (Suppl. 1), 318–334.

[Special Papers in Palaeontology 81, 2009, pp. 15–42]

AN ARTICULATED SPECIMEN OF *CHRONIOSAURUS DONGUSENSIS* AND THE MORPHOLOGY AND RELATIONSHIPS OF THE CHRONIOSUCHIDS

by JENNIFER A. CLACK* *and* JOZEF KLEMBARA†

*University Museum of Zoology, Cambridge, CB2 3EJ, UK; e-mail: j.a.clack@zoo.cam.ac.uk
†Faculty of Natural Sciences, Department of Ecology, Comenius University in Bratislava, Mlynská dolina, 84215 Bratislava, Slovakia; e-mail: klembara@fns.uniba.sk

Typescript received 10 December 2007; accepted in revised form 27 January 2008

Abstract: An almost complete, articulated specimen of the chroniosuchian *Chroniosaurus dongusensis* is described. Chroniosuchians are a little-known group of tetrapods found mainly in the Permo-Triassic of Russia that have been suggested to be late-surviving anthracosaurs. Detailed descriptions of the material have been limited, and few articulated specimens are known, though material is abundant. The new specimen allows us to provide new anatomical information on chroniosuchids, including new reconstructions of the skull, and for the first time to present a cladistic analysis that includes them. In six of the 12 most parsimonious trees they appear as basal embolomeres, lying a node above *Silvanerpeton*, supported by specific resemblances to embolomeres in the braincase region. In the remaining six trees, they appear further crownward, above *Silvanerpeton* and embolomeres, and below seymouriamorphs and other stem amniotes, though no unambiguous synapomorphies support this node. Chroniosuchians were probably semi-terrestrial, crocodile-like tetrapods but with skull specializations that suggest a very different feeding mechanism from that of crocodiles. Their dorsal osteoderms resemble those of some crocodilians and probably allowed a limited amount of both lateral and dorsoventral flexion of the trunk region.

Key words: Embolomere, anthracosaur, Permo-Triassic, Russia, braincase anatomy, crocodile-like semi-terrestrial lifestyle.

CHRONIOSUCHIANS are an enigmatic group of fossil tetrapods found mostly in the Upper Permian and Lower and Middle Triassic of Russia. First described by Vjushkov (1957), they were brought to the attention of western palaeontologists by Carroll *et al.* (1972) who briefly described and figured two genera, *Chroniosuchus* and *Bystrowiana*, placing them in the seymouriamorph section of their publication. Ivachnenko and Tverdochlebova (1980) gave the first detailed description of the group, providing reconstruction drawings though no specimen illustrations. The group has since usually been referred to as anthracosaurs (Golubev 1998*a*, *b*, 1999), and more specifically, as highly specialized embolomeres (Novikov *et al.* 2000). Chroniosuchians are divided into two families: Chroniosuchidae and Bystrowianidae. Golubev (1998*a*, 1999, 2000) recognized six species of chroniosuchid: *Chroniosaurus dongusensis*, *Chroniosaurus levis*, *Chroniosuchus paradoxus*, *Chroniosuchus licharevi*, *Jarilinus mirabilis* and *Uralerpeton tverdochlebovae*. Reviewed most recently by Novikov *et al.* (2000) chroniosuchians now include five genera and seven species of chroniosuchid and four genera and five species of bystrowianid. Novikov *et al.* (2000) added a seventh taxon, *Suchonica vladimiri* Golubev, 1999

to the six chroniosuchid species listed by Golubev (1998*a*). Golubev regarded the chroniosuchids as forming a 'single phylogenetic lineage' (Golubev 1998*a*, p. 397), and indeed very little separates the species anatomically. In most respects, their anatomy is highly conserved throughout the family and the species are diagnosed at present mainly on the ornamentation patterns of the dorsal scutes and on their stratigraphic and geographical distributions.

Most chroniosuchians are known from disarticulated remains and some only from vertebrae. The taxa are spread through a sequence of sediments in several localities in Russia, with the chroniosuchids mainly confined to the Upper Permian and the bystrowianids appearing from the Upper Permian to the Middle Triassic. Bystrowianid material is also recorded from the Upper Permian of China (Young 1979). Recently, bystrowianid remains have been reported from the Middle Triassic of Germany for the first time (Witzmann *et al.* 2008).

Chroniosuchians have a number of autapomorphies that distinguish them from any embolomeres, such as an anterior fenestra in the snout, a 'kinetic line' that runs between the skull table and the cheek but continues between the postorbital and the squamosal, lack of an intertemporal bone and the presence of dorsal scutes associated with the vertebrae. They are separated in both time and space from the latest well-known embolomere, *Archeria crassidisca*, from the Lower Permian of the southern United States, though there is a single specimen of an embolomere, *Aversor dmitrievi*, described from the Lower Permian of Russia (Gubin 1985). These two factors make assignment to the embolomeres difficult to justify, though there are other intuitively convincing anatomical similarities. No cladistic analysis has yet included chroniosuchians in investigations of the 'anthracosaur' taxa, nor even more widely among Palaeozoic tetrapods, to test their affinities.

We describe here an almost complete, articulated specimen of *C. dongusensis*, the only such specimen to represent almost an entire animal. This specimen helps clarify some of the anatomy of this poorly known group and allows us to present a new skull roof reconstruction that corrects some previous errors in the literature. This work is not intended as a full systematic review of the chroniosuchians, which is beyond the scope of this paper. Instead we seek to clarify points of the anatomy in the chroniosuchids. Finally, we present the first cladistic analysis that includes chroniosuchians in an attempt to place these animals phylogenetically. Anatomical details of chroniosuchids other than *C. dongusensis* are described and illustrated, where they amplify what is already available, and are included in the analysis.

Institutional abbreviations. PIN, Palaeontological Institute, Moscow; THU, Teikyo Heisei University, Ichihara City, Chiba Prefecture, Japan; UMZC, University Museum of Zoology, Cambridge, UK.

GEOLOGY AND MATERIAL

The specimen to be described is THUg 3308 (Text-fig. 1). It has had a checkered history, but was purchased by the Teiko Heisei University from a dealer. Thus the provenance was initially uncertain. However, comparisons by the authors with material in the collections of the PIN, have shown that there is little doubt that it derived from the Orenburg region, and probably from the right bank of the River Donga from which much chroniosuchid material has been collected. These comparisons also leave no doubt that the specimen is correctly identified as *C. dongusensis*, one of the species known from most material.

The specimen is preserved on a composite block of red sandstone, with the skeleton preserved as pinkish white bone, typical of the Orenburg region and of much other chroniosaurid material. The skeleton was initially prepared by an independent preparator in Russia, who has assembled the composite block from several pieces. Information, based on a report from a staff member of the PIN (V. V. Bulanov *in litt.* 2006), has suggested that the preparator was Dimitri Sumin, and gave the following statement: 'This is a complete skeleton only one exemplar, but both parts of this speciment was rotated to each other on 180 degree, i.e. anterior part of the skeleton was prepared in dorsal side, the posterior one in ventral side did. After preparation the first and second blocks was united to

TEXT-FIG. 1. *Chroniosaurus dongusensis*, THUg 3308. Photograph of entire specimen. Scale bar represents 10 mm.

common speciment, but they had no rigth orientation (sic)'. Some additional preparation was undertaken by Mrs Sarah Finney, in the Department of Zoology, Cambridge.

The specimen was X-rayed at the Queen's Medical Centre in Nottingham, revealing the breaks in the assembled block. Despite the above comments on the misorientation of parts of the specimen, in fact there seems to be little difference in exposure or orientation between the anterior and posterior parts of the vertebral column. Most of the vertebrae are visible in right lateral view. However, if the specimen represents a single individual, the column was presumably twisted at some point so that the skull is in dorsal view whereas the pelvis and hind limbs are in ventral view. In the middle section of the column is a disrupted region accounting for about half the reconstructed presacral length, in which some of the vertebrae are dislocated and not fully exposed. The possibility arises that the specimen is a composite of two individuals, however if so, they were two of the same species, of the same size, from the same locality and in identical matrix, such that the only real uncertainty would be the presacral vertebral count. The rest of the anatomy has been confirmed by comparison with existing collections of *C. dongusensis* in the PIN.

Chroniosaurus dongusensis is one of the earliest of the chroniosuchian genera, being commonest in the Strel'na and Purtovino Members of the Poldarsa Formation, Upper Tatarian Stage. Localities are widely spread across Russia west of the Urals from Kotlas in the north to Orenburg in the south. *Chroniosaurus* occurs with rich faunas of other tetrapods, terrestrial and aquatic, such as cynodonts, dicynodonts, pareiasaurs and temnospondyls (Golubev 2000). Golubev (2000) regarded them as members of the aquatic communities.

PIN specimens studied: *Chroniosaurus dongusensis*: 3585/4, 32, 38, 42, 79, 80, 87, 93–97, 99, 121–123, 124, 133–135 (isolated humerus, various isolated centra, no suffixes); 3713/13, 14, 17, 38, 58; and 3713 (interclavicle no suffix). *Chroniosaurus levis*: 105B/209, 1097. *Chroniosuchus* ('*Jugosuchus*') *licharevi*: 2357/2, 2005/2579. *Jarilinus* ('*Chroniosuchus*') *mirabilis*: 521/ 5, 25, 25, 27, 55, and 523/1. *Suchonica vladensis*: 4611/2.

SYSTEMATIC PALAEONTOLOGY

CHRONIOSUCHIA Tatarinov, 1972
CHRONIOSUCHIDAE Vjushkov, 1957

Genus CHRONIOSAURUS Tverdochlebova, 1972

Diagnosis. Modified from Novikov *et al.* (2000): tetrapods with narrow and elongated skull. Skull length up to about 25 cm. Ornamentation of skull pustular in juveniles and pustular and mostly pectinate in adults. Three pairs of crests on dorsal surface of skull roof: on (1) postparietal, parietal, postfrontal, frontal, prefrontal and nasal; (2) tabular and parietal; (3) tabular, supratemporal and postorbital. Ornamentation of osteoderms pustular and pectinate.

Chroniosaurus dongusensis Tverdochlebova, 1972
Text-figures 1–4, 7–9

Diagnosis. Ornamentation of scutes of pustular type.

Chroniosaurus levis Golubev, 1998*a*
Text-figures 5, 6A, H

Diagnosis. Ornamentation of scutes of pectinate type.

Remarks. The above diagnoses of *Chroniosaurus* species remain provisional, dependent upon a thorough review of the morphology of all chroniosuchids.

Description

Skull roof

Ivachnenko and Tverdochlebova (1980) and Golubev (2000) have described aspects of the skull roof of *Chroniosaurus* and other chroniosuchids. The former authors gave reconstructions of the dorsal view of the skull, the palate and both faces of the lower jaw of *C. dongusensis*. The latter author gave a revised lateral view reconstruction as a simple line drawing, but in our opinion somewhat more accurately.

The skull of THUg 3308 is preserved largely in dorsal view (Text-fig. 2A), with the left cheek folded under and inaccessible. The right side is crushed, so that the cheek is largely missing. Skull table sutures are reasonably clear, but the midline roofing bones and snout are fractured such that some sutures are difficult to trace. The following description is based not only on THUg 3308, but on comparative specimens studied in PIN.

The skull table shows the typical dermal ornament and raised ridges described by Ivachnenko and Tverdochlebova (1980), and which Golubev (2000) used as a diagnostic feature of the genus. No lateral line canals have ever been reported in chroniosuchids. However, one of the present authors has noted the existence of what may be a canal entirely within the bone, running along the ventral margin of the jugal and quadratojugal in some specimens of '*Jugosuchus*' [JAC pers. obs. 1990: specimens of the former genus *Jugosuchus* have now been split between members of the other species of chroniosuchid (Golubev 1998*a*)]. The canal may exit at a foramen in the quadratojugal, as described recently by Warren (2007) for *Ossinodus*.

The premaxillae are separated in the midline by a narrower internarial fontanelle than suggested by Ivachnenko and Tverdochlebova (1980), and there is a dorsal process that reaches as far posteriorly as half way along the external naris, seen in

TEXT-FIG. 2. *Chroniosaurus dongusensis.* Close up photographs of two skulls. A, THUg 3308. B, PIN 3585/124. Scale bar represents 10 mm.

PIN 3585/124 (Text-fig. 2B). The paired nasals are narrower than suggested by Ivachnenko and Tverdochlebova (1980), but bear the longitudinal ridges along their lateral margins that those authors described. The ridge runs posteriorly onto the prefrontal, and defines a change in angle between the midline portion of the snout and the more lateral parts. As shown by Golubev (2000), the external naris is substantially shorter than reconstructed by Ivachnenko and Tverdochlebova (1980), and most of its dorsal margin is contributed by the nasal, which sends a narrow process between the premaxillary dorsal process and the naris (PIN 3585/124) (Text-fig. 2B).

The frontals join the nasals in a strongly interdigitating suture, though this region is often poorly preserved and crushed, destroying the suture in many cases. The frontals are not as constricted between the orbits as shown by Ivachnenko and Tverdochlebova (1980), and do not contribute as much to the dorsal orbit margin. However, they do contribute a short length by means of an almost rectangular lateral lappet between pre and

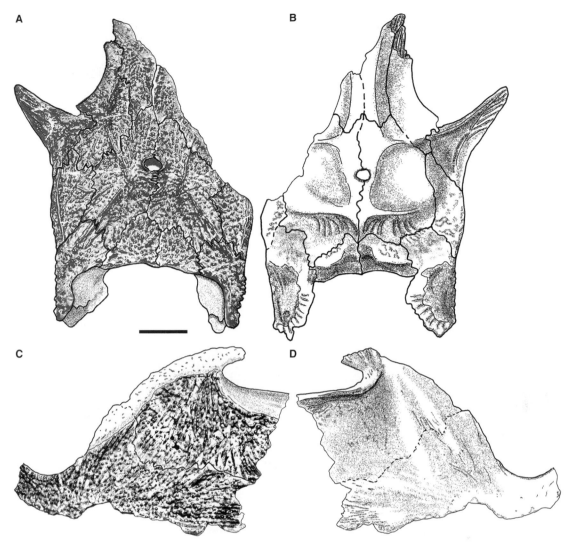

TEXT-FIG. 3. *Chroniosaurus dongusensis.* A–B, skull table PIN 3713/38 in A, dorsal (external), and B, ventral (internal) views. C–D, PIN 3713/14, left squamosal and jugal in C, lateral (external), and D, medial (internal) views. Scale bar represents 10 mm.

postfrontals. This is a feature possibly unique to Chroniosuchi-dae, seen also in *Jarilinus mirabilis* specimen PIN 523/1 (formerly *Chroniosuchus mirabilis* Golubev, 1998*a*). This feature is not well seen in THUg 3308, in which the left prefrontal and right postfrontal are missing, and the right prefrontal is displaced. The left postfrontal appears to terminate anteriorly in a pointed process, but its medial margin has been displaced under the right frontal, so that it true shape cannot be made out.

Specimen PIN 3713/38 (Text-fig. 3A–B), a well preserved skull table of *C. dongusensis* showing dorsal and ventral views, is used to supply additional information on skull table morphology. It represents a larger individual than THUg 3308 or any of the PIN 3585 series, and differences between it and these smaller individuals are almost certainly attributable to ontogeny. It is broken obliquely across the right postfrontal and left frontal.

PIN 3713/38 shows the underside of the frontals to be strengthened by longitudinal ridges, and they and the postfrontals are scooped out laterally for reception of the eyeball and associated soft tissues. In this specimen, the left postfrontal is complete and the underside shows the sutural surface for reception of the lateral lappet of the frontal that enters the orbit, and the overlap area for the prefrontal that meets it. The postfrontal is broadly crescentic in shape, and has strongly interdigitated sutures with the frontal, the parietal, the supratemporal and the postorbital.

The parietal foramen of PIN 3713/38 is set roughly at the mid-point of the interparietal suture, in contrast with the much more anteriorly located position in the skull restoration of *C. dongusensis* by Ivachnenko and Tverdochlebova (1980, fig. 1a). Just posterolateral to the parietal foramen, the conspicuous ridges marking the skull table, noted in earlier descriptions, meet and form a broad V-shape. The posterior arm of the V continues diagonally across the parietal-tabular suture and fades out on the tabular. The anterior arm of the V crosses the parietal-postorbital suture at about its mid-point and turns medially at roughly the mid-point of the postorbital. The pattern of ridges is somewhat different from that in THUg 3308, in which the ridges run approximately parallel to the midline to the posterior margin of the parietals about midway across their widths, but more anteriorly flare out gently onto the postfrontal (Text-Fig. 2A). The ridges follow a similar course to those in THUg 3308 in similarly sized specimens such as PIN 3585/124 and 3585/94. The parietals have a broad, strongly interdigitated suture with the tabulars and the postparietal-parietal suture is concave anteriorly.

The undersides of the parietals of PIN 3713/38 carry median ridges continuous with those on the frontals, excavated laterally into shallow depressions whose posterior borders are marked by transverse ridges. Along the posterior elevation, these ridges are folded and scalloped. By comparison with the embolomeres *Pholiderpeton* and *Pteroplax* (Clack 1987), the midline ridge probably bore the attachment of the sphenethmoid anteriorly.

The intertemporal is absent in all chroniosuchians. In *Chroniosaurus*, the supratemporal and postfrontal suture together and occupy its former territory. The supratemporal has a gently curved lateral margin, parallel and close to which runs a ridge in the ornament comparable to those on the parietals. In THUg 3308, these ridges continue forwards onto the postorbital and backwards to the ends of the tabulars. On the underside of PIN 3713/38, the supratemporal is roughened but does not bear an obvious sutural surface for the squamosal.

Only the right tabular is present in THUg 3308. In PIN 3713/38, the tabulars bear conspicuous ventral flanges visible in dorsal view (Text-fig. 3). In smaller specimens such as PIN 3585/124, the condition resembles that in THUg 3308, in which the flanges are not well developed and are only very slightly exposed in dorsal view. In embolomeres, the tabulars also bear ventral flanges, though they are more substantial than those of chroniosuchids. Discosauriscids also bear modest ventral flanges along the postserior margin.

In embolomeres, the tabulars carry one or two facets for attachment of the otic capsules (Clack 1987). In *C. dongusensis*, there are modest equivalents though they are not obviously double as they are in most embolomeres. However, PIN 2357/2, assigned to *Ch. licharevi* (formerly *Jugosuchus licharevi* Golubev, 1998a), shows double facets (JAC pers. obs. 1990). In *Archeria*, the anterior facet is much lower and smaller than the posterior one, and the condition approaches a single facet in some specimens (JAC pers. obs. 1980; Clack and Holmes 1988). In *Discosauriscus*, Klembara (1997) shows a small roughened patch on the tabular just mesial to a low otic flange that may also be an attachment point for part of the otic capsule.

The external surface of the tabular of *Chroniosaurus* is strongly ornamented, and drawn into a long posterior process running almost parallel to the sagittal plane of the skull. PIN 2357/2 shows a similar condition, but also shows a small pit on the posterior surface of the tabular, as well as a small process. The condition resembles that in embolomeres. In *Discosauriscus* (Klembara 1997), there is also a small process extending posteriorly from the posterolateral margin of the tabular in a similar position to the tabular horn of embolomeres, though unlike that in *Chroniosaurus*, it is not ornamented dorsally.

The postparietals are roughly D shaped in PIN 3713/38, though somewhat narrower in THUg 3308. The undersides bear roughened surfaces anteriorly for reception of the otic capsule, also clearly seen in PIN 2357/2 (*Ch. licharevi*). Their restricted anteroposterior extent may suggest a relatively much shorter otic capsule than in embolomeres, though given the poor ossification of the otic capsules (see below) this may not be a reliable guide; they may have accommodated only the opisthotics. With the posteriorly extended tabulars they form a deeply concave rear margin to the skull table. Contrary to embolomeres (e.g. Holmes 1984; Clack 1987) and seymouriamorphs (e.g. Klembara 1997; Klembara *et al.* 2005), there are no unornamented occipital flanges of the postparietals in chroniosuchians.

In chronioschids, the postorbital is incorporated into the skull table via a strongly interdigitating suture with the supratemporal, and bears some striking features. The bone is frequently preserved still bound to the supratemporal and the postfrontal as in PIN 3713/38 and THUg 3308, and in other skulls in which the rest of the cheek is poorly preserved. It is acutely triangular, with its external surface carrying a continuation of the ridge found on the supratemporal and the tabular. It carries elongate striations on its underside along the posterior margin. These striations form the contact surface with the anterior margin of the squamosal and part of the jugal, in what appears to be a continuation of the so-called kinetic line present between the supratemporal and the squamosal (e.g. Clack 1987). In embolomeres, this kinetic line sometimes separates the supratemporal and the intertemporal from the squamosal and the postorbital, though in some cases, the intertemporal bears an interdigitated suture with at least some of the postorbital dorsal margin. In *Archeria*, for example, the postorbital shows such interdigitation surfaces for attachment to the intertemporal, though it usually becomes detached during processes of preservation, so was probably not firmly held (Holmes 1989). In *Chroniosaurus*, the postorbital forms the major part of the posterior orbit margin, as it does in *Archeria*.

One of the best preserved skulls of *C. dongusensis*, PIN 3585/124, shows the snout reasonably well, and much of the description is taken from this specimen (Text-fig. 2B). Ornament on the external surface of the snout bones, though present, is reduced by comparison with those of the skull table and cheek. This specimen and THUg 3308 are similarly sized, small individuals and the limited amount of snout ornament is likely to be a condition of immaturity.

The maxilla is slender and tapering, except where a prominence rises near the anterior margin of the anterior fenestra, where it provides a smooth sigmoid surface for contact with the lacrimal. There is a short suture with the premaxilla, also gently sigmoid in PIN 3585/124; the maxilla overlaps the premaxilla. The lacrimal contributes the posterior margin of the naris, and its posterolateral margin is excavated along about half its length by the dorsal margin of the large anterior fenestra that characterizes chronioschids.

The prefrontal is acutely triangular in shape, with a narrow anterior apex running between the lacrimal and the nasal, and a posterior process forming most of the anterior margin of the orbit. With the jugal, it provides the posterior margin of the anterior fenestra, forming an antorbital pillar. The suture with the jugal along this pillar is oblique and smooth. The posterior process of the prefrontal almost mirrors the narrowly triangular postorbital contribution to the orbit. As with the lacrimal there is a distinct angle between the midline and lateral snout components, defined by the ridge as noted above. In effect, the ridge is a continuation of that which runs along the lateral margin of the skull table and in some specimens, it continues above the orbits (e.g. PIN 3585/94).

The jugal provides the ventral portion of the orbit margin. Narrow processes form straight, smooth contacts with the prefrontal and the maxilla, and together provide the curved posterior margin of the anterior fenestra. Conversely, the posterior margin of the jugal has an irregular interdigitated suture with the squamosal, seen best in PIN 3713/14, an isolated portion of cheek from a large specimen (Text-fig. 3C–D). The former specimen also reveals part of the smooth overlap area for the postorbital process. The jugal does not form part of the jaw margin in PIN 3583/124. An isolated fragment of PIN 3585 shows the internal face of the jugal, with a strengthening ridge under the orbit margin and an alary process ventrally for contact with the palate as found in embolomeres (Panchen 1972). This is not visible in ventral view as reconstructed by Ivachnenko and Tverdochlebova (1980).

The quadratojugal meets the jugal and the maxilla via smooth uninterdigitated sutures ventrally, but an interdigitated one with the squamosal in large specimens. It is a relatively narrow bone anteriorly, fitting into the posterodorsal margin of the maxilla, but broadens posteriorly where it forms the lower part of the margin of the suspensorium as a smooth unornamented flange.

The squamosal, best seen in PIN 3713/14 (Text-fig. 3C–D), provides much of the posterior margin of the suspensorium. This specimen shows both internal and external faces of the squamosal and the jugal. It shows the gently curved, unstriated nature of the contact surface for the postorbital process and the supratemporal, and confirms that the sutural contact with the supratemporal was relative short, as suggested by PIN 3585/124. The squamosal does not appear to contact the tabular, *contra* Ivachnenko and Tverdochlebova (1980, p. 42), whose lateral view reconstruction also omits the contribution of the postorbital process to the posterior orbit margin. The dorsalmost portion of the squamosal is drawn into a deeply excavated embayment, around the course of which a prominent but narrow flange runs internally. This characteristic dorsal part of the left squamosal is preserved in THUg 3308 (Text-fig. 2A).

Contact between the cheek and skull table is revealed in internal view by a specimen of the chroniosuchid PIN 2005/2579 (*Ch. licharevi*, Golubev 2000), in which the dorsal part of the squamosal-postorbital contact fits neatly into a facet on the supratemporal.

Text-figure 4 gives new reconstructions of the skull roof in lateral and dorsal views. It varies in a number of respects from previous reconstructions and is based on personal observations of a range of specimens.

Lower jaw

Lower jaws are not preserved in THUg 3308. We can confirm that there appears to be a single Meckelian fenestra, and no significant surangular crest. The reconstruction by Ivachnenko and Tverdochlebova (1980) is used as the basis for character analysis.

Palate and braincase

Few specimens show the elements of the palate well, but preliminary observations of the PIN specimens suggests that the reconstruction given by Ivachnenko and Tverdochlebova (1980) is inaccurate in several regards. However, as they showed, the pterygoids do not suture in the midline, though their medial margins run closely parallel. There are no interpterygoid vacuities. Redescription of the palate is in progress and will be provided in a future publication.

The otic capsules have not been observed by the authors in any specimen, and are not illustrated by Ivachnenko and Tverdochlebova (1980) in any chroniosuchid. We infer from this that they were poorly ossified and not preserved, even in larger, older individuals.

By contrast, the parasphenoid and the basisphenoid are often well represented, and the basioccipital occasionally remains in articulation with them. The braincase is not visible in THUg 3308, so this description is based on PIN 3585/87 (*C. dongusensis*) and PIN 105B/1097 (*C. levis*) (Text-fig. 5). The ventral view of the parasphenoid-basisphenoid complex bears close resemblance to those of embolomeres, and specifically, its proportions are most similar to those of *Archeria* (Holmes 1989) (Text-fig. 6B, J). The basipterygoid processes are bulbous posteriorly, projecting conspicuously from the body of the parasphenoid-basisphenoid complex. They are bifaceted, with subequal anterodorsal and anteroventral faces, each shaped like an inverted saddle. They meet at a smoothly rounded junction as in *Archeria* (Clack and Holmes 1988). The basal plate of the parasphenoid is more or less triangular, with the apex forming a narrow ridge between the basipterygoid processes that continues anteriorly as a long narrow cultriform process, whose complete length is still unknown. Deep carotid grooves lie either side of the ridge, and thus the basipterygoid processes lie relatively close to the midline. There are no visible carotid foramina such as are found in some early tetrapods, but which are generally absent in embolomeres including *Archeria*. The parasphenoid then widens posteriorly, where the midline ridge divides and then broadens into smooth-surfaced convex ridges. Between these lies a deep depression, whose surface sometimes bears longitudinal striations. The ridges form the parasphenoid 'wings' (basal tubera of Clack and Holmes 1988), which project substantially ventral to the basioccipital and are similar to those described in other embolomeres. They are clearly seen in dorsal view in PIN 105B/1097 (Text-fig. 5), and bear longitudinal striations indicating that hypaxial musculature originating on the cervical vertebrae probably inserted within the basal tubera (Clack and Holmes 1988).

The basisphenoid is seen in dorsal view in several specimens including PIN 105B/1097 (*C. levis*) (Text-fig. 5) and PIN 3585/4, 79–80, 99 (all *C. dongusensis*). Between the basipterygoid processes is a bowl-shaped depression, in some specimens further pierced by a small pit, the latter interpreted as the sella turcica. The arrangement is virtually identical to the condition in *Archeria*, in which the depression has been interpreted as housing the retractor eye muscles (Clack and Holmes 1988). Posteriorly, the bases of stout pillars rise from the basisphenoid, forming the processus sellares. These are joined across the

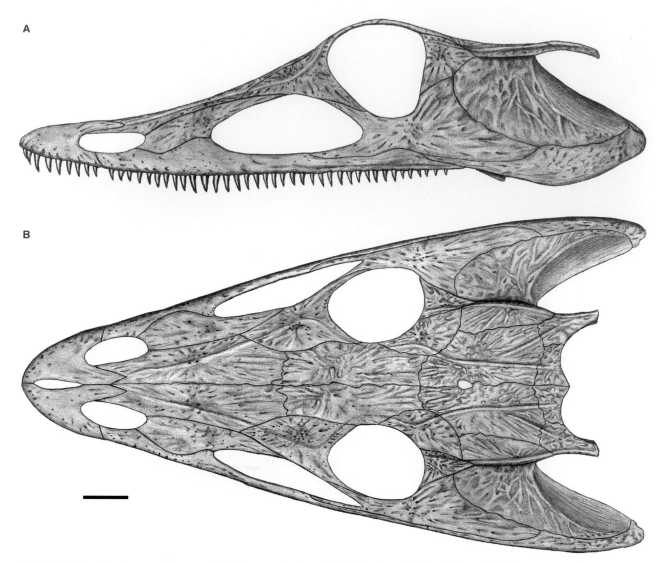

TEXT-FIG. 4. *Chroniosaurus dongusensis.* Skull reconstructions. A, left lateral and B, dorsal views. Scale bar represents 5 mm.

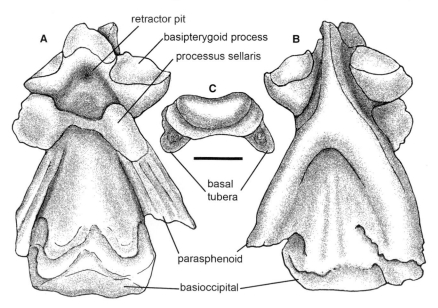

TEXT-FIG. 5. *Chroniosaurus levis*, PIN 105B/1097, parasphenoid-basisphenoid complex and basioccipital. A, dorsal, B, ventral, and C, posterior views. Scale bar represents 10 mm.

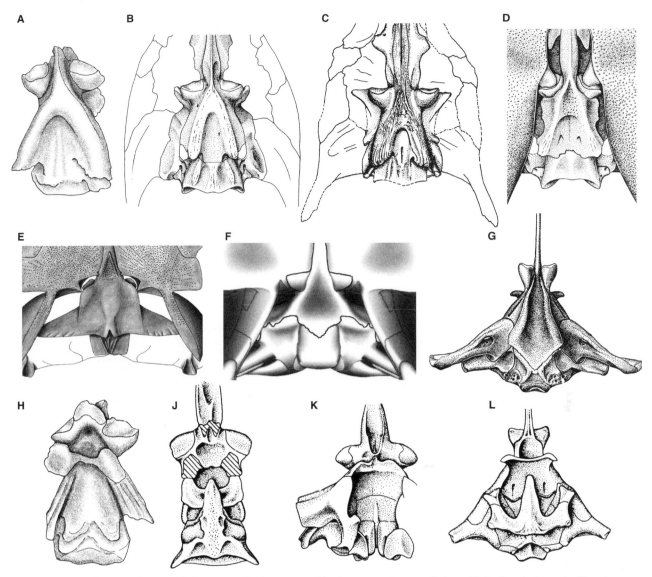

TEXT-FIG. 6. Chroniosuchian, embolomere, and other tetrapod braincases. A–G, ventral views. H–L, dorsal views. A, *Chroniosaurus levis.* B. *Archeria crassidisca* (from Clack and Holmes 1988). C, *Pholiderpeton scutigerum* (from Clack 1987). D, *Proterogyrinus scheeli* (from Holmes 1984). E, *Discosauriscus austriacus* (from Klembara *et al.* 2005). F. *Seymouria baylorensis* (from Laurin 1996). G, *Captorhinus laticeps* (from Heaton 1979). H, *Chroniosaurus dongusensis.* J, *Archeria crassidisca* (from Clack and Holmes 1988). K, *Seymouria baylorensis* (from Clack and Holmes 1988). L, *Captorhinus aguti* (from Clack and Holmes 1988). Not to scale.

midline by a narrow wall, behind which lies the broad depression accommodating the basioccipital. The appearance of this aspect of the braincase is essentially identical to that of *Archeria* (Clack and Holmes 1988).

The basioccipital is present in some specimens (e.g. PIN 105B/1097) (Text-fig. 5C). This element is one that differs most strongly from its equivalent in those embolomeres in which it is known (Text-fig. 6). It is a flattened element forming a narrow crescent, convex in dorsal and posterior views resembling that of the seymouriamorph *Discosauriscus* (Klembara 1997). In *Archeria*, by contrast, together with contributions from the exoccipital, it forms an almost complete, concave disc and the dorsal surface is divided in the midline by a tall crest of bone bearing evidence of persistant cranial segments, and also carries lateral processes to support the base of the brain (Clack and Holmes 1988).

Exoccipitals are preserved in THUg 3308, although they are only visible in one aspect, presumably dorsal, so that any contribution to the occipital face cannot be established. They are small almost rectangular bones, bearing no obvious facets for contact with the opisthotics (Text-fig. 5). They are pierced by foramina and grooves like those in *Archeria* that have been interpreted as being for cranial nerves X and/or XII (Clack and Holmes 1988). In size, shape, and in the positions of the foramina, the exoccipitals of *Chroniosaurus* resemble those of *Discosauriscus* (Klembara 1997).

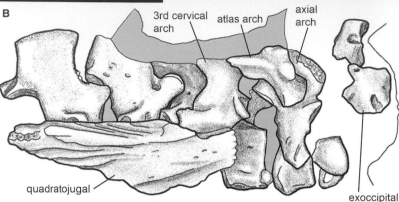

TEXT-FIG. 7. *Chroniosaurus dongusensis*, THUg 3308. A, close up photograph of cervical region. B, interpretive drawing. Grey fill represents matrix. Scale bar represents 10 mm.

Dentition

No additional details of the dentition are available from THUg 3308. According to Ivachnenko and Tverdochlebova (1980) the premaxilla of *C. dongusensis* bears about 10 teeth, and the maxilla about 36–38. We were unable to corroborate these details, though believe the figures to be roughly correct. The dentary bears a matching row of teeth to the upper jaw, according to Ivachnenko and Tverdochlebova (1980). From our observations, the tooth crowns are smooth and slightly recurved posteriorly, reaching their maximum size on the maxilla just anterior to the middle of the row.

The dentition of the palate as shown by Ivachnenko and Tverdochlebova (1980) contains some inaccuracies, and will be described in more detail in a separate paper by the current authors.

Gastralia and dorsal dermal scutes

THUg 3308 shows the dorsal dermal scutes characteristic of other small *C. dongusensis* specimens, and nothing can be added to previous descriptions. This specimen also preserves patches of articulated oval gastral scales bearing internal ridges running close to the anterior margin as in embolomeres and many other early tetrapods. Ivachnenko and Tverdochlebova (1980) portrayed them as being symmetrical about a ridge along the internal midline.

Axial skeleton

THUg 3308 preserves one of the most complete postcranial skeletons of any chroniosuchid. Of particular interest are the cervical vertebral series which includes the atlas and axis arches, described for the first time, and the pre and postsacral series. Unfortunately, no sacral vertebra is visible (Text-figs 1, 7, 8). Neural arches and pleurocentra are separate in this specimen, a juvenile condition compared with the fused condition found in larger specimens.

Text-figure 7 shows the cervical region of THUg 3308. The atlas arch is slightly displaced posteriorly, and lies over the axis arch. The atlas arch bears a modest neural spine, with dorsal and posterior components, the former effectively equivalent to the neural spine component in *Pholiderpeton*, the latter assumed to carry the postzygapophysis and similar to the posteriorly directed spine in *Archeria*: in *Archeria*, there is no dorsally directed spine. In THUg 3308 there is a facet, presumably a prezygapophysis for a pro-atlas, set somewhat below the dorsal component of the spine. Overall, the element resembles those of the embolomeres *Archeria*, *Pholiderpeton* and *Proterogyrinus*, with the closest resemblance to *Archeria*.

What is presumed to be the atlas intercentrum lies close to its life position. It is somewhat bulbous as in *Archeria*. The axis arch, in articulation with its pleurocentrum, lies immediately behind. Neither the atlas pleurocentrum nor axis intercentrum can be identified. A small element lying below the putative atlas intercentrum may represent a fragment of one of these elements

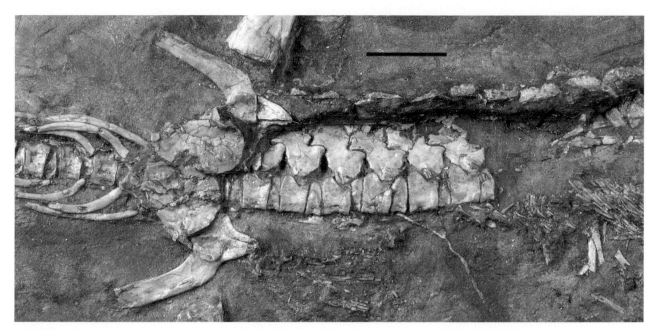

TEXT-FIG. 8. *Chroniosaurus dongusensis*, THUg 3308. Close up photograph of presacral column and pelvis. Scale bar represents 10 mm.

or possibly a broken portion of the posterolateral process of the atlantal intercentrum present in *Archeria*.

The axis arch bears a relatively small prezygapophysis, below the posteriorly directed and apparently unfinished, posteriorly sloping anterior margin of the neural spine, similar to that of *Archeria*. The spine itself is largely obscured, but appears to be less tall than at least some of the succeeding cervical neural arches, as in many early tetrapods including embolomeres. The postzygapophysis is well developed. The relatively narrow transverse process bears an unfinished facet for the diapophysis. The ventral half of the axial pleurocentrum is seen in lateral view; its perichondral surface is less extensive than in succeeding pleurocentra, leaving unfinished bone visible posteriorly. It bears a lateroventral depression such as is also found in most succeeding pleurocentra. A tiny, almost spherical piece of bone, which lies between this and the next pleurocentrum, may represent the intercentrum of cervical three.

Cervical arch three also appears to have a relatively narrow neural spine, though the zygapophyses are robust and rounded and borne on prominent conical processes. The pleurocentrum of this vertebra is rectangular in lateral view, and bears a deep ventrolateral depression and a somewhat shallower one more dorsally. The centrum is narrower than in most succeeding vertebrae.

Cervical arch four appears to have a spine very much narrower anteroposteriorly than the others, though this appearance is probably exaggerated by the slight posterior displacement of cervical arch three. The fourth cervical spine is not reduced in *Archeria* (Holmes 1989), and the condition is unknown in *Pholiderpeton*, *Proterogyrinus* and other embolomeres (Panchen 1966; Holmes 1984; Clack 1987). However, reduction of the fourth or fifth cervical neural spine is commonly found among early tetrapods, such as *Eryops* (Moulton 1974), *Greererpeton* (Godfrey 1989) and *Pederpes* (Clack and Finney 2005).

Cervical neural arches 5 through eight are reasonably well preserved with eight and nine remaining in articulation with their respective centra; however, a portion of the right quadratojugal, exposed in internal view, overlies and obscures several of the centra (Text-fig. 8). The column is then disrupted until about six vertebrae anterior to the sacrum. The tops of all visible neural arches are obscured by the dorsal dermal osteoderms or by matrix that supports the osteoderms. All neural spines nonetheless are taller and more rectangular than indicated by Ivachnenko and Tverdochlebova (1980), who reconstructed them as low and rounded. In the less disrupted presacral region, the neural arches are tightly packed, in contact for most of their anterior and posterior surfaces dorsal to the zygapophyses (Text-fig. 8).

Transverse processes are short and more or less triangular, bearing small, unfinished diapophyses. Zygapophyseal surfaces appear to meet in a horizontal plane, though for the most part, only their lateral edges are visible. Where the neural arches have become slightly detached, deeply excavated sockets are seen in the dorsal parts of the pleurocentra to receive the pedicels. All pleurocentra are about twice as long anteroposteriorly as the intercentra, and are sheathed in perichondrium. The intercentra, as in other embolomeres, show perichondral surface covering the ventral half in lateral view, with unfinished bone exposed in the more dorsal parts. They are about as tall as the pleurocentra. A small parapophysis is set at the apex of the perichondral covering.

About 11 vertebrae are represented in the tail region, exposed mainly in right ventrolateral view. Long, curved postsacral ribs articulate with the first three postsacral vertebrae via their intercentra. The bases of haemal arches are still in contact with some of the more posterior intercentra. Intercentra become relatively a little narrower in the tail region, and the neural arches of the last six visible vertebrae appear to have lower arches than those in the

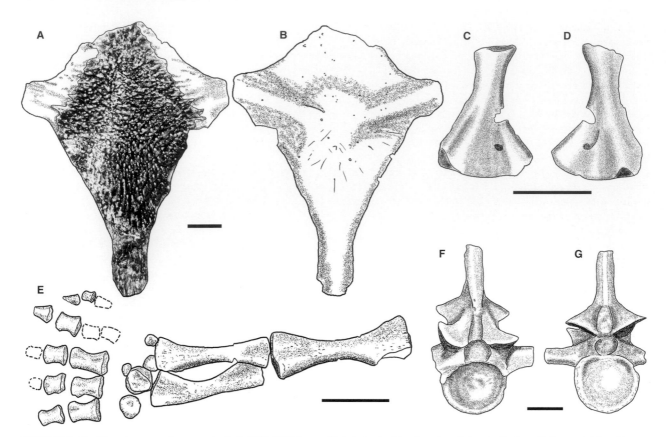

TEXT-FIG. 9. *Chroniosaurus dongusensis.* A–B, PIN 3713 (no suffix), interclavicle. A, external (ventral) view. B, internal (dorsal) view. C–D, PIN 3585 (no suffix), humerus. C, ventral view. D, dorsal view. E, PIN 3585/42, hind limb. F–G, *Suchonica vladensis*, PIN 4611/2, neural arch and pleurocentrum. F, anterior view. G, posterior view. Scale bars represent 10 mm.

trunk, though they are not completely exposed. Thus it remains possible that they bore elongate neural spines as in *Archeria*.

Only a few fragments of trunk ribs are visible on THUg 3308: ribs do not seem to be well preserved in any chroniosuchid specimen, which contrasts with the condition in most embolomeres, where long curved ribs are usually conspicuous features of the anatomy. What can be seen in *Chroniosaurus* suggests they were double-headed, narrow, delicate and slightly curved (PIN 3585/124), and not very securely attached to their respective vertebrae. An exception to this generality is the occurrence of three pairs of more robust, curved postsacral ribs, evident in THUg 3308 (Text-fig. 8), a feature also present in *Silvanerpeton* (Ruta and Clack 2006). *Archeria* has five pairs in this region (Holmes 1989).

Isolated vertebral elements are preserved in the PIN collections, with intercentra and pleurocentra from differently sized individuals and from different parts of the column: both elements are complete discs in most cases (a few intercentra are incomplete dorsally), and resemble those of other embolomeres. Intercentra have convex articulating surfaces, that fit into the matching concave surfaces of the pleurocentra. PIN 4611/2 (*Suchonica vladimiri*) is a pleurocentrum with neural arch still in articulation, and the unit is more or less indistinguishable from a comparable set of *Archeria* vertebrae (UMZC T.93) (Text-fig. 9F–G).

Assuming that the collectors and preparators of THUg 3308 maintained the articulated parts of the skeleton in roughly their correct relationship, an estimate of the presacral count is possible. Approximately 11 or 12 trunk vertebrae could be accommodated in the gap between the cervical and immediately presacral regions, which would produce a total of about 27, though this estimate should be treated with caution. It would be somewhat shorter than the presacral column of *Proterogyrinus* at 32 (Holmes 1984), and considerably shorter than 40 as found in *Archeria* (Holmes 1989). However, it is longer than that of more terrestrially adapted forms such as *Gephyrostegus* (Carroll 1970) and seymouriamorphs (Klembara and Bartík 2000), which have 24.

Appendicular skeleton

Very little of the pectoral skeleton is preserved in THUg 3308. A possible partial clavicle, exposed just posterior to the skull, shows part of the dorsal stem, and its (broken) junction with the ventral plate. However, it supplies no details additional to those in the PIN collection. An almost complete isolated clavicle is represented in the series PIN 3585/121–123, 133–135. It shows part of the broken dorsal stem, which is approximately parallel-sided and slightly twisted. The clavicular plate is approximately oval, convex ventrally and with a slightly thickened posterior edge. Both clavicles are in articulation with an interclavicle in PIN 3585/32, showing that the clavicular blades do not quite meet in the midline [as in figure 12a but not as in figure 12v of Ivachnenko and Tverdochlebova (1980)]. Clavicular plates and the interclavicle bear low-profile ornament of pits, ridges and grooves similar to that on the snout. The interclavicle of PIN 3585/32 has a strongly tapering stem whose posterolateral margins are concave. Posteriorly they form a narrow parasternal process. In larger animals, represented in PIN 3713, the interclavicle is kite-shaped, with more nearly straight posterolateral margins except for the posteriormost region, where they are parallel (Text-fig. 9A–B). The exposed surface bears strongly marked ornament, and resembles that of embolomeres including *Archeria* (Romer 1957) and *Pholiderpeton* (Clack 1987). The clavicular overlap surfaces are well seen in PIN 3713, and although the anterior portion of the bone is broken off, it is likely that here too, the clavicles did not meet in the midline. The underside of this bone shows a smooth surface pierced by a few nutrient foramina, and strengthened by thickened regions running in a broad Y pattern from the centre of ossification.

No scapulocoracoid has been identified by us in any of the material. There are a few examples of the humerus in the PIN 3585 series. PIN 3585/38 shows one associated with some articulated dorsal scutes, and also with slender elements that may represent an ulna or radius and a few phalanges or metacarpals. The phalangeal formula of the manus remains unknown.

In the PIN 3585 series is also an isolated humerus, on which the following description is based (Text-fig. 9C–D). The articular surfaces are rather poorly developed, not unexpected in a juvenile animal, but in great contrast to the primitive 'L' shaped humerus of embolomeres and other early tetrapods, even from a juvenile stage (e.g. *Greererpeton*: Godfrey 1989). From the proximal end, which is a flattened oval in cross section, the bone narrows slightly, and then flares out smoothly, somewhat more strongly along the posterior edge than the anterior, forming the entepicondyle. This is thickened slightly along the edge, and just inboard of the thickened region is pierced by the entepicondylar foramen. The whole bone is strengthened by a ridge running from the more posterior region of the proximal end towards the anterior region of the distal end and distally, this bears part of what is taken to be the radial facet. The thickened ridge is interpreted as in part, the ectepicondyle, identifying this surface as dorsal. More of the radial condyle is seen on the underside at the distal anteroventral corner. This bone is entirely unlike the robust humeri of embolomeres, and indeed seems to have no obvious close parallels among other early tetrapods. The authors saw no examples of humeri from larger individuals in the PIN collection, so are unable to say whether there were significant ontogenetic changes to the humerus in life, or whether their morphology approached that of larger embolomeres such as *Archeria* (Romer 1957) or *Proterogyrinus* with age (Holmes 1984).

The pelvis is represented in THUg 3308 by both ilia and ischia, exposed in external view approximately *in situ* (Text-fig. 8). The ilium has a very short neck in contrast to the longer neck illustrated by Ivachnenko and Tverdochlebova (1980), and the robust parallel-sided posterior process is set about 30 degrees to the plane of the acetabulum. There is no dorsal process like that present in other embolomeres, and in overall appearance the pelvis more closely resembles that of a temnospondyl such as *Dendrerpeton* (Holmes *et al.* 1998). The ischia are approxi-

mately oval, thickened in the region of the dorsal margin forming the lower portion of the acetabulum. Pubes are missing, and were probably unossified, as in many early tetrapods, such as the anthracosaur *Silvanerpeton* (Ruta and Clack 2006). They may have been ossified in older individuals, as noted by Schoch, (1999) for *Mastodonsaurus giganteus*.

There are two reasonably complete hind limbs known for *C. dongusensis*. PIN 3585/42, illustrated in a restoration by Ivachnenko and Tverdochlebova (1980), and THUg 3308 (Text-figs 1, 9E). The two specimens provide slightly differing information, and also appear to differ a little in the proportions of the metatarsals and phalanges. In THUg 3308, the right (assuming the general view of the hind quarters of the specimen is ventral) hind limb is less complete than the left, with, in addition to femur, tibia and fibula, only two tarsals and three partial metatarsals. The left includes three tarsals, five metatarsals and the phalanges from four digits.

The femora are slender and waisted and their general appearance is like that of a temnospondyl such as *Trimerorhachis* (Pawley 2007) rather than that of an embolomere such as *Archeria* (Romer 1957). Neither articular end can be seen clearly in any specimen. The dorsal surface, seen on the right in THUg 3308, shows the intercondylar fossa distally. The left femur is exposed in ventrolateral view with a low adductor crest and relatively deep popliteal area (though this may have been exaggerated by compression). Tibiae are roughly equally expanded at proximal and distal ends; in *Archeria* and most early tetrapods, the proximal end is broader than the distal. Otherwise they are unremarkable. The fibulae are more deeply concave along their inner (anterior) edges, as in most other early tetrapods, though again, both ends appear roughly equally expanded. There is a posterior fibular ridge visible on the right fibula of THUg 3308, exposed edge on in posterior view. Overall, both tibia and fibula are slender in proportions.

Tarsals are rounded, and ossification is obviously incomplete: only the larger elements display perichondral surfaces.

Metatarsals in THUg 3308 are considerably more slender and are proportionately longer than in PIN 3585/42. Grooves at the proximal and distal ends of the left series probably accommodated the flexor tendons. Digit one is complete and has two phalanges, one of which is an ungual, seen also in PIN 3585/42. Digit two in THUg 3308 has three phalanges represented, though the third is poorly preserved; it may or may not be an ungual. Two other digits are represented, one by four phalanges, the last of which may or may not be an ungual, and the other by two phalanges. The former is probably digit four, the latter digit three. Digit five is not represented in THUg 3308, but there two phalanges are present in PIN 3585/42. The phalangeal formula of the pes therefore remains unknown and cannot be compared with that of embolomeres and *Silvanerpeton*, which have the synapomorphous formula of 2-3-4-5-5 (Ruta and Clack 2006).

PHYLOGENETIC ANALYSIS

The data matrix employed in this analysis is based on that of Ruta and Clack (2006), which was originally designed to establish the phylogenetic position of *Silvanerpeton miripedes*, an Early Carboniferous anthracosaur-grade stem amniote (*sensu* Ruta *et al.* 2003). *Chroniosaurus* was added to the original list of 33 taxa. Two characters from the original dataset were restated (LAC 2 and TAB 2) to accommodate conditions in *Chroniosaurus* more clearly; one character from the original set was rescored (*Paleothyris* character 46, SUTEMP 2, changed from 0 to ?); three characters (275, vomerine shagreen field; 277 size relations of upper to lower marginal dentition; 278, mandibular oral sulcus) were regarded as duplicates and deleted; and a further 15 were added. The additional characters describe several features of the braincase, of which some regions are very well preserved in *Chroniosaurus* and in some embolomeres. Other characters relate to the skull roof and postcranial skeleton and reflect characters of *Chroniosaurus* not covered in the original set. The total number of characters is 327 (see Appendix).

When 54 phylogenetically uninformative characters were omitted, the analysis retrieved 12 most parsimonious trees (MPTs) of tree length = 998, consistency index (CI) = 0.37, homoplasy index (HI) = 0.63, CI excluding uninformative characters = 0.34, HI excluding uninformative characters = 0.66, retention index = 0.55 and rescaled consistency index = 0.21. The strict consensus is shown in Text-figure 10 with Bremer support and bootstrap values indicated at the nodes. It places *Chroniosaurus* in a polytomy with embolomeres, *Silvanerpeton* and all other stem amniotes. This result corroborates the suggestion that the 'traditional' embolomeres form a natural

group with *Eoherpeton*, as found by Ruta and Clack (2006). That analysis placed *Silvanerpeton* as basal not only to embolomeres but also to all other stem amniotes. In six of the MPTs, *Chroniosaurus* emerged as a basal embolomere, whereas in the other six, it appeared above *Silvanerpeton* and embolomeres, and below the seymouriamorphs and other stem amniotes. The remaining tree topology is almost identical to that obtained by Ruta and Clack (2006), except that this new analysis produces a polytomy of the three discosauriscids.

Using only Ruta and Clack (2006) data set and adding *Chroniosaurus* the data set yielded a consensus topology from three MPTs that placed *Silvanerpeton* crownward to the embolomeres, succeeded by *Chroniosaurus* and *Gephyrostegus*, a result similar to that found in six of the 12 MPTs in this study. The additional characters in this study thus provided some support for the relationship between *Chroniosaurus* and the embolomeres, but the results were ambiguous.

Comments on character statements, definitions and distribution

In formulating some of the new characters we come up against a number of problems. One of the most striking is the inadequacy of language for describing shapes, especially three dimensional ones and in defining exactly why objects that we see appear similar to each other and different from others. For example, as illustrated above, many

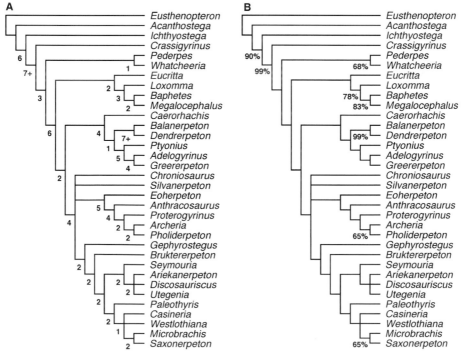

TEXT-FIG. 10. A, strict consensus of 12 MPTs, with decay indices (Bremer support) at nodes. B, same, showing bootstrap support.

aspects of the parasphenoid-basisphenoid complex are virtually identical between *Chroniosaurus* and embolomeres, specifically *Archeria*, whereas they differ strongly where known from those of seymouriamorphs or early amniotes as exemplified by *Captorhinus* (Text-fig. 6). Some of the new characters included in this analysis have been explicitly designed to try to encapsulate some of these observed similarities. These indeed emerge as the majority of synapomorphies supporting the clade of embolomeres plus *Chroniosaurus* in six of the 12 MPTs. Of these, several are unique to this clade and are unambiguously synapomorphic. These are: 310, basisphenoid width between basipterygoid processes, state 1; 311, basisphenoid retractor pit shape, state 1; 313, basipterygoid process shape, state 1; 315, parasphenoid shape 1, state 1 (unknown and equivocal in *Anthracosaurus*, reversed in *Eoherpeton*). Character 323, posterior margin of the postparietal shape is a further synapomorphy of the clade, though it is also found in some of the lepospondyl taxa and in *Whatcheeria*. Character 312, dorsum sellae shape state 1 is also found in *Silvanerpeton*. In contrast to these six trees, those in which *Chroniosaurus* is placed with *Silvanerpeton* and other stem amniotes share only characters that are found elsewhere as homoplasies, are reversals, or are plesiomorphies.

To express these similarities in words, let alone in itemized character states that have to be tailored to distinguish them from other states, is almost impossible. Further problems arise in that many of the characters, such as the shape of the faces on the basipterygoid processes, or the type of suture that one bone makes with another, are not adequately described in the literature, even in the most detailed descriptions. This may be either because they are not preserved, or this feature was not significant to, and thus overlooked by, the original describer. As our need for discrimination becomes ever more demanding, so the detail required becomes ever greater. It may be that in fact, several of these such characters could probably be scored could the original material be re-examined, though for the present study this has proved impractical. Furthermore, even with apparently clearly defined characters, one researcher's assessment of how it should be scored may differ from another who comes from a different viewpoint. Much of this has been rehearsed before, but the current study emphasizes the point, sometimes denied, that drawing up a data matrix is a subjective exercise, strongly dependent on the interests of the observer.

DISCUSSION

Morphology and lifestyle of chroniosuchians

As noted above, Golubev (2000) suggested that chroniosuchians formed part of an aquatic community, the Sokolki Assemblage, alongside brachyopids and kotlassiids. However, it is not clear on what basis this assignment was made, since terrestrial taxa including pareiasaurs, dicynodonts and gorgonopsids are also found there. A number of factors suggest that chroniosuchians were terrestrial. First, the absence of lateral line grooves on the skull of even juvenile specimens suggests that they were likely to have been terrestrial. One feature that could imply an aquatic habit is that the limb bones so far known have poorly developed articular surfaces, and rather simple morphology, lacking many of the processes associated with muscle insertions. Such apparent reduction in complexity is sometimes associated with aquatic taxa, but it is equally associated with juvenility. Since no large, presumably adult, limb elements have been recovered, the case for an aquatic lifestyle in the adults is hard to make on this evidence. However, as in crocodiles, it remains possible that both the juveniles and adults manifested some aquatic abilities.

Possession of dorsal osteoderms is a feature often associated with terrestrial or semi-terrestrial tetrapods including crocodilians and dissorophid temnospondyls. Chroniosuchian osteoderm morphology differs from that of both these taxa. Dissorophids are the most terrestrially adapted temnospondyls. In some genera, such as *Cacops* and *Dissorophus*, osteoderms in the anterior part of the column are fused to the tips of the neural spines and are overlain by an alternating set of more external osteoderms with which they articulate by sliding joints. In *Dissorophus* there is a complex joint between these external osteoderms and the neural spines, and in both genera the articulations permit only dorsoventral flexion of the front part of the column (Dilkes and Brown 2006; Dilkes 2007). Chroniosuchids differ from dissorophids in having laterally broad osteoderms, one per neural spine, articulating via more or less horizontal overlap surfaces to each other plus interarticulations to the neural spine and to each other.

Crocodilian dermal osteoderms and their function in bracing the vertebral column and musculature during locomotion have been studied by Salisbury and Frey (2000). They showed two types of osteoderm covering: an 'open' system in which the osteoderms are linked only by more or less horizontal facets, and a 'closed' system in which the distal end of each osteoderm is downturned by up to 90 degrees. The closed system is associated with anteriorly directed processes at the anterodistal corners of the osteoderm, fitting into a slot in the next anterior osteoderm. While both systems allow a certain degree of both lateral and dorsoventral flexion, lateral flexion is more restricted in the 'closed' system.

Chroniosuchids differ from crocodilians in having a single rather than paired midline series of osteoderms, and they also differ in having the interarticulations between osteoderms situated close to the midline rather than laterally. On balance, and without knowing the axial muscle anatomy of the chroniosuchians, their system seems to be more comparable with the 'closed' arrangement in crocodilians such as *Goniopholis*, allowing both dorsoventral and limited lateral flexion. Salisbury and Frey (2000, p. 113) suggest that the 'closed' osteoderm covering allowed 'a suite of terrestrial locomotor modes equally as diverse as that of extant crocodiles'. It may thus be possible to infer that chroniosuchians also had a range of locomotor modes that included both swimming and walking. Long bone microstructure also supports the suggestion that chroniosuchians were terrestrial (Laurin *et al.* 2004). Thus they may not have been permanent or exclusive members of the aquatic community.

Though there are some evident parallels between the construction of chroniosuchid postcranial skeletons and in the overall shape of the skull to those of crocodilians, some contrasts are also evident. Crocodilians are noted for their solidly fused 'akinetic' skulls, even having lost the synapomorphous archosaurian antorbital fenestra. By contrast, chroniosuchians have developed an antorbital fenestra, in combination with an apparently kinetic skull structure. This suggests a very different feeding mechanism from crocodiles, which employ the 'twisting' method especially when feeding in water and which is facilitated by the solidly constructed skull. The structure in chroniosuchians is considered further below.

Skull morphology reveals several ontogenetic changes of the kind that have been documented elsewhere. Sutures become more fully interdigitated, and ornament becomes deeper and more regular. However, the structure of the sutural connexion between the skull table plus postorbital and the cheek does not seem to change with growth. This 'kinetic line' is retained not only in larger individuals of *Chroniosaurus*, but throughout the two chroniosuchian families, and so is likely to be functionally significant. The arrangement appears fairly obviously to be a development of the 'kinetic line' in other embolomeres, but extended down the cheek via the suture between the postorbital and the squamosal. In embolomeres, this loose junction between skull table and cheek has been considered to be simply the result of having a broad compression-resisting 'butt joint' at the abrupt angled junction between these two skull elements (Clack 1987), but given the condition in chroniosuchians, such an explanation is probably oversimplified. At least in the juvenile specimens studied, the 'kinetic line' is possibly continuous with the line of uninterdigitated sutures that runs between the jugal-prefrontal and lacrimal-nasal-prefrontal sutures. The elaboration of the basipterygoid processes may be related, though no biomechanical studies have so far been conducted on the possible range of movement that such processes allow.

In *Allosaurus* (Rayfield *et al.* 2001) it was shown that compressive vectors were routed in loops around the cranial fenestrae, and this route serves to minimize stress and strain in response to applied forces. It was suggested that nonmechanical functions for the fenestrae, such as to house a gland, may have been less important than mechanical ones, such as stress management devices. A similar function may have applied to the skull of chroniosuchids, complemented by the development of strengthening ridges on the nasals, the frontals and the skull table. In *Allosaurus*, those bones are thickened and absorb some of the stresses during feeding. However, the state of the ridges is variable among chroniosuchid genera, and in *Chroniosuchus*, the ridges are less prominently developed or sometimes absent. (This may constitute a consistent and diagnostic difference between *Chroniosaurus* and *Chroniosuchus*, but that requires further study to confirm.) Loss of the intertemporal in chroniosuchians may similarly be associated with strengthening of the skull table. On the other hand, the antorbital bar, formed by the jugal and the prefrontal in chroniosuchians, is relatively slender and would tend to concentrate the stresses through an apparently weak component of the skull, tending to reduce the overall strength of the skull (E. Rayfield pers. comm.). Unfortunately, at the present time, complete skulls are not available for computed tomography scanning and biomechanical analysis that could test any of these hypotheses.

The detailed similarities of the parabasisphenoid between chroniosuchians and the derived Permian embolomere *Archeria*, though compelling, may be more widely distributed than first suspected. The braincases of *Eoherpeton* and *Pholiderpeton* (Text-fig. 6) show many similar features, but others are less well-known or are unknown in the relevant regions: features such as the bulbous basipterygoid processes, the narrow basisphenoid region between them, the bowl-shaped retractor pit and the shape of the processus sellares may be general embolomere characters. *Chroniosaurus* apparently shows none of the synapomorphies of the parasphenoid found in seymouriamorphs (e.g. *Seymouria*: Laurin 1996; *Discosauriscus*: Klembara 1997) (Text-fig. 6).

The fully embolomerous centra appear convergent upon those of the later embolomeres. Equally, the similarities between the basioccipital and the exoccipital of *Chroniosaurus* and *Discosauriscus* also appear convergently derived. In this case there are differences between the atlantal centra of the two taxa, related to the fact that *Chroniosaurus* has developed embolomerous centra, with enlarged intercentra, whereas in *Discosauriscus* the intercentra are small ventral wedges (Klembara and Bartík 2000). This has the effect of giving a better match between the faces of the occipital condyle and the atlantal intercentrum in *Discosauriscus*, compared with what appears to be that element in *Chroniosaurus*. There, a flattened, narrow condyle (Text-fig. 5C) appears to articulate with a dorsoventrally deep and anteriorly rounded atlantal intercentrum that is similar to that of *Archeria*. In *Discosauriscus*, centra remain fully notochordal, so that the morphology of the atlantal joint is explicable. However, centra of even small chroniosuchians retain relatively smaller notochordal canals, and in larger specimens they are completely occluded. The difference is hard to explain.

Biogeographical implications

If chroniosuchians are derived relatives of embolomeres, it suggests that they split from the line leading to other embolomeres at some time in the Early Carboniferous. At present, *Silvanerpeton* from the Viséan represents the earliest member of the clade encompassing that genus, embolomeres and all other stem amniotes (Ruta and Clack 2006), with the next two earliest embolomeres being *Eoherpeton* (Smithson 1985) and *Proterogyrinus* (Smithson 1986). All three taxa are recorded from Scotland, with *Proterogyrinus* also found in the eastern United States (Holmes 1984). *Archeria*, from the Lower Permian of Texas and Oklahoma, is the latest embolomere to be found in the United States (Holmes 1989). On this basis, embolomeres may have originated in Europe and spread both east and west during the Carboniferous, surviving longer, in a highly derived form, as chroniosuchians, in the east. If chroniosuchians are not derived embolomeres, they remain an enigmatic group of stem amniotes whose biogeographic and phylogenetic origins are unresolved.

Acknowledgements. We dedicate this paper to Andrew Milner, who has been a close colleague and friend for many years, and in the case of JAC, throughout her career. This project was partly supported by the Scientific Grant Agency, Ministry of Education of the Slovak Republic and the Slovak Academy of Sciences, Grant Nr. 1/D197/09. We thank our Russian colleagues Valery Golubev and Oleg Lebedev for access to collections, discussion of specimens, and hospitality during our stay. The drawings in Text-figure 4 were made by Andrej Čerňanský (Comenius University in Bratislava). The photograph of PIN 3585/124 in Text-figure 2B was taken by Július Kotus (Bratislava) and that in Text-figure 7 by Russell Stebbings (UMZC). Alex Foss facilitated the CT scanning at the Queen's Medical Centre in Nottingham. We thank Marcello Ruta for rerunning our PAUP analysis when computer problems struck, and for comments on the manuscript. Steve Salisbury and Emily Rayfield made helpful comments on functional morphology and pointed us to relevant literature including 'in press' articles. Robert Holmes, Florian Witzmann and Sean Modesto made helpful comments that improved the manuscript.

REFERENCES

CARROLL, R. L. 1970. The ancestry of reptiles. *Philosphical Transactions of the Royal Society of London B*, **257**, 267–308.

—— KUHN, O. and TATARINOV, L. P. 1972. Batrachosauria (Anthracosauria), Gephyrostegida-Chroniosuchida. *In* KUHN, O. (ed.). *Handbuch der Paläoherpetologie, Teil 5B.* Gustav Fischer Verlag, Stuttgart, 80 pp.

CLACK, J. A. 1987. *Pholiderpeton scutigerum* Huxley, an amphibian from the Yorkshire coal measures. *Philosphical Transactions of the Royal Society of London B*, **318**, 1–107.

—— and FINNEY, S. M. 2005. *Pederpes finneyae*, an articulated tetrapod from the Tournaisian of western Scotland. *Journal of Systematic Palaeontology*, **2**, 311–346.

—— and HOLMES, R. 1988. The braincase of the anthracosaur *Archeria crassidisca* with comments on the interrelationships of primitive tetrapods. *Palaeontology*, **31**, 85–107.

DILKES, D. W. 2007. New data on the vertebrae and osteoderms of *Cacops aspidephorus* and *Dissorophus multicinctus* (Temnospondyli: Dissorophidae). *Journal of Vertebrate Paleontology*, **27** (Suppl. 3), 68A.

—— and BROWN, L. E. 2006. Biomechanics of the vertebrae and associated osteoderms of the Early Permian amphibian *Cacops aspidephorus*. *Journal of Zoology*, **271**, 396–407.

GODFREY, S. J. 1989. The postcranial skeletal anatomy of the Carboniferous tetrapod *Greererepton burkemorani* Romer, 1969. *Philosophical Transactions of the Royal Society of London B*, **323**, 75–33.

GOLUBEV, V. K. 1998a. Revision of the Late Permian chroniosuchians (Amphibia: Anthracosauria) from Eastern Europe. *Paleontological Journal*, **32**, 390–401.

—— 1998b. Narrow-armoured chroniosuchians (Amphibia: Anthracosauria) from the Late Permian of Eastern Europe. *Paleontological Journal*, **32**, 278–287.

—— 1999. A new narrow-armoured chroniosuchian (Amphibia: Anthracosauromorpha) from the Upper Permian of Eastern Europe. *Paleontological Journal*, **33**, 166–173.

—— 2000. The faunal assemblages of Permian terrestrial vertebrates from Eastern Europe. *Paleontological Journal*, **34**, s211–s224.

GUBIN, Y. M. 1985. The first anthracosaur from the Permian of the East European platform. *Paleontological Journal*, **3**, 118–122.

HEATON, M. J. 1979. Cranial anatomy of the primitive captorhinid reptiles from the Late Pennsylvania and Early Permian of Oklahoma and Texas. *Bulletin of the Oklahoma Geological Survey*, **127**, 1–84.

HOLMES, R. 1984. The Carboniferous amphibian *Proterogyrinus scheelei* Romer, and the early evolution of tetrapods. *Philosophical Transactions of the Royal Society of London B*, **306**, 431–524.

—— 1989. The skull and axial skeleton of the Lower Permian anthracosauroid amphibian *Archeria crassidisca* Cope. *Palaeontographica*, **207**, 161–206.

—— CARROLL, R. L. and REISZ, R. R. 1998. The first articulated skeleton of *Dendrerpeton acadianum* (Temnospondyli: Dendrerpetontidae) from the Lower Pennsylvanian locality of

Joggins, Nova Scotia and a review of its relationships. *Journal of Vertebrate Paleontology*, **18**, 64–79.

IVACHNENKO, M. F. and TVERDOCHLEBOVA, G. I. 1980. *Systematics, morphology, and stratigraphic significance of the Upper Permian chroniosuchians from the east of the European part of the USSR.* Izdatel'stvo Saratovkogo Universiteta, 69 pp., Saratov, Russia [In Russian].

KLEMBARA, J. 1997. The cranial anatomy of *Discosauriscus* Kuhn, a seymouriamorph tetrapod from the Lower Permian of the Boskovice Furrow (Czech Republic). *Philosophical Transactions of the Royal Society of London B*, **352**, 257–302.

—— and BARTÍK, I. 2000. The postcranial skeleton of *Discosauriscus* Kuhn, a seymouriamorph tetrapod from the Lower Permian of the Boskovice Furrow (Czech Republic). *Transactions of the Royal Society of Edinburgh: Earth Sciences*, **90**, 287–316.

——BERMAN, D. S., HENRICI, A. and ČERŇANSKÝ, A. 2005. New structures and reconstructions of the skull of the seymouriamorph *Seymouria sanjuanensis* Vaughn. *Annals of the Carnegie Museum*, **74**, 217–224.

LAURIN, M. 1996. A redescription of the cranial anatomy of *Seymouria baylorensis*, the best known seymouriamorph (Vertebrata: Seymouriamorpha). *Paleobios*, **17**, 1–16.

—— GIRONDOT, M. and LOTH, M.-M. 2004. The evolution of long bone microstructure and lifestyle in lissamphibians. *Paleobiology*, **30**, 589–613.

MOULTON, J. M. 1974. A description of the vertebral column of *Eryops* based on the notes and drawings of A. S. Romer. *Breviora*, **428**, 1–44.

NOVIKOV, I. V., SHISHKIN, M. A. and GOLUBEV, V. K. 2000. Permian and Triassic anthracosaurs from Eastern Europe. 60–70. *In* BENTON, M. J., KUROCHKIN, E. N., SHISHKIN, M. A. and UNWIN, D. M. (eds). *The age of dinosaurs in Russia and Mongolia*. Cambridge University Press, Cambridge, 696 pp.

PANCHEN, A. L. 1966. The axial skeleton of the labyrinthodont *Eogyrinus attheyi*. *Journal of Zoology*, **150**, 199–222.

—— 1972. The skull and skeleton of *Eogyrinus attheyi* Watson (Amphibia: Labyrinthodontia). *Philosophical Transactions of the Royal Society of London B*, **263**, 279–326.

PAWLEY, K. 2007. The postcranial skeleton of *Trimerorhachis insignis* Cope, 1878 (Temnospondyli: Trimerorhachidae): a plesiomorphic temnospondyl from the Lower Permian of North America. *Journal of Paleontology*, **81**, 873–894.

RAYFIELD, E. J., NORMAN, D. B., HORNER, C. C., HORNER, J. R., SMITH, P. M., THOMASON, J. J. and UPCHURCH, P. 2001. Cranial design and function in a large theropod dinosaur. *Nature*, **409**, 1033–1037.

ROMER, A. S. 1957. The appendicular skeleton of the Permian embolomerous amphibian *Archeria*. *University of Michigan Contributions from the Museum of Paleontology*, **13**, 103–159.

RUTA, M. and CLACK, J. A. 2006. A review of *Silvanerpeton miripedes*, a stem amniote from the Lower Carboniferous of East Kirkton, West Lothian, Scotland. *Transactions of the Royal Society of Edinburgh: Earth Sciences*, **97**, 31–63.

—— COATES, M. I. and QUICKE, D. L. J. 2003. Early tetrapod relationships revisited. *Biological Reviews*, **78**, 251–345.

SALISBURY, S. W. and FREY, E. 2000. A biomechanical transformational model for the evolution of semi-spheroidal articulations between adjoining vertebral bodies in crocodilians. 85–134. *In* GRIGG, G. C., SEEBACHER, F. and FRANKLIN, C. E. (eds). *Crocodilian biology and evolution*. Surrey Beatty and Sons, Chipping Norton, NSW, 446 pp.

SCHOCH, R. R. 1999. Comparative osteology of *Mastodonsaurus giganteus* (Jaeger, 1828) from the Middle Triassic (Lettenkeurper: Longobardian) of Germany (Baden-Württemberg, Bayern, Thüringen). *Stuttgarter Beiträge zur Naturkunde B*, **278**, 1–175.

SMITHSON, T. R. 1985. The morphology and relationships of the Carboniferous amphibian *Eoherpeton watsoni* Panchen. *Zoological Journal of the Linnean Society*, **85**, 317–410.

—— 1986. A new anthracosaur from the Carboniferous of Scotland. *Palaeontology*, **29**, 603–628.

TATARINOV, L. P. 1972. Seymouriamorphe aus der Fauna der UdSSR. 70–80. *In* KUHN, O. (ed.). *Handbuch der Paläoherpetologie, Part 5B*. Gustav Fischer, Stuttgart, 80 pp.

TVERDOCHLEBOVA, G. I. 1972. A new batrachosaur genus from the Upper Permian of the south Urals. *Paleontological Journal*, **1**, 95–103.

VJUSHKOV, B. P. 1957. New unusual animals from the Tatarian deposits of the European part of the USSR. *Doklady Academia Nauk SSSR*, **113**, 183–186.

WARREN, A. A. 2007. New data on *Ossinodus pueri*, a stem tetrapod from the Early Carboniferous of Australia. *Journal of Vertebrate Paleontology*, **27**, 850–862.

WITZMANN, F., SCHOCH, R. R. and MAISCH, M. W. 2008. A relict basal tetrapod from Germany: first evidence of a Triassic chroniosuchian outside Russia. *Naturwissenschaften*, **95**, 67–72.

YOUNG, C. C. 1979. A new Late Permian fauna from Jiyuan, Honan. *Vertebrata Palasiatica*, **17**, 99–113. [In Chinese].

APPENDIX

Characters 1–238 are taken from Ruta *et al.* (2003) and Ruta and Clack (2006); character abbreviations follow those authors. Characters 239–309 are taken from Clack and Finney (2005) and Ruta and Clack (2006). Characters 310–327 are newly created for this analysis.

Character list

1. Absence (0) or presence (1) of alary process. PREMAX 1.
2. Alary process less than (0) or at least one-third as wide as premaxillae (1). PREMAX 3.

3. Premaxillae more (0) or less than (1) two-thirds as wide as skull. PREMAX 7.

4. Absence (0) or presence (1) of shelf-like premaxilla-maxilla contact mesial to tooth row on palate. PREMAX 9.

5. Presence (0) or absence (1) of anterior tectal. TEC 1.

6. Presence (0) or absence (1) of lateral rostral. LAT ROS 1.

7. Absence (0) or presence (1) of septomaxilla. SPTMAX 1.

8. Septomaxilla not a detached ossification inside nostril (0); detached (1). SPTMAX 2.

9. Absence (0) or presence (1) of paired dorsal nasals. NAS 1.

10. Nasals more (0) or less than (1) one-third as long as frontals. NAS 2.

11. Absence (0) or presence (1) of condition: nasals broad plates delimiting most of nostril posterodorsal and mesial margins, with lateral margins diverging abruptly anteriorly. NAS 5.

12. Parietal-nasal length ratio less than (0) or greater than 1.45 (1). NAS 6.

13. Prefrontal less than (0) or more than (1) three times longer than wide. PREFRO 2.

14. Prefrontal not sutured with premaxilla (0) or sutured (1). PREFRO 6.

15. Prefrontal without (0) or with (1) stout, lateral outgrowth. PREFRO 7.

16. Absence (0) or presence (1) of condition: prefrontal entering nostril margin. PREFRO 8.

17. Prefrontal not sutured with maxilla (0) or sutured (1). PREFRO 9.

18. Lacrimal without (0) or with (1) dorsomesial digitiform process. LAC 4.

19. Lacrimal without (0) or with (1) V-shaped emargination along posterior margin. LAC 5.

20. Absence (0) or presence (1) of condition: portion of lacrimal lying anteroventral to orbit abbreviated. LAC 6.

21. Total length of lacrimal less than (0) or more than (1) two and a quarter times its maximum preorbital depth. LAC 7.

22. Maxilla extending behind level of posterior margin of orbit (0); terminating anterior to it (1) MAX 3.

23. Maxilla not entering (0) or entering (1) orbit margin. MAX 5.

24. Frontal unpaired (0) or paired (1). FRO 1.

25. Frontal shorter than (0), longer than (1), or subequal to (2) parietals. FRO 2.

26. Absence (0) or presence (1) of condition: anterior margin of frontals deeply wedged between posterolateral margins of nasals. FRO 6.

27. Absence (0) or presence (1) of parietal-tabular suture. PAR 1.

28. Absence (0) or presence (1) of parietal-postorbital suture. PAR 2.

29. Anterior margin of parietal lying in front of (0), level with (1), or behind (2) orbit midlength. PAR 4.

30. Anteriormost third of parietals not wider than frontals (0); at least marginally wider (1). PAR 5.

31. Parietals more than two and a half times as long as wide (0) or less (1). PAR 6.

32. Parietal-frontal suture not strongly interdigitating (0); strongly interdigitating (1). PAR 8.

33. Parietal-postparietal suture not strongly interdigitating (0); strongly interdigitating (1). PAR 9.

34. Postparietals paired (0) or unpaired (1). POSPAR 2.

35. Postparietal less than (0) or more than (1) four times wider than long. POSPAR 3.

36. Postparietals without (0) or with (1) median lappets. POSPAR 4.

37. Absence (0) or presence (1) of postparietal-exoccipital suture. POSPAR 5.

38. Nasals not smaller than postparietals (0) or smaller (1). POSPAR 10

39. Postfrontal not contacting tabular (0) or contacting it (1). POSFRO 3.

40. Absence (0) or presence (1) of condition: posterior margin of postfrontal lying flush with posterior jugal margin. POSFRO 4.

41. Intertemporal present (0) or absent (1) as a separate ossification. INTEMP 1.

42. Intertemporal not interdigitating with cheek (0) or interdigitating (1). INTEMP 2.

43. Intertemporal not contacting squamosal (0) or contacting it (1). INTEMP 3.

44. Absence (0) or presence (1) of condition: intertemporal shaped like a small, subquadrangular bone, less than half as broad as the supratemporal. INTEMP 4.

45. Presence (0) or absence (1) of supratemporal. SUTEMP 1.

46. Absence (0) or presence (1) of condition: supratemporal forming anterior edge of temporal notch. SUTEMP 2.

47. Absence (0) or presence (1) of condition: supratemoral narrow and strap-like, at least three times as long as wide. SUTEMP 3.

48. Supratemoral contact with squamosal smooth (0) or interdigitating (1). SUTEMP 4.

49. Tabular present (0) or absent (1) as separate ossification. TAB 1.

50. Absence (0) or presence (1) of rounded, subdermal, button-like posterior process of tabular. TAB 3.

51. Tabular contacts squamosal on dorsal surface of skull table (0) or not (1). TAB 4.

52. Tabular contact with squamosal smooth (0) or interdigitating (1). TAB 5.

53. Parietal-parietal width smaller than (0) or greater than (1) distance between posterior margin of skull table and posterior margin of orbits measured along the skull midline. TAB 7.

54. Postorbital without (0) or with (1) ventrolateral digitiform process fitting into deep vertical jugal groove. POSORB 2.

55. Postorbital contributing to (0) or excluded from (1) margin of orbit. POSORB 3.

56. Postorbital irregularly polygonal (0) or broadly crescentic and narrowing to a posterior point (1). POSORB 4.

57. Postorbital not contacting tabular (0) or contacting it (1). POSORB 5.

58. Postorbital not wider than orbit (0) or wider (1). POSORB 6.

59. Absence (0) or presence (1) of condition: postorbital at least one-fourth the width of the skull table at the same transverse level. POSORB 7.

60. Anteriormost part of dorsal margin of postorbital with sigmoid profile absent (0) or present (1). POSORB 8.

61. Anterior part of squamosal lying behind (0) or in front (1) of parietal midlength. SQU 1.
62. Squamosal without (0) or with (1) broad, concave semi-circular embayment. SQU 3.
63. Absence (0) or presence (1) of 'squamosotabular' in place of squamosal and tabular. SQU 4.
64. Jugal not contributing (0) or contributing (1) to ventral margin of skull roof. JUG 2.
65. Jugal not contacting (0) or contacting (1) pterygoid. JUG 3.
66. Jugal depth below orbit greater (0) or smaller (1) than half orbit diameter. JUG 4.
67. Jugal without (0) or with (1) V-shaped indentation of dorsal margin. JUG 7.
68. Jugal not extending (0) or extending (1) anterior to anterior orbit margin. JUG 8.
69. Absence (0) or presence (1) of condition: quadratojugal much smaller than squamosal. QUAJUG 2.
70. Absence (0) or presence (1) of condition: quadratojugal an anteroposteriorly elongate and dorsoventrally narrow splinter of bone. QUAJUG 3.
71. Quadrate without (0) or with (1) dorsal process. QUA 1.
72. Absence (1) or presence (0) of preopercular. PREOPE 1.
73. Absence (0) or presence (1) of internarial fenestra. INT FEN 1.
74. Interorbital distance greater than (0), smaller than (1), or subequal to (2) half skull table width. ORB 1.
75. Interorbital distance greater than (0), smaller than (1) or subequal to (2) maximum orbit diameter. ORB 2.
76. Absence (0) or presence (1) of angle at anteroventral orbit corner. ORB 3.
77. Absence (0) or presence (1) of condition: in lateral view, orbit deeper than long. ORB 4.
78. Position of pineal foramen behind (0), at the level of (1) or anterior to (2) interparietal suture mid length. PIN FOR 2.
79. Fossa at dorsolateral corner of occiput, not bordered laterally, roofed over by skull table and floored by dorsolateral extension of opisthotic (0); fossa near dorsolateral corner of occiput, roofed over by occipital flanges of tabular and postparietal and bordered laterally and ventrally by dorsolateral extension of opisthotic meeting ventromedial flange of tabular (1); small fossa near ventrolateral corner of occiput bordered laterally by ventromedial flange of tabular, roofed over by dorsal portion of lateral margin of supraoccipital-opisthotic complex and floored by lateral extension of opisthotic (2); absence of fossa (3). PTF 1.
80. Absence (0) or presence (1) of abbreviated skull roof postorbital region. SKU TAB 1.
81. Lateral line system on skull roof totally enclosed (0), mostly enclosed with short sections in grooves (1), mostly in grooves with short sections enclosed (2), entirely in grooves (3), absent (4). SC 1.
82. Mandibular canal totally enclosed (0), mostly enclosed, short sections in grooves (1), mostly in grooves, short sections enclosed (2), entirely in grooves (3), absent (4). SC 2.
83. Absence (0) or presence (1) of condition: ventral, exposed surface of vomers (i.e. excluding areas of overlap with surrounding bones) narrow, elongate, and strip-like, without extensions anterolateral or posterolateral to choana and two and a half to three times longer than wide. VOM 1.
84. Vomer without (0) or with (1) denticles. VOM 4.
85. Vomer excluded from (0) or contributing to (1) interpterygoid vacuities. VOM 5.
86. Vomer not forming (0) or forming (1) suture with maxilla anterior to choana. VOM 7.
87. Vomer with (0) or without (1) toothed lateral crest. VOM 8.
88. Vomer with (0) or without (1) anterior crest. VOM 9.
89. Palatine with (0) or without (1) fangs. PAL 1.
90. Palatine without (0) or with (1) denticles. PAL 2.
91. Palatine with (0) or without (1) tooth row (3 or more teeth). PAL 4.
92. Ectopterygoid with (0) or without (1) fangs. ECT 2.
93. Ectopterygoid without (0) or with (1) denticles. ECT 3.
94. Ectopterygoid longer than/as long as palatines (0) or not (1). ECT 4.
95. Ectopterygoid with (0) or without (1) tooth row (3 or more teeth). ECT 5.
96. Absence (0) or presence (1) of pterygoid flange oriented transversely. PTE 3.
97. Pterygoid without (0) or with (1) posterolateral flange. PTE 9.
98. Pterygoids not sutured with each other (0) or sutured (1). PTE 10.
99. Pterygoid without (0) or with (1) distinct, mesially directed process for the basipterygoid recess. PTE 13.
100. Palatal ramus of pterygoid without (0) or with (1) distinct, anterior, unornamented digitiform process. PTE 16.
101. Presence (0) or absence (1) of interpterygoid vacuities. INT VAC 1.
102. Absence (0) or presence (1) of condition: interpterygoid vacuities occupying at least half of palatal width. INT VAC 2.
103. Absence (0) or presence (1) of condition: interpterygoid vacuities concave along their whole margins. INT VAC 3.
104. Absence (0) or presence (1) of condition: interpterygoid vacuities together broader than long. INT VAC 4.
105. Presence (0) or absence (1) of anterior palatal vacuity. ANT VAC 1.
106. Anterior palatal vacuity single (0) or double (1). ANT VAC 2.
107. Supraoccipital absent (0) or present (1) as separate ossification. SUPOCC 1.
108. Absence (0) or presence (1) of condition: exoccipitals enlarged, about as broad as high and forming stout, double occipital condyles. EXOCC 3.
109. Absence (0) or presence (1) of condition: exoccipitals forming continuous, concave, strap-shaped articular surfaces with basioccipital. EXOCC 4.
110. Basioccipital notochordal (0) or not (1). BASOCC 1.
111. Absence (0) or presence (1) of condition: basioccipital circular and recessed. BASOCC 6.
112. Absence (0) or presence (1) of condition: opisthotic forming thick plate with supraoccipital, separating exoccipitals from skull table. OPI 2.

113. Parasphenoid without (0) or with (1) elongate, strut-like cultriform process. PASPHE 1.
114. Parasphenoid without (0) or with (1) posterolaterally directed, ventral thickenings (ridges ending in basal tubera). PASPHE 3.
115. Parasphenoid without (0) or with (1) single median depression. PASPHE 6.
116. Parasphenoid without (0) or with (1) paired lateral depressions. PASPHE 7.
117. Ventral cranial fissure not sutured (0), sutured but traceable (1), absent (2). PASPHE 9.
118. Parasphenoid without (0) or with (1) triangular denticle patch with raised margins at base of cultriform process. PASPHE 12.
119. Jaw articulation lying behind (0), level with (1) or anterior to (2) occiput. JAW ART 1.
120. Presence (0) or absence (1) of parasymphysial plate. PSYM 1.
121. Parasymphysial plate without (0) or with (1) paired fangs. PSYM 2.
122. Parasymphysial plate without (0) or with (1) tooth row. PSYM 3.
123. Parasymphysial plate with (0) or without (1) denticles. PSYM 4.
124. Dentary with (0) or without (1) anterior fang pair. DEN 2.
125. Dentary with (1) or without (0) chamfered ventral margin. DEN 3.
126. Dentary without (0) or with (1) U-shaped notch for premaxillary tusks. DEN 4.
127. Absence (0) or presence (1) of condition: rearmost extension of mesial lamina of splenial closer to anterior margin of adductor fossa than to anterior end of jaw. SPL 2.
128. Absence (0) or presence (1) of suture between splenial and anterior coronoid. SPL 3.
129. Absence (0) or presence (1) of suture between splenial and middle coronoid. SPL 4.
130. Postsplenial without (0) or with (1) mesial lamina. POSPL 2.
131. Postsplenial with (0) without (1) pit line. POSPL 3.
132. Angular without (0) or with (1) mesial lamina. ANG 2.
133. Angular contacting prearticular (0) or not (1). ANG 3.
134. Angular not reaching (0) or reaching (1) posterior end of lower jaw. ANG 4.
135. Surangular with (0) or without (1) pit line. SURANG 3.
136. Prearticular sutured with splenial (0) or not (1). PREART 5.
137. Anterior coronoid with (0) or without (1) fangs. ANT COR 2.
138. Anterior coronoid with (0) or without (1) denticles. ANT COR 3.
139. Anterior coronoid with (0) or without (1) tooth row. ANT COR 4.
140. Middle coronoid present (0) or absent (1). MID COR 1.
141. Middle coronoid with (0) or without (1) fangs. MID COR 2.
142. Middle coronoid with (0) or without (1) denticles. MID COR 3.
143. Middle coronoid with (0) or without (1) marginal tooth row. MID COR 4.
144. Posterior coronoid with (0) or without (1) fangs. POST COR 2.
145. Posterior coronoid with (0) or without (1) denticles. POST COR 3.
146. Posterior coronoid with (0) or without (1) tooth row. POST COR 4.
147. Posterior coronoid without (0) or with (1) posterodorsal process. POST COR 5.
148. Posterior coronoid not exposed (0) or exposed (1) in lateral view. POST COR 6.
149. Posterodorsal process of posterior coronoid not contributing (0) or contributing (1) to tallest point of lateral margin of adductor fossa ('surangular' crest). POST COR 7.
150. Adductor fossa facing dorsally (0) or mesially (1). ADD FOS 1.
151. Dentary teeth not larger (0) or larger (1) than maxillary teeth. TEETH 5.
152. Marginal tooth crowns not chisel-tipped (0) or chisel-tipped (1). TEETH 6.
153. Marginal tooth crowns without (0) or with (1) dimple. TEETH 7.
154. Cleithrum with (0) or without (1) postbranchial lamina. CLE 2.
155. Cleithrum co-ossified with (0) or separate from (1) scapulocoracoid. CLE 3.
156. Clavicles meet anteriorly (0) or not (1). CLA 3.
157. Absence (0) or presence (1) of condition: posterior margin of interclavicle drawn out into parasternal process. INTCLA 1.
158. Absence (0) or presence (1) of condition: parasternal process elongate and parallel-sided for most of its length. INTCLA 2.
159. Absence (0) or presence (1) of condition: interclavicle wider than long. INTCLA 3.
160. Interclavicle rhomboidal with posterior half longer (0) or shorter (1) than anterior half. INTCLA 4.
161. Absence (0) or presence (1) of separate scapular ossifications. SCACOR 1.
162. Glenoid subterminal (0) or not (1) (scapulocoracoid extending ventral to posteroventral margin of glenoid). SCACOR 2.
163. Presence (0) or absence (1) of enlarged glenoid foramen. SCACOR 3.
164. Absence (0) or presence (1) of ventromesially extended infraglenoid buttress. SCACOR 4.
165. Presence (0) or absence (1) of anocleithrum. ANOCLE 1.
166. Latissimus dorsi process offset anteriorly (0) or aligned with ectepicondyle (1). HUM 1.
167. Absence (0) or presence (1) of distinct supinator process projecting anteriorly. HUM 2.
168. Presence (0) or absence (1) of ventral humeral ridge. HUM 3.
169. Latissimus dorsi process confluent with (0) or distinct from (1) deltopectoral crest. HUM 4.
170. Presence (0) or absence (1) of entepicondylar foramen. HUM 5.
171. Presence (0) or absence (1) of ectepicondylar foramen. HUM 6.

172. Presence (0) or absence (1) of distinct cctepicondyle. HUM 7.

173. Absence (0) or presence (1) of condition: ectepicondylar ridge extending distally to reach distal humeral end. HUM 8.

174. Distal extremity of ectepicondylar ridge aligned with ulnar condyle (0), between ulnar and radial condyles (1), or aligned with radial condyle (2). HUM 9.

175. Humerus without (0) or with (1) expanded extremities (waisted). HUM 10.

176. Radial condyle terminal (0) or ventral (1). HUM 11.

177. Posterolateral margin of entepicondyle lying distal with respect to plane of radial-ulnar facets (0) or not (1). HUM 13.

178. Posterolateral margin of entepicondyle markedly concave (0) or not (1). HUM 14.

179. Width of entepicondyle greater (0) or smaller (1) than half humeral length. HUM 15.

180. Portion of humeral shaft length proximal to entepicondyle smaller (0) or greater (1) than humeral head width. HUM 16.

181. Presence (0) or absence (1) of accessory foramina on humerus. HUM 17.

182. Humerus length greater (0) or smaller (1) than the length of two and a half mid-trunk vertebrae. HUM 18.

183. Radius longer (0) or shorter (1) than humerus. RAD 1.

184. Radius longer than (0), as long as (1), or shorter than (2) ulna. RAD 2.

185. Absence (0) or presence (1) of olecranon process. ULNA 1.

186. Absence (0) or presence (1) of dorsal iliac process. ILI 3.

187. Supraacetabular iliac buttress less (0) or more (1) prominent than postacetabular buttress. ILI 6.

188. Absence (0) or presence (1) of transverse pelvic ridge. ILI 7.

189. Acetabulum directed posteriorly (0) or laterally (1). ILI 10.

190. Ischium not contributing (0) or contributing (1) to pelvic symphysis. ISC 1.

191. Absence (0) or presence (1) of distinct process on internal trochanter. FEM 1.

192. Absence (0) or presence (1) of condition: internal trochanter separated from femur by distinct trough-like space. FEM 2.

193. Absence (0) or presence (1) of distinct rugose area on fourth trochanter. FEM 3.

194. Proximal end of adductor crest of femur not reaching (0) or reaching (1) midshaft length. FEM 4.

195. Femur shorter than (0), as long as (1), or longer than humerus (2). FEM 5.

196. Without (0) or with (1) flange on posterior edge. TIB 7.

197. Fibula not waisted (0) or waisted (1). FIB 1.

198. Absence (0) or presence (1) of ridge near posterior edge of fibula flexor surface. FIB 3.

199. Absence (0) or presence (1) of rows of tubercles near posterior edge of flexor surface of fibula. FIB 4.

200. Absence (0) or presence (1) of ossified tarsus. TAR 1.

201. Absence (0) or presence (1) of one proximal tarsal ossification, or presence of more than two ossifications (2). TAR 2.

202. Tarsus without (0) or with (1) L-shaped proximal tarsal element. TAR 3.

203. Absence (0) or presence (1) of distal tarsals between fibulare and digits. TAR 4.

204. Absence (0) or presence (1) of distal tarsals between tibiale and digits. TAR 5.

205. Cervical ribs with (0) or without (1) flattened distal ends. RIB 2.

206. Ribs mostly straight (0) or ventrally curved (1) in at least part of the trunk. RIB 3.

207. Absence (0) or presence (1) of triangular spur-like posterodorsal process in at least some trunk ribs. RIB 5.

208. Absence (0) or presence (1) of condition: elongate posterodorsal flange in mid-trunk ribs. RIB 6.

209. Axial arch not fused (0) or fused (1) to axial (pleuro)centrum. CER VER 3.

210. Absence (0) or presence (1) of extra articulations above zygapophyses in at least some trunk and caudal vertebrae. TRU VER 1.

211. Absence (0) or presence (1) of condition: neural and haemal spines rectangular to fan-shaped in lateral view. TRU VER 2.

212. Absence (0) or presence (1) of condition: neural and haemal spines facing each other dorsoventrally. TRU VER 3.

213. Haemal spines not fused (0) or fused (1) to caudal centra. TRU VER 4.

214. Absence (0) or presence (1) of extra articulations on haemal spines. TRU VER 5.

215. Absence (0) or presence (1) of ossified pleurocentra. TRU VER 7.

216. Trunk pleurocentra not fused midventrally (0) or fused (1). TRU VER 8.

217. Trunk pleurocentra not fused middorsally (0) or fused (1). TRU VER 9

218. Neural spines without (0) or with (1) distinct convex lateral surfaces. TRU VER 10.

219. Neural spines of trunk vertebrae not fused to centra (0) or fused (1). TRU VER 11.

220. Presence (0) or absence (1) of trunk intercentra. TRU VER 13.

221. Trunk intercentra not fused middorsally (0) or fused (1). TRU VER 14.

222. Absence (0) or presence (1) of lateral and ventral carinae on trunk centra. TRU VER 15.

223. Absence (0) or presence (1) of condition: tallest ossified part of neural arch in posterior trunk vertebrae lying above posterior half of vertebral centrum. TRU VER 19.

224. Absence (0) or presence (1) of of prezygapophyses on trunk vertebrae. TRU VER 20.

225. Absence (0) or presence (1) of postzygapophyses on trunk vertebrae. TRU VER 21.

226. Absence (0) or presence (1) of prezygapophyses on proximal tail vertebrae. TRU VER 22.

227. Absence (0) or presence (1) of postzygapophyses on proximal tail vertebrae. TRU VER 23.

228. Absence (0) or presence (1) of prezygapophyses on distal tail vertebrae. TRU VER 24.

229. Absence (0) or presence (1) of postzygapophyses on distal tail vertebrae. TRU VER 25.
230. Absence (0) or presence (1) of capitular facets on posterior rim of vertebral mid-trunk centra. TRU VER 26.
231. Height of neural arch in mid-trunk vertebrae greater (0) or smaller (1) than distance between pre- and postzygapophyses. TRU VER 27.
232. Absence (0) or presence (1) of digits. DIG 1.
233. Absence (0) or presence (1) of no more than four digits in manus. DIG 2.
234. Absence (0) or presence (1) of no more than five digits in manus. DIG 3.
235. Absence (0) or presence (1) of no more than three digits in manus. DIG 4.
236. Presence (0) or absence (1) of dorsal fin. DOR FIN 1.
237. Presence (0) or absence (1) of caudal fin. CAU FIN 1.
238. Presence (0) or absence (1) of basal scutes. BAS SCU 1.
239. Anterior tectal: narial opening ventral to it (0); narial opening anterior to it (1).
240. Basioccipital: indistinguishable from exoccipitals (0); separated by suture (1).
241. Basioccipital: ventrally exposed portion longer than wide (0); shorter than wide (1).
242. Lacrimal contributes to narial margin: no, excluded by anterior tectal (0); yes (1); no, excluded by nasal/maxillary or prefrontal/maxillary suture (2).
243. Maxilla external contact with premaxilla: narrow contact point not interdigitated (0); interdigitating suture (1).
244. Median rostral (= internasal): mosaic (0); paired (1); single (2); absent (3).
245. Nasals contribute to narial margin: no (0); yes (1).
246. Opisthotic paroccipital process ossified and contacts tabular below posttemporal fossa: no (0); yes (1); posttemporal fossa absent (2).
247. Postparietal occipital flange exposure: absent (0); present (1).
248. Prefrontal-postfrontal suture: anterior half of orbit (0); middle or posterior half of orbit (1); absent (2).
249. Premaxilla forms part of choanal margin: broadly (0); point (1); excluded by vomer (2).
250. Squamosal suture with supratemporal position: at apex of temporal embayment (0); dorsal to apex (1); ventral to apex (2).
251. Tabular emarginated lateral margin: no (0); yes (1).
252. Tabular facets/buttresses for braincase ventrally: no (0); single (1); double (2).
253. Tabular occipital flange exposure: absent (0); extends as far ventrally as does postparietal (1); extends further ventrally than does postparietal (2).
254. Ectopterygoid reaches adductor fossa: no (0); yes (1).
255. Palatine-ectopterygoid exposure: more or less confined to tooth row (0); broad mesial exposure additional to tooth row (1).
256. Pterygoids flank parasphenoid: for most of length of cultriform process (0); not so (1).
257. Pterygoid junction with squamosal along cheek margin: unsutured (0); half and half (1); sutured entirely (2).

258. Parasphenoid wings: separate (0); joined by web of bone (1).
259. Parasphenoid sutures to vomers: yes (0); no (1).
260. Parasphenoid carotid grooves: curve round basipterygoid process (0); lie posteromedial to basipterygoid process (1).
261. Vomers separated by parasphenoid for more than half length: yes (0); no (1).
262. Vomers separated by pterygoids: for more than half length (0); for less than half length (1); not separated (2).
263. Ectopterygoid denticle row: present (0); absent (1).
264. Maxilla tooth number: more than 40 (0); 30–40 (1); less than 30 (2).
265. Maxillary caniniform teeth (about twice the size of neighbouring teeth): absent (0); present (1).
266. Palatine row of smaller teeth: present (0); absent (1).
267. Palatine denticle row: present (0); absent (1).
268. Parasphenoid shagreen field: present (0); absent (1).
269. Parasphenoid shagreen field location: anterior and posterior to basal articulation (0); posterior to basal articulation only (1); anterior to basal articulation only (2).
270. Pterygoid shagreen: dense (0); a few discontinuous patches or absent (1).
271. Prearticular denticulated field: defined edges (0); scattered patches (1); absent (2).
272. Premaxillary teeth with conspicuous peak: absent (0); present (1).
273. Vomer fang pairs: present (0); absent (1).
274. Vomerine fang pairs noticeably smaller than other palatal fang pairs: no (0); yes (1).
275. Vomerine row of small teeth: present (0); absent (1).
276. Vomerine denticle row lateral to tooth row: present (0); absent (1).
277. Meckelian bone visible between prearticular and infradentaries: present (0); absent (1).
278. Naris position: ventral rim closer to jaw margin than height of naris (0); distance to jaw margin similar to or greater than height of naris (1).
279. Naris shape: slit-like (0); round or oval (1); upper margin ragged (2).
280. Naris orientation: ventrally facing (0); dorsolaterally facing (1).
281. Naris size relative to choana: less than 50 per cent (0); same or larger (1).
282. Suspensorium proportions: distance from quadrate to temporal embayment anterior margin about equal to maximum orbit width (discounting any anterior extensions) (0); distance less than maximum orbit width (1); distance more than maximum orbit width (2).
283. Ornament character: fairly regular pit and ridge with starburst pattern at regions of growth (0); irregular but deep (1); irregular but shallow (2); absent or almost absent (3).
284. Centra: rhachitomous (0); gastrocentrous (1); holospondylous (2).
285. Centrum (sacral): distinguishable by size and shape from pre- and postsacrals (1); not so distinguishable (0).
286. Cleithrum dorsal end: smoothly broadening to spatulate dorsal end (0); distal expansion marked from narrow stem

by notch or process or decrease in thickness (1); tapering (2).

287. Cleithrum stem cross section at mid section: flattened oval (0); complex (1); single concave face (2).

288. Femur adductor blade: distinguished distally from shaft by angle or notch (0); fades into shaft distally (1).

289. Humerus shape: ends more or less in line, little torsion apparent (0); ends offset by more than 60 degrees (1).

290. Humerus latissimus dorsi process: part of ridge (0); distinct but low process (1); spike (2).

291. Humerus anterior margin: smooth finished bone convex margin (0); anterior keel with finished margin (1); cartilage-finished (2); smooth concave margin (3).

292. Humerus radial and ulnar facets: confluent (0); separated by perichondral strip (1).

293. Neural arch ossification: paired in adult (0); single in adult (1).

294. Neural arch (sacral): distinguishable by spine morphology (1); not so distinguishable (0).

295. Pelvis: single ossification (0); at least two ossifications per side (1).

296. Pelvis obturator foramina: multiple (0); single or absent (1).

297. Ribs (trunk): no longer than height of neural arch plus centrum (0); less than two and a half times this height (1); more than two and a half times this height (2).

298. Ribs (trunk): tapered distally or parallel-sided (0); expanded distally into overlapping posterior flanges (1).

299. Ribs (trunk) bear proximodorsal (uncinate) processes: absent (0); present (1).

300. Ribs (trunk) differ strongly in morphology in 'thoracic' region: absent (0); present (1).

301. Rib (sacral) distinguishable by size: shorter than trunk ribs, longer than presacrals (1); same length as presacrals (0).

302. Rib (sacral) distinguishable by shape: broader than immediate presacrals but not broader than mid-trunk proximal shafts (0); broader than mid-trunk proximal shafts (1).

303. Scapulocoracoid dorsal blade: absent (0); present (1).

304. Scutes: tapered and elongate, four times or greater than four times longer than broad (0); ovoid, no more than three times longer than broad (1).

305. Tibia and fibula width at narrowest point: 50 per cent of length (0); less than 30 per cent of length (1).

306. Tibia and fibula meeting along their length (0); separated by interepipodial space (1).

307. Number of pes digits: more than five (0); five (1); fewer than five (2).

308. Posterior process of ilium a slender, subhorizontal rod, with parallel dorsal and ventral margins, more than five times longer than deep: absent (0); present (1).

309. Process '2' of humerus: absent (0); present (1).

310. Basisphenoid: very narrow between basipterygoid processes, latter separated by about width of one process or less (1); basipterygoid processes set further apart than width of one process (0).

311. Basisphenoid in dorsal view: smooth concave retractor pit between basipterygoid processes (1); retractor pit paired or absent (0).

312. Basisphenoid in dorsal view: robust, broad bases for processi sellares, latter form backwall to retractor pit (1); processi sellares not forming stout buttresses (narrow wall or absent) (0).

313. Basipterygoid processes: processes rounded in ventral view, but not hemispherical (0); conspicuously bulbous, rounded posteriorly, nearly hemispherical (1); essentially simple triangular extensions from parasphenoid with little three dimensional morphology in ventral view (2).

314. Basipterygoid processes: faces of processes more or less concave throughout (0); faces reverse toroidal (ie convex centrally, concave laterally) (1).

315. Parasphenoid: forms narrow crest between basipterygoid processes (1); area between basipterygoid processes not a crest, but flat or depressed (0).

316. Parasphenoid cultriform process: flat, depressed, or rounded in section anteriorly (0); cultriform process a sharp keel, V-shaped in cross section from basiptergoid processes anteriorly (1).

317. Centra trunk: ossified portions of pleuocentra and intercentra differ in height by more than 25% (0); pleurocentra and intercentra about equal in height (1).

318. Ribs postsacral: elongate curved, at least three pairs (1); such pairs fewer or absent (0).

319. LAC 2 restated: lacrimal enters orbit margin (0); lacrimal excluded from orbit margin, with jugal/prefrontal contact (1).

320. Postorbital: with ventral component an acutely triangular extension forming posterior margin of orbit: absent (0); present (1).

321. Postorbital/intertemporal contact: interdigitating (0); or noninterdigitating (1).

322. Postorbital/squamosal contact: interdigitating (0); or smooth and noninterdigitating (1).

323. Postparietal margin: produced into posterior peak at midline ('widow's peak): absent = 0, present = 1.

324. Supratemporal margin: contributes to posterior emargination of cheek (temporal notch): absent (0); present (1).

325. Supratemporal/squamosal contact: interdigitating (0); noninterdigitating butt joint (1).

326. TAB 2 restated: Tabular horn projection: absent (0); elongate blade-like unornamented (1); blade-like but ornamented (2).

327. Supratemporal/intertemporal margin: irregular (0); smoothly convex (1).

Data matrix

Missing data are coded as ? Polymorphisms for states 0 and 1 are coded as A, for states 1 and 2 as B, for states 2 and 3 as C. D represents polymorphisms for states 0, 1 and 2.

Acanthostega

0? 00010? 10	0000000000	0001200111	1000000000	1??? 000100	0000000110	0000000100
0011100000	1100000000	0100000100	0000010000	0000001000	1100100000	1010101100
11011000? 0	0000011000	0010000000	0010001100	0110010011	1110100101	0000000000
0000100000	0001111000	0100010100	0001000010	0001000? 10	10000001? 0	0100001000
0200? 00001	0100000000	1000000010	?? 00000? 10	? 000010		

Adelogyrinus

0? 0? 11?? 11	01000? 1001	0011? 1? 1? 1	011000? 1? 0	1??? 1??? 1?	?? 0010? 000	? 01?? 10000
01? 12000? 0	? 3????????	??????????	??????? 001	1??????? 1?	??? 100??? 0	1?? 11?????
??????????	01111? 0? 01	???? 1?????	??????????	??????????	??????????	????? 111? 0
???? 111? 01	? 1? 11???? 1	0???? 1?????	? 22? 31? 11??	??? 110011?	?? 11011?? 0	?? 1? 0?????
? 1120?????	?? 10?? 0010	??? 0??????	?????? 00??	???????		

Anthracosaurus

0? 10110? 10	0010000000	1001101021	1000010000	0100000100	1? 10000000	1000011100
0101100231	B410? 01100	1000000100	00001? 0001	010110200?	??? 100? 10?	1? 0? 1? 1???
??????????	000??????	??????????	??????????	??????????	??????????	??????????
??????????	??????????	????????? 1	1113121001	122100? 11?	1? 120? 1?? 1	201? 11? 021
122???????	??????????	????????? 1	???????? 00	0011010		

Archeria

0? 10110? 10	0010000000	1101101021	1010010000	0000000000	1? 00010000	1001110100
0101100230	33??????	?? 0? 000100	0000?? 0001	0001102001	0011001111	1101101010
1011011101	0101111100	0001110010	1012001101	1112111111	1011201101	2111010000
0000111000	1001111110	01010111? 1	11031210? 1	1? 2000? 1? 0	?? 100? 11? 0	21???? 1021
? 0A1112101	1010112000	111? 111011	1111111111	1? 11111		

Ariekanerpeton

0? 10111110	0000000000	0101201021	100010? 000	0010000000	1? 10010010	1001010000
? 102000211	3C11001101	1110111? 00	A0001?? 00?	0? 01002011	??? 100? 1? 1	11011? 1010
1011011011	0001101110	1? 0?? 0110	1012101100	11110?? 111	???? 2? 1???	???? 0100? 0
000011A100	00011???? 0	01010111? 1	11131A1101	0121102110	? 0120111? 0	2000111011
01B1? 22? 11	300? 111000	1111111000	?? 0? 010? 00	0001? 00		

Balanerpeton

110011?? 10	1000000000	0001100021	1AA000? 000	0101000100	1? 00010000	1100010000
01011002? 1	4401101101	1010101000	01111?? 001	0? 10002001	??? 0000101	1100111011
??? 101??? 1	1001110? 00	0??? 1? 01? 0	11?? 101100	1111101011	11? 0201001	2011000000
0000100000	0001111?? 0	01100111??	11? 31? 0110	1001112? 01	1210011010	200? 111111
110001? 01?	3? 101? 0000	1011111000	?? 0? 000001	0001000		

Baphetes

0? 00111010	0000100010	0001100021	1?? 0000000	0101010101	1? 00000000	1001000000
0111200010	3? 01001101	1010100100	10001? 000?	0000012000	101? 00???	1?????????
??????????	000??????	????? 00010	1012001100	1???? 11011	????? 0110?	????????? 0
0000????? 0	0??????????	??????? 11?	1101110012	1121102110	1211011000	2000111011
1200? 11? 12	20?? 1?? 1??	??? 011?? 00	?? 0001?? 00	0001000		

Bruktererpeton

0??? 11?? 10	00? 0? 00?? 0	1?? 1101021	100010? 0??	00? 000000?	?? 000?????	?? 0???????
?? 0???? 2? 1	???????? 0?	? 11???? 1??	??????????	??????????	??????????	??????????
??????????	? 0011?????	1? 01? 10110	1012011101	11121?????	1101201001	21?? 0100? 0
0000110000	0001111110	01??? 111??	? 1? 31?? 2A?	0?? 110????	12110? 1?? 0	?? 0011? 111
?? C1? B2101	0? 00?? 2000	?? 11111???	?????? 00??	?? 00101		

Caerorhachis

0??????? 10	00????????	00? 1? 00021	111000? 00?	01? 0000? 0?	0? 00010000	100? 0?? 00
? 10??? B? 0	??? 10? 1101	1010100100	00001?? 00?	??? 0102000	1000001101	? 100101010
1011011001	000??????	??????????	??????????	???? 111111	11? 1? 01?? 1	2111010000
0000110000	0001111110	01??? 111??	?? 031?? 1? 0	1?? 0???? 1?	1110011010	0? 01111? 11
1201??? 0??	?? 1??? 1000	??? 01110? 0	???? 0? 01? 1	000?? 00		

Casineria
```
??????????   ??????????   ??????????   ??????????   ??????????   ??????????   ??????????
??????????   ??????????   ??????????   ??????????   ??????????   ??????????   ??????????
??????????   ???11?????   1?????0??0   1???1?1111   101??????1   ????2?????   ????0100?0
0???11?000   ???1??????   110101111?   ??????????   ??????????   ??????????   ??????????
???1?B????   3???1?1000   ???111?00?   ??????0???   ??????
```

Chroniosaurus
```
??1?11??10   0000000000   1001201021   100000?000   1???000?01   1?00010000   110011?011
?1112002?0   4?01?1?101   001011?000   1???00?001   0?0110200?   ???100?1?1   1100101110
101111???1   000?110000   ??????01?0   110?1?110?   10??0??11    0?????1??1   20???1?000
0000111000   1001111110   ?1???111?1   11131?02?0   0121?0?110   ??110111?0   ?01?111111
1011??????   301?1??000   ???1111101   1111111111   1111121
```

Crassigyrinus
```
0?01111110   0000010000   0001100021   1??001?000   00?0000001   1?00000000   1000000000
01111100?0   1300010000   0000000100   ?00001???0   0?01101000   1010001000   1100101010
101??????0   000?1?1000   ?????10010   001B001100   001001?011   111121110?   ?????100?0
00000??0?0   00?10????0   01???1?1??   11131?1120   1210001010   1011001020   1000011121
121??11101   200?1?2000   ???011?000   ??0000?100   1?01101
```

Dendrerpeton
```
1000111110   0000000000   0001000021   111010?000   0100000101   1?00010000   1100010A00
11020002?0   4?011?1101   1010101010   01101??101   0?1000210?   ????00??1    11001????0
?????????1   0001110?00   00011?01?0   1012101100   1111101111   11?1201??1   20110000?0
0000100000   00011????0   01100111?1   11131111?1   1?0?112101   1210011010   ?001111111
1000?1?01?   3?1?111000   ??11111000   0000010000   0011000
```

Discosauriscus
```
0?10111110   0000000000   0101201021   100010?000   0010000001   1?10010000   1101010000
1101100211   3311001101   1110111100   A0001??00?   0?01002011   ???1001111   1101101010
1011011111   0001101110   1001000110   1012101100   111101?111   100020100?   ????000000
0000111100   0001111110   010101?1?1   11131A1101   0121102110   ?012011A20   2000111011
0001022111   300?111000   1111111000   00??000000   0001001
```

Eoherpeton
```
0?10110?10   0000000000   1101101021   100000?000   00?0000000   1?000?0000   ?000010100
?1020?0230   441?????00   ?0?0000?00   ?000??????   ?0?110?0??   ????001111   1100101010
101101?0?1   000?1?????   0101?10010   1011001101   1????11111   ???1?0110?   ?????100?0
0000110000   0001111??0   0??????1??   11?31????0   1120???000   ???2001??0   00????1?11
1121???101   101?01200?   ??1?11?011   1110010?10   1?10101
```

Eucritta
```
0?0?11??10   0000000000   0001200021   1000?0?000   010101010?   1?00010000   000??10000
?1011100?0   ???1001101   10??00?100   0000??????   ??0110200?   ????00????   ??????????
??????????   000?11AA00   0???1?0??0   101?001100   1?1111??11   ????201??1   ????0000??
??????????   ??????????   ?1?????1??   ?1?31?01?2   100?102?1?   11?2?11000   ??00?1????
100??11?0?   2???1?1100   ??10111000   ????01??00   0001000
```

Eusthenopteron
```
0?00000?0?   ??00000000   00000?0000   0000000?01   0000000000   0000000100   0000000000
0000000200   0000000000   0000000000   0000000000   0000000010   0000000000   0000000100
0100?0??0    0000000?00   00000?0000   0000000000   0000000000   0?00000000   ?????00000
0000100000   0000000000   00??000000   00?000000?   0000000?0?   0000?00000   ?0000000??
0??0?0??0?   0?0?0?0???   ??0????000   ?00?000000   1?01?01
```

Gephyrostegus
```
0?1011??10   0000000000   1101111021   100010?000   00?000000?   1?00010000   1000010000
01011002?1   441100??01   1011110100   00001??00?   0?0???2??1   ???0001110   1000100010
0011011101   0001111100   0101110110   1012011101   1112111111   ??01201101   2111010000
0000110000   00011?????   010101111??  ?1?31?02??   1?21101???   ??10011??0   20001110?1
122111210?   001?112000   ??11111000   ??0?010?0?   1???101
```

Greererpeton
0? 01110? 11   0111011000   0001110121   0110000100   A??? 000101   0100000110   0000000000
0111000010   2200001100   0000000101   0000010001   0110012000   1010011100   1100100100
10110100? 0   1000100? 01   0001110010   1011001100   1012101011   1111201001   20? 1101100
0000100000   0001111?? 0   010101? 1? 1   1203011010   1110112110   1110001010   211? 011011
1100011001   1010111010   0111111001   0000010000   001? 000

Ichthyostega
0? 01000? 10   0000000000   0001200111   1001? 0? 000   1??? 000100   0100000000   0000000000
0011000030   0000010110   0100000100   000000? 000   0000000000   1010000100   0000101100
11011000? 0   0000011110   0101? 10000   0010011100   0112111011   1010? 10001   10001100? 0
0000100000   0001111000   01??? 1010?   01020? 102?   1? 21100? 1?   1012001021   101? 010000
0000? 00000   ? 11? 002101   ?? 00000000   ?? 0000? 010   ? 000000

Loxomma
0??? 111010   0000000010   0001200021   1? 1000? 000   0101010101   1? 00000000   000?? 00000
01? 2? 00010   3? 01001101   1010100100   ? 000?? 0???   ? 00001200?   ?????????   ? 1001?????
?????????1   ? 00???????   ?????????   ?????????   ?????????   ?????????   ?????????
?????????   ?????????   ???????? 11   11?? 110012   112? 1? 2110   ?? 1201100?   1000111011
120??? ????   ?????????   ????????? 0   ?? 0001?? 00   0001000

Megalocephalus
0? 00111010   0010100010   0001100021   1000000000   1??? 010101   ?? 00000000   1000100000
0111200010   3301001101   1010100100   0000000001   0000012000   1010001111   1100101110
11111100? 1   000???????   ?????????   ?????????   ?????????   ?????????   ?????????
?????????   ?????????   ???????? 11   12? 1110012   1? 21102110   1211011000   2000111011
120??? ????   ?????????   ????????? 1   ?? 0001?? 00   ? 001000

Microbrachis
0? 10011?? 11   0100000001   1101001121   1100101000   1??? 1??? 00   001000? 001   1001010000
010? 0002? 1   C311001111   1111101? 01   00001?? 011   0010002011   ??? 0001101   1101101010
1011011011   0001111110   1????? 01? 0   1??? 011111   1011011011   1001201001   ???? 010010
0000111001   ? 011111111   11001111? 1   10? 31? 110?   0? 11110110   101? 0111? 0   201? 111011
0? 1202? 11?   3? 101? 2000   11?? 111000   ?? 0? 01? 000   ?? 0?? 0?

Paleothyris
0? 10111? 10   00? 0? 00000   11? 1101121   1100100001   1??? 0? 1000   1? 10010000   1001010000
0101100021   4410001111   1????? 10100   00001? 1001   0? 0110201?   ??? 1001???   ? 10110????
??????? 00?   000?? 01???   ? 10??? 1110   1012111111   101110? 011   1101201001   2111110010
0000111010   0011111?? 0   010101 11? 1   11031? 12??   0? 2??? 1011?   12? 11? 0001   ? 01? 111011
01C21?? 11?   3010112000   1110111000   ???? 01? 100   ? 00?? 00

Pederpes
??????? 1?   0? 00000000   0001? 000??   ?? 10000000   0110010101   1? 10000000   ? 00? 01? 100
? 0? 2101?? 0   1?? 1???? 01   001? 00? 100   00000?????   ?? 0100100?   ????????? 0   00? 0??????
0????????0   0?? 0111?? 0   ?? 0?? 10100   1011001100   101101????   000020110?   ???? 000? 00
00?? 10? 000   00011???? 0   01??? 111??   ? 10? 1? 10? 2   1? 10? 000??   ???? 101000   1? 000? 1021
1220? 11102   200? 1? 21? 1   ?? 1111?? 01   ?? 0? 000? 00   0? 01000

Pholiderpeton
0? 10110? 10   000? 00? 000   11011? 1?? 1   1?? 0010000   00? 0000000   1?? 00? 0000   ? 000100000
0101100030   3310001100   ? 000000100   00001? 0001   0001102000   0010001111   1100101010
1011011101   01011? 1???   00010?? 0? 0   ? 01B?? 1111   1? 121?????   ?????????   ???? 0100? 0
0000111000   10011?????   01????? 1? 1   12031210? 0   1220001010   11100111? 0   201? 111021
1221111101   ? 0111? 2000   1111????? 1   ?? 10111? 10   1? 10111

Proterogyrinus
0? 10110? 10   0010000000   1001101021   1000010000   0000000000   1? 00010000   1001110? 00
0101100230   B41? 0??? 00   101? 000100   0000?? 0001   000110200?   ??? 00011??   11001?? 0? 0
1011010?? 1   0101111000   0001110010   1012011100   1111111111   1011201101   2111010000
0000110000   0001111110   01010111? 1   11030210? 0   122000? 010   ?? 10011020   201??? 10? 1
122101? 111   2010102000   1111111011   ?? 0? 110000   1011111

Ptyonius
```
0? 00110? 10    0000010101    0001111121    0110010000    1??? 00100?    1? 01000110    1001010000
? 1011002? 0    44000? 1110    0100000100    01001?? 0? 1    0? 1000201?    ??? 100????    ??? 11?????
??????????    0001110? 01    0?? 11? 0?? 1    1??? 011111    101110? 011    ???? 201???    ????? 10011
1111??? 01?    ? 001111110    01100111??    ? 1? 311? 10?    0??? 01? 11?    1012001???    201? 01? 111
1? 3200?? 1?    3? 10111000    0?? 0112000    ?? 2? 0?? 101    ? 01?? 0?
```

Saxonerpeton
```
0? 1011?? 10    0000000000    11? 1201021    100000? 010    1??? 1??? 00    001000100?    0001010011
? 1000002? 1    ? A11? 01111    ????? 01? 00    00001?????    ?? 0000201?    ??? 100????    ??? 1??????
??????????    0001111110    01?? 1? 01? 0    11?? 111111    1112101011    1001B01001    20110100? 0
0?????? 0??    ? 0111????? 1    111001? 1? 1    11? 312110?    0? 1110011?    10110111? 0    ? 01? 11? 011
0? 3202?? 1?    3? 101? 2000    111? 111000    ?? 2? 00? 000    ?? 0????
```

Seymouria
```
0? 10111110    0000000000    1101101021    1000A01000    0110000100    1? 10010000    1001010100
110200? 211    4411001101    1111111100    00001? 0001    0001002011    ??? 1001111    1101101010
1011011111    000011011? 0    1101111110    1012111100    1112111111    0? 10201101    2011000000
0000111110    0011111110    0101010111? 1    1113111101    ?? 21102110    ? 0121111? 0    2000111111
120111? 011    3111111000    111? 111000    00AA010000    0? 01000
```

Silvanerpeton
```
0? 1????? 10    0000000000    0001101021    100000? 000    0000000000    1000010010    1001? 10000
? 1? 210? 2? 0    4? 11?? 1101    ????? 00100    0000?????? 1    0? 0110200?    ???? 001?? 1    1?? 01?????
??? 10100? 1    000? 111000    0?????? 0?? 0    101? 001111    101201?? 11    ???? 2? 1?? 1    20?? 0100? 0
0000110000    00? 11?????    ? 1010111??    ? 1031?? 0? 0    0? B? 10? 1??    ? 0? 10? 1000    ? 00011? 011
? 1D1? 1?? 0?    ???? 1? B000    1111111100    010? 000100    1? 00111
```

Utegenia
```
0? 10111010    0000000000    01011A1021    100010? 000    0000000000    1? 10010000    1001010000
? 102100211    3C11001101    0111011? 00    ? 0001?? 00?    0? 01002001    ??? 100?? 1    11011? 1010
1011011001    00011? 2110    1???????? 0    1????? 11??    11110?? 111    ???? 2? 1???    ???? 0100? 0
000001A000    0001111?? 0    010101011? 1    11131A1100    0111102110    ? 012001000    2000111011
0121? 22?? B    300? 111000    ?? 11011000    ?? 0? 000000    ?? 01? 11
```

Westlothiana
```
0? 10111011    00? 00? 0? 0?    11? 1101121    100000? 00?    1??? 00000?    1? 10010001    ? 00?? 10000
0102? 0? 2? 1    ????????? 1    111? 100? 00    0000?? 1???    ?? 0010201?    ??? 100????    ??? 11?????
??????????    000? 1?????    0???? 11110    10121? 1111    1012111011    1101211001    2111? 100? 0
0000111110    0011111110    11??? 111??    11? 31?? 0??    0? 0? 10? 11?    11? 10111? 0    ? 0????? 011
0? 310?? 110    301? 112000    ?? 1? 111000    ???? 000001    ? 1?? 10?
```

Whatcheeria
```
0? 0? 11?? 10    0000000000    0001110011    1000010000    0110000100    1?? 0000000    100?? 11? 00
? 001101010    22? 10??? 01    001? 00? 1??    00001?????    ?? 0? 101? 00    011000????    1?? 0?? 1000
110? 1001? 0    0000101110    1? 01100? 10    1011001100    10?? 111011    0?? 0?? 101?    ???? 1100? 0
0000101000    00011????? 0    01??? 1? 1??    ? 10? 1010? 2    111? 1000? 0    ?? 121011? 0    2100010021
103011? 101    ?? 11? 021? 1    111?? 1? 010    ?? 00000? 00    ? 11000
```

[Special Papers in Palaeontology 81, 2009, pp. 43–59]

THE EARLY CRETACEOUS LIZARDS OF EASTERN ASIA: NEW MATERIAL OF *SAKURASAURUS* FROM JAPAN

by SUSAN E. EVANS* *and* MAKOTO MANABE†

*Research Department of Cell and Developmental Biology, UCL, University College London, Gower Street, London WC1E 6BT, UK; e-mail: ucgasue@ucl.ac.uk
†National Science Museum, 3-23-1 Hyakunin-cho, Shinjuku-ku, Tokyo 169-0073 Japan; e-mail: manabe@kahaku.go.jp

Typescript received 21 April, 2008; accepted in revised form 5 June 2008

Abstract: New material of the Early Cretaceous Japanese lizard *Sakurasaurus* has been recovered from the locality of Kaseki-kabe, Kuwajima District, Hakusan City. Remains of a single disarticulated individual permit a more comprehensive description, reconstruction and phylogenetic analysis of this taxon. Although the original study, based largely on jaw material, suggested scincomorph affinity, a new cladistic analysis places *Sakurasaurus* on the scleroglossan stem as a sister taxon of the Chinese genus *Yabeinosaurus*. *Sakurasaurus* and *Yabeinosaurus* provide another link between the faunal assemblage of the Japanese Tetori Group and those of the roughly contemporaneous and geographically adjacent Jehol Group of China, although there are also many differences. A similar pattern is revealed by broader comparisons between the Sino-Japanese lizard faunas and those of other roughly contemporaneous Asian horizons.

Key words: Asia, Japan, Reptilia, Squamata, *Sakurasaurus*, Tetori Group, *Yabeinosaurus*.

THE LIZARD genus *Sakurasaurus* was described from the Early Cretaceous (Valanginian–Hauterivian) Okuradani Formation at Shokawa (now part of Takayama City), Gifu Prefecture, Japan, on the basis of jaw material and a few skull roofing bones (Evans and Manabe 1999a). The lower jaw morphology suggested scincomorphan affinity (robust dentition, heterodonty, deep subdental shelf), and the form of the frontal and parietal bones (sculpture, paired frontals) was consistent with this, as were the results of a preliminary phylogenetic analysis. However, given the fragmentary nature of the material, this result was considered provisional. Conrad (2008) included *Sakurasaurus* in his comprehensive review of squamate phylogeny and also recovered a placement for *Sakurasaurus* within Scincomorpha, at the base of Cordyliformes (Cordylidae + Gerrhosauridae).

More recently, new material of *Sakurasaurus* has been recovered from the Kuwajima Formation of neighbouring Ishikawa Prefecture. The Kaseki-Kabe, or fossil cliff, of Kuwajima District, Hakusan City (formerly Shiramine village) has yielded a rich and diverse plant and animal assemblage including fish (Yabumoto 2000), amphibians (Matsuoka 2000a), mammals and tritylodont synapsids (Rougier *et al.* 2007; Setoguchi *et al.* 1999a, b; Matsuoka 2000b; Manabe *et al.* 2000a, b; Takada *et al.* 2001), dinosaurs (Azuma and Tomida 1995; Hasegawa *et al.* 1995;

Manabe 1999; Manabe and Barrett 2000; Manabe *et al.* 2000a, b), pterosaurs (Unwin *et al.* 1996, 1997; Unwin and Matsuoka 2000), turtles (Hirayama 1996, 1999, 2000), choristoderes (Evans and Manabe 1999b; Matsumoto *et al.* 2007) and a diversity of Early Cretaceous squamates (Evans *et al.* 1998; Evans and Manabe 1999a, 2000, 2008; Evans *et al.* 2006), including material of *Sakurasaurus*.

The Kuwajima material of *Sakurasaurus* includes isolated jaw and skull roofing elements. More important, however, is a small block, SBEI 199 (Text-fig. 1), preserving jaw elements (maxillae and dentaries) attributable to *Sakurasaurus* in association with other bones representing much of the skull and some parts of the postcranial skeleton. Although the posterior teeth are less robust than those of the holotype, the Kuwajima skull is considerably smaller than the original material from Shokawa and the specimen is immature (separation of vertebral neural arches and condyles, disarticulation of skull elements). It is referred to *Sakurasaurus* on the basis of maxillary shape (inflated facial process, slender slightly upturned premaxillary process, medial recess, structure of medial facets), the shape of the paired frontals, the heterodont dentition and the general morphology of the jaw (protuberant anterior dentary process, shape of subdental ridge, bifurcate posterior dentary, long prearticular). It is possible that

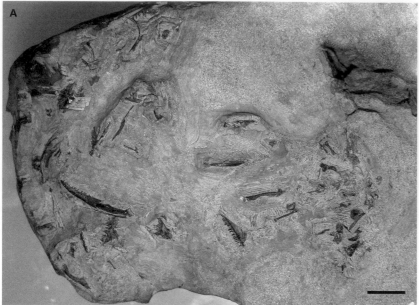

TEXT-FIG. 1. *Sakurasaurus* sp. Kuwajima Formation, Japan. SBEI 199. A, original specimen. B, map of major elements. Scale bar represents 10 mm.

the Kuwajima specimens belong to a different species of *Sakurasaurus* from those of Shokawa but, given the immaturity of the material, it is here referred to *Sakurasaurus* sp. More detailed knowledge of *Sakurasaurus* and of the roughly contemporaneous Chinese (Jehol Biota) lizard *Yabeinosaurus* (Evans *et al.* 2005) necessitates a reassessment of their phylogenetic positions.

GEOLOGY

Rocks of the Tetori Group outcrop in central Honshu, Japan and span an interval from the Middle Jurassic to the Early Cretaceous (Maeda 1961; Matsukawa and Obata 1994; Kusuhashi *et al.* 2002; Fujita 2003; Isaji *et al.* 2005). The fossil material described in this paper was collected from Kuwajima District, Hakusan City, Ishikawa Prefecture. In this region, the 'Kaseki-Kabe' (or fossil cliff) exposes the upper part of the Kuwajima Formation, Itoshiro Subgroup. This horizon is generally referred to the Lower Cretaceous (Kusuhashi *et al.* 2002; Isaji *et al.* 2005; Evans *et al.* 2006) and is probably either Valanginian (e.g. Isaji *et al.* 2005) or Hauterivian (Fujita 2003) in age. A Zircon fission-track date of 135 ± 7 Ma recorded (Gifu-ken Dinosaur Research Group 1992) for the laterally equivalent Okuradani

Formation at Shokawa, Gifu Prefecture (Maeda 1961) is consistent with this.

Palaeoenvironmental analysis of the Kuwajima Formation and of the Okuradani Formation at Shokawa (Isaji *et al.* 2005) suggests that the fossiliferous beds were deposited on a well-vegetated inland floodplain. Facies representative of peat marsh and shallow lake environments yield a predominantly aquatic assemblage (invertebrates, fish, turtles, choristoderes). In Kuwajima, however, there is a rarer third facies, interpreted as a periodically flooded subaerial swamp (Isaji *et al.* 2005), that has produced a higher proportion of terrestrial taxa (e.g. tritylodont synapsids, mammals, lizards, dinosaurs). The new material of *Sakurasaurus* comes from this facies.

Institutional abbreviations. IBEF, formerly Izumi Board of Education, Izumi Village; SBEG, formerly Board of Education, Shokawa Village; SBEI, formerly Shiramine Board of Education, Kuwajima Village.

Abbreviations used in the text-figures. an.p, angular process; ant.p, anterior protrusion; a.p, alar process; a.zy, anterior zygapophysis; ch, choana; con, condyle; co.p, coronoid process; cr.P, edge of crista prootica; D, dentary; E.ft, ectopterygoid facet; Ep, epipterygoid; Ex.o, exoccipital; f.12, hypoglossal nerve foramina; Fr, frontal; Fr.ft, frontal facet; Fr.P.ft,, frontoparietal articulation; f.v, fenestra vestibulae; i.P, incisura prootica; l.Co, left coronoid; l.D, left dentary; L.ft, possible lacrimal facet; l.J, left jugal; l.Jw, left jaw; l.Mx, left maxilla; l.Ot, left otooccipital; l.Pra, left prearticular; l.Pof, postorbitofrontal; l.Pro, left prootic; l.Sp, left splenial; mk.fs, meckelian fossa; my.f, mylohyoid foramina; Mt, metatarsal; Mx.ft; N, nasals; N.ft, nasal facet; pa, recess for processus ascendens of supraoccipital; Pal.pr, palatine process; P.ft, parietal facet; Pmx, premaxilla; Pmx.ft, premaxillary facet; Pra, prearticular; ppr, postparietal process; Prf.ft, prefrontal facet; Pt, pterygoid; rap, retroarticular process; r.D, right dentary; r.Fe, femur; r.Fr, right frontal; r.il, right ilium; r.J, right jugal; r.Mx, right maxilla; r.Pal, right palatine; r.Pra, right prearticular; r.Prf, right prefrontal; r.Q, right quadrate; rst, lateral opening of the recessus scala tympani; r.Vo, vomer; Sp.Ft, splenial facet; Sq.n, squamosal notch; Sr.ft, sacral rib facets; St, supratemporal; Su, surangular; V, vertebra; v.f, recess leading to vagus foramen; vno, vomeronasal opening.

SYSTEMATIC PALAEONTOLOGY

SQUAMATA Oppel, 1811

Genus SAKURASAURUS Evans and Manabe, 1999*a*

Type species. Sakurasaurus shokawensis Evans and Manabe, 1999*a*.

Type specimen. IBEF VP 17, a left mandible (Evans and Manabe 1999*a*).

Material. IBEF VP 18 (a right dentary); IBEF VP 19 (a right maxilla); IBEF VP 20 (a parietal); SBEG VP 001 (Paratype left maxilla); SBEG VP 002 (a left maxilla); SBEG VP 003 (a right maxilla); SBEI 199 (associated skeletal elements), SBEI 566 (right dentary); SBEI 1523 (left maxilla); SBEI 1524 (left dentary); SBEI 1605 (partial left maxilla); SBEI 1727 (jaw fragments); SBEI 1798 (left frontal); SBEI 1802 (partial left frontal).

Generic diagnosis (revised from Evans and Manabe 1999a). Limbed squamate characterized by the following combination of characters: heterodont dentition in mature animal with slender acuminate anterior teeth and more robust posterior teeth; slender curved process of maxilla fits into lateral recess on premaxilla; palate incompletely neochoanate; prearticular with triangular angular flange; mylohyoid foramina fully enclosed within splenial; protuberant anterior dentary process.

Occurrence. Currently limited to beds of the Tetori Group (Okuradani Formation, Kuwajima Formation) of Central Honshu, Japan, dated to Early Cretaceous (Valanginian–Hauterivian).

Remarks. Sakurasaurus resembles *Yabeinosaurus* and differs from all other known Late Jurassic and Early Cretaceous lizard taxa in the combination of the following features: large, inflated facial process of maxilla; interdigitated frontoparietal suture; small lateral parietal wings; postparietal processes with broad and horizontal bases; retention of parietal foramen; long paired frontals with strong median suture and anteroventral flanges; narrow nasal bones; single median premaxilla; quadrate with broad but shallow lateral conch, notch for squamosal peg; postorbital and postfrontal apparently fused into a postorbitofrontal that closes or restricts upper temporal fenestra; angular flange on prearticular; palatal teeth present on palatine and pterygoid; dermal skull ornamentation but no osteoderms; robust hind limbs. *Sakurasaurus* differs from *Yabeinosaurus* in its smaller adult size (based on comparisons of small immature *Yabeinosaurus*, Evans *et al.* 2005), the shorter parietal body with more laterally directed post-parietal processes, broad-based triangular angular flange rather than finger-like process, and the heterodonty of mature adult specimens.

Description of sbei 199

SBEI 199 is a small block bearing a bone association that covers an area of approximately 105 × 80 mm (Text-fig. 1A–B). One edge of the bone cluster meets the edge of the block and at least part of the postcranial skeleton originally extended beyond this point. The association includes representatives of many of the skull bones, parts of the vertebral column and rib cage and parts of the right hind limb and pelvis, but no identifiable elements of the forelimbs or pectoral girdle. The vertebrae are represented by separated neural arches and centra suggesting immaturity, and this would also explain why the skeletal components disarticulated completely but without obvious damage prior to burial. There are no duplicated elements and the bones are consistent in terms of size, structure, left-right pairing and level of maturity. We are therefore confident that SBEI 199 bears the remains of one individual.

Skull. The skull is represented by bones of the skull roof, jaws, suspensorium, palate and braincase. In comparison with modern lizards (e.g. *Dipsosaurus, Cnemidophorus, Chalcides*), the snout-pelvis length was *c.* 80–100 mm with the skull *c.* 20 mm in length. The skull bones are well formed and have their facet surfaces fully developed. Thus although the animal may have been immature, it was probably not a hatchling. Comparison with *Yabeinosaurus*, where a growth series of specimens is available from juvenile to large adult, suggests that *Sakurasaurus* reached maturity at a smaller size.

Text-figure 2 presents a reconstruction of the skull in dorsal, lateral and partial palatal views based mainly on the elements from SBEI 199, with the addition of the parietal IBEF VP 20 (Evans and Manabe 1999*a*). Allowance must be made for the fact that the individual bones cannot be removed from the block so that their precise orientations must be estimated, especially where articulations are not fully visible.

The maxillae are both preserved in medial aspect and show features consistent with those of the original Shokawa specimens, although they are considerably smaller (Text-fig. 3A–B, D), have a lower tooth count, and show more rapid replacement (and hence differences in spacing). The anterior premaxillary process is small and bifurcated, with short tapering medial and lateral processes that clasped the margins of the premaxilla. The posterior, orbital process is deep, overlapping the lateral surface of the jugal and excluding it, at least partially, from the orbital margin. The facial process is large and somewhat inflated. It bears a narrow nasal facet along the anterodorsal margin. Posteroventral to this is a larger overlap surface for the prefrontal and, below that, an elongated recess. A similar recess in some modern lizards (e.g. teiids, SEE, pers. obs.) accommodates part of the nasal sac. There is no obvious facet for a lacrimal on the maxilla, although facets on the prefrontal suggest that a lacrimal may have been

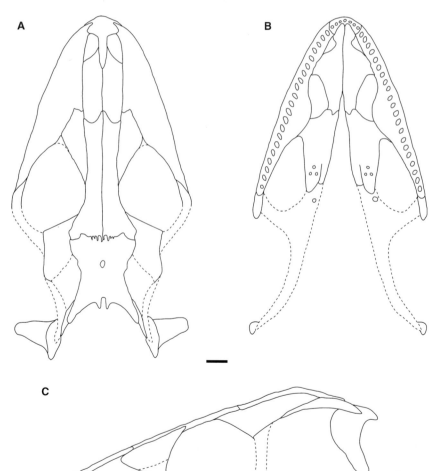

TEXT-FIG. 2. *Sakurasaurus,* reconstruction of the skull. A, dorsal view. B, partial ventral view. C, left lateral view. Scale bar represents 1 mm.

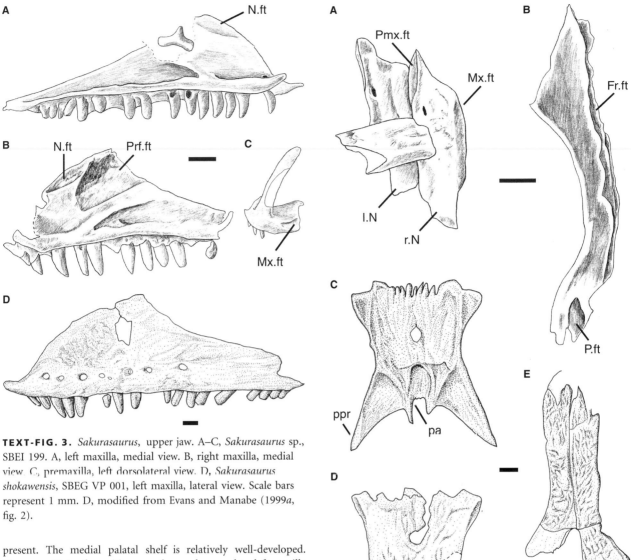

TEXT-FIG. 3. *Sakurasaurus*, upper jaw. A–C, *Sakurasaurus* sp., SBEI 199. A, left maxilla, medial view. B, right maxilla, medial view. C, premaxilla, left dorsolateral view. D, *Sakurasaurus shokawensis*, SBEG VP 001, left maxilla, lateral view. Scale bars represent 1 mm. D, modified from Evans and Manabe (1999*a*, fig. 2).

TEXT-FIG. 4. *Sakurasaurus*, skull roof. A–B, *Sakurasaurus* sp., SBEI 199. A, nasals in dorsal view. B, right frontal in ventromedial view. C–D, *Sakurasaurus shokawensis*, IBEF VP 20, parietal, in C, ventral, and D, partial dorsal views. E, IBEF VP 18, paired frontals in dorsal view. Scale bars represent 1 mm. C–E, modified from Evans and Manabe (1999*a*, fig. 5).

present. The medial palatal shelf is relatively well-developed. There are 18 tooth positions on the nearly complete left maxilla and 13 on the right (against 24 in more mature specimens from the Okuradani Formation, Evans and Manabe 1999*a*). The teeth are fully pleurodont, acuminate and slightly recurved posteriorly.

The premaxilla (Text-fig. 3C) is a single, median element with a bilaterally flattened nasal process that becomes dorsoventrally flattened towards the tip. The alveolar margin bears three small teeth, one in the midline and two on the right. However, these are well spaced and the likely tooth count would be 5–7. The premaxillary teeth are smaller than those on the maxilla. The alveolar margin extends posterolaterally to form a large, bifurcate, palatal shelf that met the vomers. Curved dorsolateral facets accommodated the tips of the maxillae.

The two nasals are preserved together (Text-fig. 4A), with the right bone exposed in dorsal view and the left in ventral view. They are overlain by an oblong bone fragment that is the broken posterior end of one or other nasal. The nasal is rectangular and relatively narrow. The anterodorsal surface is slightly ornamented but this ornamentation becomes weaker posteriorly where the nasal would have overlapped the frontal. The anteromedial margin is extended into a short, facetted premaxillary

process, whereas the anterolateral margin is incised by the narial margin. The lateral edge of the right bone is well-preserved and bears a subvertical facet that matches the corresponding facet on the maxilla.

Only the right frontal is preserved, exposed in ventromedial aspect (Text-fig. 4B). It is long and narrow, although as preserved it appears wider than it actually is because of the anteroventral expansion (see below). In its proportions (length/width) and deep anteroventral process, it resembles frontals described from Shokawa (Evans and Manabe 1999*a*,

Text-fig. 4E) and is also similar to the frontals of the Chinese *Yabeinosaurus*. The interfrontal suture is relatively deep, with several horizontally interlocking laminae that held the bones securely together, especially in the interorbital region. The anterior margin of the frontal tapers slightly towards the midline. Laterally, each bone is expanded into a deep anteroventral flange that braced the orbitonasal margin. The posteromedial margin diverges slightly from the midline, but the posterolateral edge is interdigitated, with a ventrolateral facet that accommodated an anterior parietal lappet. Mesokinetic bending is unlikely to have been possible at the frontoparietal suture. In mature Shokawa specimens, the frontal is sculptured and similar sculptured frontal fragments have been recovered from the Kaseki-kabe locality. A parietal referred to *Sakurasaurus* from Shokawa (IBEF VP 20, Text-fig. 4C–D) has a morphology consistent with that of the frontal in terms of ornamentation, width and the structure of the anterior joint surface.

The right prefrontal is preserved in posterolateral view (Text-fig. 5A). It is a large bone with a small, posteriorly tapering frontal process and a large triangular preorbital body. The anterolateral surface bears a large facet for the maxilla, while the posterolateral edge has a recessed area that probably accommodated a lacrimal. Judging from the position of facets and the orientation of the bone, the prefrontal would have been exposed as a dorsolateral rectangle flanking the frontal. The orbitonasal flange is of moderate width and is extended ventrally into a short pillar-like palatine process with an anterolateral groove for a lacrimal duct that presumably passed between it and the lacrimal. The ventral articulation with the palatine does not appear to have been tightly sutured.

Close to the left jugal and right quadrate is an irregular element (Text-fig. 5B). As interpreted, the dorsal margin bears a narrow anterior facet, ventral to which is a short thickened orbital margin. The ventrolateral edge is almost straight and the posterior margin is broken. Judging from the shape, size and fit of this element with other bones, it is probably a left postorbitofrontal (conceivably a postfrontal), the dorsal facet fitting the frontal and parietal and the jugal contacting the anteroventral edge. Under this interpretation, the bone is similar to the postorbitofrontal of *Yabeinosaurus* (Evans *et al.* 2005, see Discussion). As preserved, it is not clear how much of the posterior margin is damaged, but the upper temporal fenestra would have been at least partially closed by this bone (Text-Fig. 2A).

Both jugals are preserved in part. The anterior, suborbital ramus of the left jugal is preserved in lateral aspect (Text-fig. 5C), close to the postorbitofrontal. Most of its anterior region bears a large facet for the maxilla. Lying adjacent to it is a twisted bar of bone (?, Text-fig. 5C) that cannot be fully exposed because of surrounding elements. It may be the detached postorbital ramus of the left jugal or possibly the squamosal. The right jugal lies in the angle between the right ilium and femur and is exposed in a partial medial view (Text-fig. 5D) that shows the junction between the suborbital and postorbital rami. This region bears a medial facet for the ectopterygoid, but lacks any trace of a posterior process.

A small, anteriorly tapering element close to the left jugal and prootic is, by its size and shape, interpreted as a supratemporal as it seems too small to be the squamosal. It is probably, but not certainly, the left bone (Text-fig. 5E).

The right quadrate (Text-fig. 5F) is a semicircle of thick bone, forming a rounded, but mediolaterally shallow conch. The anterior margin is raised into a low tympanic crest notched dorsally by a laterally open squamosal pit. The posterior margin is curved, with the dorsal head slightly overhanging. The bone narrows ventrally, but the mandibular condyle is not exposed. One epipterygoid overlies the prootic. It is columnar with a slight expansion of its dorsal and ventral ends (Text-fig. 5J).

The right vomer is complete (Text-fig. 5G) and is preserved in palatal view. It is a roughly triangular bone with a flat ventral surface except for a sharp dorsomedial angulation. As reconstructed (Text-fig. 2B), the adjacent vomers would have been in contact along their medial margins anteriorly, but were separated posteriorly by the narrow anterior limb of the interpterygoid vacuity. There is a tapering anterior premaxillary process and a blunt facetted posterior process into which the palatine locked. The lateral margin is excavated by two embayments. A small triangular anterior notch marks the opening of the vomeronasal duct; posterior to this the bone flares out laterally and would have overlapped the maxillary shelf closing off the vomeronasal opening (incomplete neochoanate condition). The remainder of the lateral margin formed the choanal margin.

The right palatine is complete and also exposed in palatal aspect (Text-fig. 5H). The anterior margin bears a large choanal embayment but the choanal groove on the palatine itself is relatively shallow. A forked maxillary process separates the choanal margin from the suborbital fenestra. The main body of the bone is relatively slender and bears a small cluster of teeth towards the posterior margin. A facet immediately medial to this would have accommodated the pterygoid. Another tooth-bearing palatal fragment lies at right angles to the right palatine (Text-fig. 5I). It is probably part of a pterygoid.

The braincase is represented by the left prootic and left otooccipital. The prootic is preserved in lateral view, but is partially obscured and crushed by the overlying epipterygoid (Text-fig. 5J) and an adjacent cervical vertebra. This view shows a large anterior notch for the trigeminal nerve (incisura prootica), framed dorsally by a broad anterodorsal alary process and ventrally by a strong, deep, anterior inferior process. Further posteroventrally, the bone is developed into a crista prootica, but the overlying elements make it impossible to prepare the region ventromedial to the crest or to gauge its depth. The left otooccipital is separated from the remaining skull elements at one edge of the block. It is preserved in posterolateral view (Text-fig. 5K–L). The anterodorsal part of the bone is broken and the paroccipital process is not preserved. The exoccipital component is visible already fully fused to the opisthotic and contributes posteriorly to the occipital condyle. It is pierced by two large hypoglossal foramina. Between the exoccipital body and the opisthotic is an elongated concavity that led to the vagus foramen. This is separated from the lateral opening of the recessus scala tympani by a distinct opisthotic/exoccipital flange. A more anterior flange separates the recessus from the fenestra vestibuli. As preserved, the basioccipital seems to have made little contribution to the ventrolateral border of the recessus scala tympani.

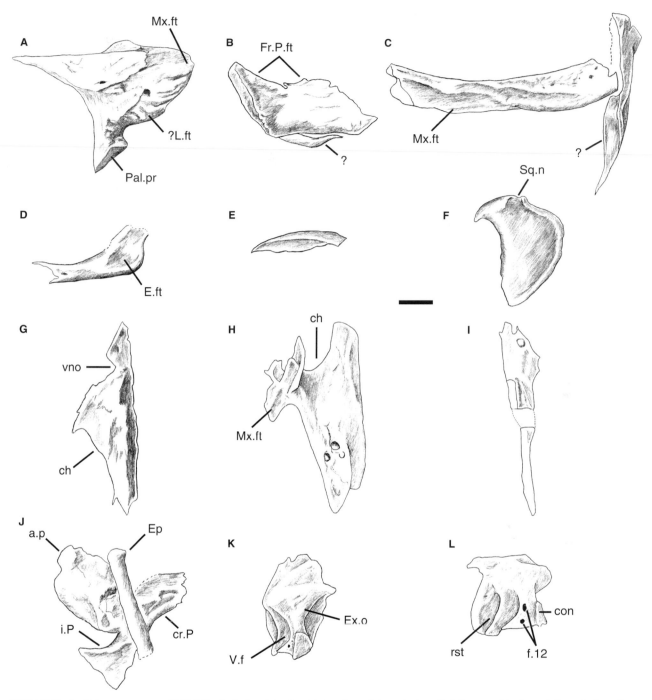

TEXT-FIG. 5. *Sakurasaurus* sp., SBEI 199, skull elements. A, right prefrontal, lateral view. B, left postorbitofrontal, lateral view. C, left jugal, lateral view. D, right jugal, medial view. E, possible supratemporal. F, right quadrate, lateral view. G, right vomer, palatal view. H, right palatine, palatal view. I, possible pterygoid, palatal view. J, left prootic and epipterygoid, lateral view. K–L, left otooccipital, in K, posterior view, and L, posterolateral view. Scale bar represents 1 mm.

The mandible

The left dentary and surangular are preserved in association in lateral view (Text-figs 6, 7); but the remaining parts of the left mandible are disarticulated and, though generally well-preserved, are scattered across the block. The dentary is rather shallow with a tapering symphysial region and a line of six lateral neurovascular foramina of which the posterior two are slit-like. The posterior margin is bifurcate and accommodates the surangular. There are 24–26 tooth positions filled anteriorly with relatively simple

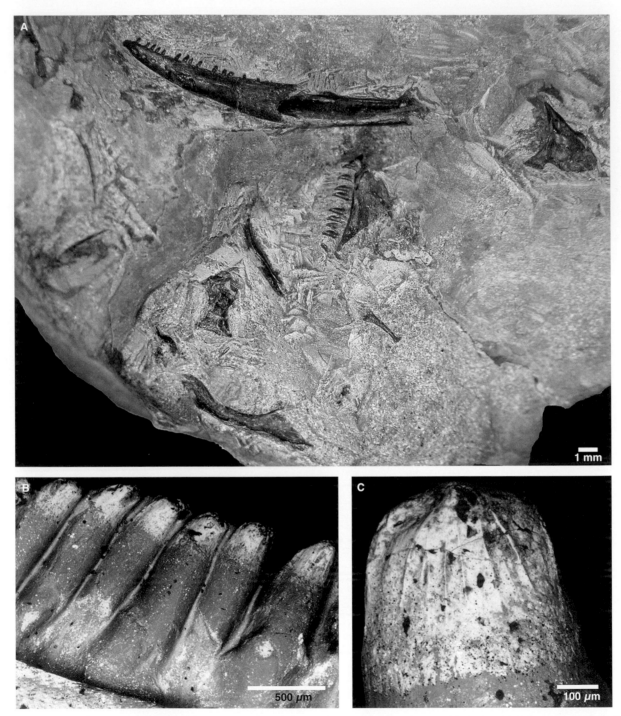

TEXT-FIG. 6. *Sakurasaurus*, jaw and dentition. A, *Sakurasaurus* sp., SBEI 199, enlargement of block showing left mandible and right maxilla. B–C, *Sakurasaurus shokawensis*, IBEF VP 18, scanning electron micrographs of the dentition, B, anterior dentary dentition. C, mature posterior tooth. B–C, modified from Evans and Manabe (1999a, fig. 3).

columnar teeth and posteriorly with more conical teeth, but they do not show the marked heterodonty found in the type material of *Sakurasaurus shokawensis*, where the posterior teeth are enlarged and robust, with expanded tips (Text-fig. 6B–C). This could reflect the immaturity of the SBEI 199 individual (compare its size with that of the original specimens, Text-fig. 7G–I).

The right dentary is preserved in medial view at one edge of the block, but can only be partially prepared. It presents an open meckelian fossa below a moderately developed subdental ridge that extends forward into a small symphysial projection (Text-fig. 7D), like that of the original material from Shokawa (Text-fig. 7G–I). The teeth are pleurodont with a typical lingual

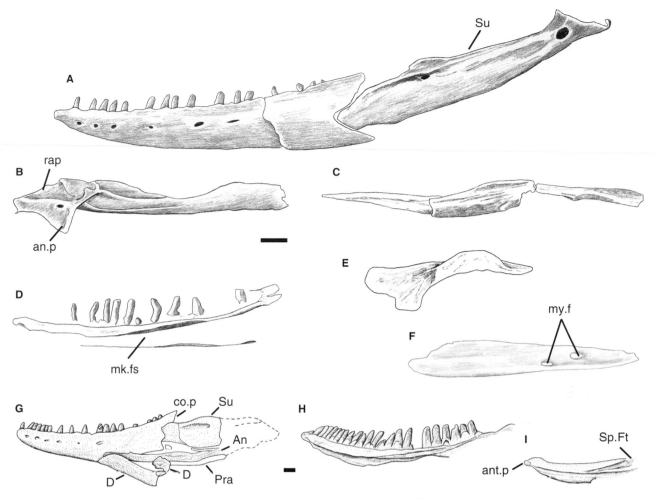

TEXT-FIG. 7. *Sakurasaurus*, lower jaw. A–F, *Sakurasaurus* sp., SBEI 199. A, left mandible, lateral view. B, left prearticular, medial view. C, right prearticular, medial view. D, right dentary, partial medial view. E, left coronoid, medial view. F, left splenial, medial view. G–I, *Sakurasaurus shokawensis*. G, IBEF VP 17, left mandible, type specimen, in lateral view. H–I, IBEF VP 18, right dentary, in H, medial, and I, partial ventral, views. Scale bars represent 1 mm. G–I, modified from Evans and Manabe (1999*a*, fig. 1).

replacement pattern. The left splenial is separated and lies some distance from the left dentary. It is preserved in medial view as a long, anteriorly tapering element perforated by an anteroinferior mylohyoid foramen and a posterior mylohyoid foramen, both completely enclosed by the splenial itself (Text-fig. 7F). Judging from the size of the splenial and dentary, most of the meckelian fossa was covered by the bone. The left surangular (Text-fig. 7A) is shorter than the dentary and has been slightly disarticulated posteriorly. Its dorsal margin bears a small prominence, below which the bone is perforated by a small surangular foramen. Posteriorly, the surangular tapers into its articular process and is again perforated, this time by a large lateral foramen for part of the mandibular nerve. The left prearticular-articular is also disarticulated and is preserved in a roughly dorsolateral view (Text-fig. 7B), although the anterior part of the bone is broken away. The adductor fossa was relatively large and open dorsally. There is a small, but damaged, articular surface and a slender, posteriorly directed retroarticular process with a dorsal concavity. Medially, the process is extended into a triangular angular flange, separated from the dorsal concavity by a distinct ridge. The foramen for the chorda tympani nerve lies medial to the ridge so that it perforates the flange rather than the concavity of the retroarticular process. The right prearticular is also preserved, in medial view (Text-fig. 7C). Although the anterior part is broken and slightly displaced, it shows that the prearticular had a long articulation with the dentary. A U-shaped bone (Text-fig. 7E) separated from the remaining jaw elements may be the left coronoid, but no details of its articular surfaces are visible.

Postcranial skeleton

The postcranial skeleton is represented by isolated vertebrae and ribs and parts of the right hind limb. The vertebrae have open neurocentral sutures denoting immaturity (Text-fig. 8A). Limited details are preserved, but the vertebrae are procoelous

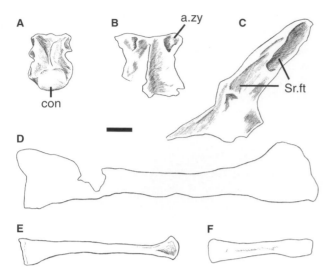

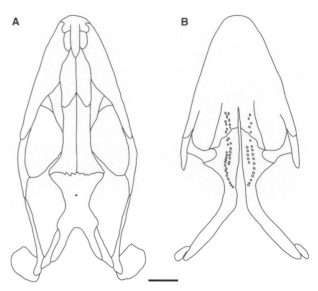

TEXT-FIG. 8. *Sakurasaurus* sp., SBEI 199, postcranial skeleton. A, vertebral centrum, dorsal view. B, vertebral neural arch, dorsal view. C, right ilium, medial view. D, outline of right femur. E, a metatarsal. F, a phalanx. In D–F, proximal end is to the left. Scale bar represents 1 mm.

TEXT-FIG. 9. *Yabeinosaurus tenuis*, reconstruction of the skull. A, dorsal view. B, partial ventral view; based mainly on IVPP V13285 and IVPP V13284 (Evans *et al.* 2005). Scale bar represents 10 mm.

with small rounded condyles. Cervical vertebrae are small with short centra bearing a distinct midventral crest. Overlying the left maxilla is a single isolated hypapophyseal element forming a small dorsal bar and a slender, ventrally directed process (Text-fig. 3A). It is not known whether this is a typical cervical element or the hypapophysis of the atlas. Only one neural arch is clearly preserved, again at the edge of the block (Text-fig. 8B). It is roughly square (probably cervical) with a low medial spine and a slight development of the zygosphene-zygantral system, as small raised edges to the zygapophyses themselves. Several more broken vertebrae, as well as ribs, are scattered across the block (not all shown in Text-fig. 1).

The right pelvis and hind limb are represented by an ilium and femur respectively. The ilium is preserved in medial view (Text-fig. 8C). It shows a well-developed pubic process, a small anterior angle and pronounced sacral rib facets, the second of which is positioned somewhat distally. The blade is short and tapering, angled posterodorsally at about 45 degrees to the horizontal. Adjacent to it is the femur, a long robust bone with a strong shaft (crushed) and expanded proximal and distal ends (Text-fig. 8D), but without ossified epiphyses. The bone is large in relation to the pelvis. Isolated pedal elements are also preserved, including a long slender metapodial (Text-fig. 8E) that is probably a fourth metatarsal and several robust phalanges (Text-fig. 8F). From this limited evidence, the hind limbs seem to have been strongly built with long feet.

DISCUSSION

Phylogenetic position

On the basis of original jaw and roofing elements from Shokawa, Evans and Manabe (1999*a*) ran a preliminary cladistic analysis that suggested a basal scincomorph position for *Sakurasaurus*. More recently, Conrad (2008) included *Sakurasaurus* in his comprehensive review of squamate relationships, getting a placement at the base of the scincomorph group, Cordyliformes. In the same cladogram, the Chinese lizard *Yabeinosaurus* was placed at the base of Anguimorpha, although Evans *et al.* (2005) obtained more basal positions for this taxon (either at the base of Scleroglossa or just outside Iguania + Scleroglossa, but see Townsend *et al.* 2004).

The new material of *Sakurasaurus* from Kuwajima allows it to be more fully coded as well as revealing similarities to *Yabeinosaurus* (Evans *et al.* 2005) in skull, jaw and dental morphology that are suggestive of a relationship.

Sakurasaurus (Text-fig. 2A–C) and *Yabeinosaurus* (Text-fig. 9A–B) resemble one another in the features of the anterior dentition (mostly conical, striated); strongly built maxilla with inflated facial process and slender premaxillary process; teeth on the palatine and pterygoid; the paired elongate frontals with a strong median suture and anteroventral flanges; the shape of the prefrontal (inflated facial part, slender frontal process), quadrate (broad but shallow lateral conch, notch for squamosal peg) and parietal (interdigitated central part to the fronto-parietal suture, dorsoventrally flattened, but wide postparietal processes with deeply placed pit for the processus ascendens, extended but thin lateral margins; the strong angular process on the articular and dermal skull ornamentation, but no cranial or postcranial osteoderms. *Yabeinosaurus* has a large, irregularly shaped bone in the posterolateral margin of the orbit that restricts the upper temporal opening and is interpreted as a combined postorbital and postfrontal (Evans *et al.* 2005). As preserved, this resembles the element here interpreted as the left postorbitofrontal. The two genera differ in some aspects of parietal shape (in *Yabeinosaurus* the postparietal processes are broader and meet across the midline), the shape of the angular process (triangular in *Sakurasaurus*, elongate and recurved in *Yabeinosaurus*), in body size (the adult *Yabeinosaurus* is larger) and in the heterodonty of mature *Sakurasaurus shokawensis*. However, parietal shape changes with size/age in *Yabeinosaurus* (Evans *et al.* 2005), with younger specimens showing a greater resemblance to *Sakurasaurus*. The same may also be true of the angular flange. Nonetheless, mature specimens of *Yabeinosaurus* and *Sakurasaurus* differ in the robusticity of the posterior dentition. Retention of separate generic status for the Japanese lizard is supported.

We initially recoded *Sakurasaurus* into a data matrix (a modification of Evans and Barbadillo 1998) incorporating a number of Jurassic and Early Cretaceous taxa and ran an analysis using PAUP version 3.0 (Swofford 1993). Beginning with a heuristic search because of the very large data set, we obtained 32 equally parsimonious trees [Tree length (TL) = 643, Consistency Index (CI) = 0.42, Rescaled Consistency Index (RC) = 0.24]. In all trees, *Sakurasaurus* and *Yabeinosaurus* emerged as sister taxa, but the strict and semi-strict consensus placed them in an unresolved trichotomy with Iguania and Scleroglossa, very similar to the position with *Yabeinosaurus* alone (Evans *et al.* 2005). We therefore tried coding the more complete data set for *Sakurasaurus* (see Appendix) into the matrix of Conrad (2008). This matrix is very large (222 taxa, 363 characters) and analysis with PAUP version 4.06b10 (Swofford 2001) took us to the limits of disc space. We therefore ran a first heuristic analysis with *Sakurasaurus*, *Yabeinosaurus* and all extant taxa, as well as the fossil outgroups Kuehneosauridae (Triassic lepidosauromorphs, Europe and North America, Robinson 1962; Colbert 1966), *Marmoretta* (Middle–Late Jurassic lepidosauromorph, Europe, Evans 1991), Rhynchocephalia, and *Huehuecuetzpalli* (Early Cretaceous basal squamate, Mexico, Reynoso 1998). *Sakurasaurus* and *Yabeinosaurus* emerged as sister taxa, lying at the base of Scleroglossa (non-iguanian squamates). As this was different from Conrad (2008) result, we re-ran the analysis with *Yabeinosaurus* and *Sakurasaurus* taken separately. In this situation, *Yabeinosaurus* emerged at the base of Anguimorpha, as in Conrad's analysis, and *Sakurasaurus* at the base of Scleroglossa. The three above analyses were then rerun with a reduced modern data set (using single species representatives for all genera, and eliminating some of the more derived representatives of major clades). The results for each analysis were the same with regard to placement of *Yabeinosaurus* and

Sakurasaurus. We then ran a final analysis in which we included all of the more basal fossil taxa from Conrad (2008) analysis. These are mostly Jurassic to Early Cretaceous genera (*Meyasaurus*, Early Cretaceous, Spain, Evans and Barbadillo 1997; *Dorsetisaurus*, Late Jurassic/Early Cretaceous, North America and Europe, Hoffstetter 1967; Prothero and Estes 1980; *Tepexisaurus*, Early Cretaceous, Mexico, Reynoso and Callison 2000; *Paramacellodus*, Late Jurassic/Early Cretaceous, Europe, North America, ?North Africa, Hoffstetter 1967; Evans and Chure 1998; Richter 1994; *Pseudosaurillus*, Early Cretaceous, Europe, Hoffstetter 1967; *Becklesius*, Early Cretaceous Europe, Hoffstetter 1967; *Dalinghosaurus*, Early Cretaceous, China, Evans and Wang 2005; *Hoyalacerta*, Early Cretaceous, Spain, Evans and Barbadillo 1999; *Eichstaettisaurus*, Late Jurassic/Early Cretaceous, Europe, Broili 1938, Evans *et al.* 2004; *Ardeosaurus*, Late Jurassic, Europe, Mateer 1982; *Scandensia*, Early Cretaceous, Spain, Evans and Barbadillo 1998; *Parviraptor*, Late Jurassic/Early Cretaceous, Europe and North America, Evans 1994*a*, 1996; *Bavarisaurus*, Late Jurassic, Germany, Evans 1994*b*) and token representatives of more derived clades like the macrocephalosaur boreoteiioids (*Macrocephalosaurus*, Late Cretaceous, Mongolia, Gilmore 1943), carusiids (*Carusia*, Late Cretaceous, Mongolia, Borsuk-Białynicka 1985) and mosasauroids (*Aigialosaurus*, Late Cretaceous, Croatia, Kramberger 1892; *Platecarpus*, Late Cretaceous, North America, Russell 1967) (a total of 82 taxa). The consensus of 160 equally parsimonious trees (TL = 2006; CI = 0.239; RC = 0.144) placed *Yabeinosaurus* and *Sakurasaurus* as sister taxa in a small clade with *Ardeosaurus* (Late Jurassic, Germany, Mateer 1982) and *Scandensia* (Early Cretaceous, Spain, Evans and Barbadillo 1998). This clade lies at the base of Scleroglossa (Text-fig. 10) but, as a whole, is not united by any unequivocal synapomorphies.

Biogeography of Asian lizards

The fossiliferous beds of the Tetori Group are dated, on several criteria, as of Neocomian (Berriasian–Hauterivian) age (Isaji *et al.* 2005). The ages of the horizons yielding the Chinese Jehol Biota are somewhat more controversial, but have been suggested to span a period from Hauterivian to Barremian–Albian age (e.g. Smith *et al.* 1995; Wang and Zhou 2003; Zhou *et al.* 2003; Wang *et al.* 2000). Adding the more problematic Daohougou Bed (Daohougou, Inner Mongolia), which is interpreted to contain a pre-Jehol biota (Late Jurassic or earliest Cretaceous age, Wang *et al.* 2005), there would undoubtedly have been temporal overlap between Chinese and Japanese assemblages. Furthermore, as the Sea of Japan was not yet open, the Tetori and Jehol biotas lived in geographical proximity (Text-fig. 11). The

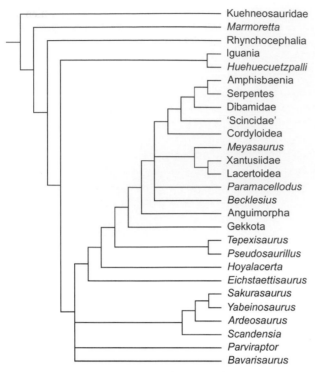

TEXT-FIG. 10. Consensus tree of 160 equally parsimonious trees (Length = 2006; Consistency Index, = 0.239; Rescaled Consistency Index = 0.144). Monophyletic clades of living taxa have been collapsed to single nodes.

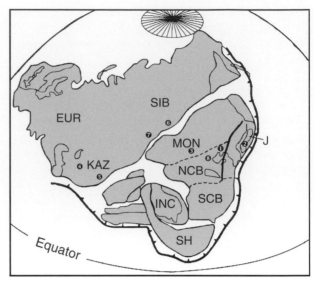

TEXT-FIG. 11. Map showing Asia in the Early Cretaceous (redrawn and emended from Zhou *et al.* 2003, fig. 1) with major lizard localities numbered as follows: 1, Jehol Biota localities, China; 2, Tetori Group, Japan; 3, Höövor, Öösh, and other Early Cretaceous localities, Mongolia; 4, Kyzylkum, Uzbeckistan; 5, Fergana Valley, Kyrgyzstan; 6, Mogoito, Transbaikalia; 7, Shestakovo, West Siberia; 8, Shandung, China. Tectonic divisions: EUR, Europe; INC, Indo-China; KAZ, Kazakhstan; MON, Mongolia; NCB, north China; SCB, south China; SIB, Siberia; SH, Shan Thai.

palaeoenvironment would have been broadly similar – freshwater wetlands with lakes and swamps supporting a mix of aquatic, semi-aquatic and terrestrial taxa (Chang *et al.* 2003; Isaji *et al.* 2005). Previous works have shown that these biotas share one fish taxon (the amiiform *Sinamia*, Yabumoto 2000) and have a comparable complement of freshwater reptiles attributed to the group Choristodera (the short-necked *Monjurosuchus*, and closely related long-necked genera *Hyphalosaurus* (China) and *Shokawa* (Japan), Matsumoto *et al.* 2007). Both regions have yielded a diverse lizard assemblage which is still under study. From the Jehol Biota localities (and their lateral equivalents in Gansu and Shandong, Wang and Evans 2006*a*, *b*), seven lizard genera have been described to date: *Yabeinosaurus* (Young 1958; Ji *et al.* 2001; Evans *et al.* 2005); the anguimorph *Dalinghosaurus* (Ji 1998; Ji and Ji 2004; Evans and Wang 2005); the remarkable long-ribbed glider *Xianglong* (Li *et al.* 2007); the problematic *Jeholacerta* (Ji and Ren 1999) and *Liaoningolacerta* (Ji 2005), both based on undiagnostic juvenile specimens; the paramacellodid *Mimobecklesisaurus* (Li 1985); and the possible scincomorph *Pachygenys* (Gao and Cheng 1999). There are also unnamed juvenile lizards from the older Daohugou Bed at Daohugou, Inner Mongolia (Evans and Wang 2007, in progress). In Japan, the Okuradani Formation (Shokawa, Gifu Precture) and the

Kuwajima Formation (Kuwajima District, Hakusan City, Ishikawa Prefecture) have together yielded at least six taxa, of which three have been formally described: *Sakurasaurus* (Evans and Manabe 1999*a*, this paper); the long-bodied anguimorph *Kaganaias* (Evans *et al.* 2006) and a herbivorous boreoteioid *Kuwajimalla* (Evans and Manabe 2008). There is also the problematic *Tedorisaurus* (Shikama 1969) from Fukui Prefecture, based on a specimen in a private collection that is unavailable for study. It may not even be squamate. Of these, only *Sakurasaurus* is recognized to have a close Jehol relative (*Yabeinosaurus*). However, *Mimobecklesisaurus* (Li 1985) from Gansu Province, China, has been attributed to the Jurassic/Cretaceous group Paramacellodidae due to the presence of rectangular osteoderms and similar rectangular osteoderms are common in the Tetori deposits. Some of the apparent faunal differences between China and Japan may result from preservational differences, in that isolated jaws are difficult to compare with fully articulated skeletons. Others (e.g. the absence of the herbivorous *Kuwajimalla* from the Jehol Biota) are real, but this may change as more material is described from each region.

Lizard fossils are also known from other Asian and Siberian localities. Jurassic records are rare. Middle Jurassic (Callovian) lizards have been briefly described from Kyrgyzstan (Callovian, Nessov 1988; Fedorov and Nessov

1992) and have been recorded, but not yet described, from Xinjiang, China (Clark *et al.* 2006). Late Jurassic lizards are known only from Kazakhstan (the paramacellodid *Sharovisaurus*, Hecht and Hecht 1984). Early Cretaceous lizard bearing localities are somewhat more numerous and include Mongolia (Alifanov 1993*a*, 2000*a*; Conrad and Norell 2006), Kazakhstan, Kyrgyzstan and Uzbekistan (Nessov 1988; Nessov and Gao 1993; Gao and Nessov 1998) and Siberia (e.g. Averianov and Skutschas 1999; Averianov and Fayngertz 2001; Leshchinskiy *et al.* 2001) (Text-fig. 11). If correctly interpreted, these lizard faunas include priscagamids, pleurodont iguanians, gekkotans, paramacellodids and other scincomorphs and anguimorphs including varanoids and xenosaurs/shinisaurs. However, many of these records, like the Tetori fossils, are based on disarticulated jaws and other elements that can be difficult both to interpret correctly and to compare with articulated specimens. Nonetheless, there are some similarities with the Sino-Japanese biotas. 'Xenosaurs' [under current understanding (Conrad 2004) these could be either xenosaur or shinisaur relatives] have been recorded from Höövor, Mongolia (Alifanov 1993*a*), Uzbekistan (Nessov and Gao 1993; Gao and Nessov 1998) and western Siberia (Averianov and Fayngertz 2001; Leshchinskiy *et al.* 2001), but these need to be compared with *Dalinghosaurus* from the Jehol Biota and with Late Cretaceous carusiids of Mongolia (Borsuk-Bialynicka 1985). Scincoid paramacellodids have been reported from western Siberia (Averianov and Fayngertz 2001; Leshchinskiy *et al.* 2001) and Transbaikalia (Averianov and Skutschas 1999) based on the presence of rectangular osteoderms and, less securely, on relatively simple dentitions. Gekkonomorphs have been recorded from Höövor (Alifanov 1993*a*) and Öösh (Conrad and Norell 2006) in Mongolia and possibly from Kuwajima (Evans and Manabe 2000). Early agamids are also recorded from Höövor (Alifanov 1993*a*) and also from Central Asia (Nessov 1988); determining whether the Jehol *Xianglong* is genuinely acrodont (Li *et al.* 2007) requires a better preserved adult skull. A fully herbivorous dentition like that of the Tetori *Kuwajimalla* (Evans and Manabe 2008) has not been described in any other Jurassic or Early Cretaceous lizard (only in the Late Cretaceous macrocephalosaurs of Mongolia, e.g. Gilmore 1943), but Alifanov (1993*a*, 2000*a*, *b*) reported mongolochamopines from Höövor and these have been posited to be basal members of the boreoteiioid (macrocephalosaur) line, although with a much simpler dentition. Generally, therefore, the lizard fauna of Asia fits the pattern described by other authors for this and other groups (e.g. Leshchinskiy *et al.* 2001; Manabe *et al.* 2000*a*, *b*; Zhou *et al.* 2003) in combining 'relicts' [in this context, survivors of Jurassic lineages that were extinct or nearly extinct elsewhere (e.g. paramacellodids, possibly the *Yabeinosaurus/Sakurasaurus* clade)]

with more crownward lineages (e.g. shinisaurs, agamids, macrocephalosaurs) that may have arisen during East Asia's isolation from the rest of Laurasia.

The exceptional diversity of the Late Cretaceous lizard assemblages of Mongolia and Chinese Inner Mongolia (e.g. Gilmore 1943; Sulimski 1975; Borsuk-Białynicka 1984; Alifanov 1993*a*, *b*, 2000*a*, *b*; Gao and Hou 1995, 1996; Gao and Norell 2000) attests to the presence of thriving Jurassic and Early Cretaceous terrestrial squamate communities from which these lineages arose. As yet, these earlier communities remain relatively poorly known despite recent progress and further work on Jurassic and Early Cretaceous squamates is needed to trace the history of East Asian lineages through the various stages of geographical isolation (Mongol-Okhotsk Sea, Turgai Sea, Barrett *et al.* 2002; Zhou *et al.* 2003) and reconnection (to western Eurasia and then to western North America).

Acknowledgements. This paper is dedicated to Andrew Milner in recognition of his outstanding contributions to the field of palaeoherpetology. The work presented here began under funding from a Royal Society-Japan Society for the Promotion of Science grant (SE and MM) and continued under a grant-in-aid for Scientific Research (No. 15340179) from the Japan Society for the Promotion of Science (to MM) and grants from the Prefectural Government of Ishikawa and the village of Shiramine (now Kuwajima District, Hakusan City). We are deeply grateful to the latter two bodies for their support of the Kuwajima project; to Dr Yuan Wang (Institute of Vertebrate Paleontology and Paleoanthropology, Beijing, China) for his continued collaboration (with SEE) on Jehol lizards and to Dr Jack Conrad (American Museum of Natural History, New York) for a prepublication copy of his squamate phylogeny review and permission to use his data matrix to re-explore the relationships of *Sakurasaurus* and to Professor Magdalena Borsuk-Białynicka (Institut Paleobiologii, Warsaw) and Dr Randall Nydam (Midwestern University, Arizona) for comments on an earlier draft of the manuscript. SBEI 199 was mainly prepared by Mikiko Yamaguchi under the supervision of Ichio Yamaguchi (Kuwajima). Jane Pendjiky (UCL) assisted in the preparation of the figures and Ryoko Matsumoto took the photographs.

REFERENCES

ALIFANOV, V. R. 1993*a*. Some peculiarities of the Cretaceous and Palaeogene lizard faunas of the Mongolian People's Republic. *Kaupia*, **3**, 9–13.

—— 1993*b*. The Upper Cretaceous lizard fauna of Mongolia, and the problems of the first interamerican contact. *Paleontological Journal*, **27**, 79–85.

—— 2000*a*. The fossil record of Cretaceous lizards from Mongolia. 368–389. *In* BENTON, M. J., SHISHKIN, M. A., UNWIN, D. M. and KUROCHKIN, E. N. (eds). *The age of dinosaurs in Russia and Mongolia*. Cambridge University Press, Cambridge, 696 pp.

—— 2000b. Macrocephalosaurs and the early evolution of lizards of Central Asia. *Trudi Paleontologicheskoyo Instituta*, **272**, 1–126 [in Russian].

AVERIANOV, A. O. and FAYNGERTZ, A. V. 2001. Lizards of the Early Cretaceous of Western Siberia. 6–8. *In* ANANJEVA, N. B. (ed.). *Questions of herpetology. Materials of the First Congress of the A.M. Nikol'skii Herpetological Society*. Pushchino-Moskva MGU, Moscow [In Russian with English summary].

—— and SKUTSCHAS, P. P. 1999. Paramacellodid lizard (Squamata, Scincomorpha) from the Early Cretaceous of Transbaikalia. *Russian Journal of Herpetology*, **6**, 115–117.

AZUMA, Y. and TOMIDA, Y. 1995. Early Cretaceous dinosaur fauna of the Tetori Group in Japan. 125–131. *In* SUN, AILING and WANG, YUANQING (eds). *Sixth Symposium on Mesozoic Terrestrial Ecosystems and Biota, Short papers*. China Ocean Press, Beijing, 250 pp.

BARRETT, P. M., HASEGAWA, Y., MANABE, M., ISAJI, S. and MATSUOKA, H. 2002. Sauropod dinosaurs from the Lower Cretaceous of eastern Asia: taxonomic and biogeographical implications. *Palaeontology*, **45**, 1197–1218.

BORSUK-BIAŁYNICKA, M. 1984. Anguimorphans and related lizards from the Late Cretaceous of the Gobi Desert, Mongolia. *Palaeontologia Polonica*, **46**, 5–105.

—— 1985. Carolinidae, a new family of xenosaurid-like lizards from the Upper Cretaceous of Mongolia. *Acta Palaeontologica Polonica*, **30**, 151–176.

BROILI, F. 1938. Ein neuer fund von ?*Ardeosaurus* H. von Meyer. *Sitzungsberichte der Bayerischen Akademie der Wissenschaften, München*, **1938**, 97–114.

CHANG MEEMANN, CHEN PEIJI, WANG YUANQING and WANG YUAN (eds) 2003. *The Jehol Biota: emergence of feathered dinosaurs and beaked birds*. Shanghai Scientific and Technical Publishers, Shanghai, China, 208 pp.

CLARK, J. M., XU XING, EBERTH, D. E., FORSTER, C. A., MACHLUS, M., HEMMING, S., WANG YUAN and HERNANDEZ, R. 2006. The Middle-to-Late Jurassic terrestrial transition: new discoveries from the Shishugou Formation, Xinjiang, China. 26–28. *In* BARRETT, P. M. and EVANS, S. E. (eds). *Ninth International Symposium on Mesozoic Terrestrial Ecosystems and Biota, Abstracts and Proceedings*. Natural History Museum, London. 187 pp.

COLBERT, E. C. 1966. *Icarosaurus* – a gliding reptile from the Triassic of New Jersey. *American Museum Novitates*, **2246**, 1–23.

CONRAD, J. L. 2004. Skull, mandible, and hyoid of *Shinisaurus crocodilurus* Ahl (Squamata, Anguimorpha). *Zoological Journal of the Linnean Society*, **141**, 399–434.

—— 2008. Phylogeny and systematics of Squamata (Reptilia) based on morphology. *Bulletin of the American Museum of Natural History*, **310**, 1–182.

—— and NORELL, M. A. 2006. High-resolution X-ray computed tomography of an Early Cretaceous gekkonomorph (Squamata) from Öösh (Övörkhangai, Mongolia). *Historical Biology*, **18**, 405–431.

EVANS, S. E. 1991. A new lizard-like reptile (Diapsida: Lepidosauromorpha) from the Middle Jurassic of Oxfordshire. *Zoological Journal of the Linnean Society*, **103**, 391–412.

—— 1994a. A new anguimorph lizard from the Jurassic and Lower Cretaceous of England. *Palaeontology*, **37**, 33–49.

—— 1994b. The Solnhofen (Jurassic, Tithonian) lizard genus *Bavarisaurus*: new skull material and a reinterpretation. *Neues Jahrbuch fur Geologie und Paläontologie, Abhandlungen*, **192**, 37–52.

—— 1996. *Parviraptor* (Squamata: Anguimorpha) and other lizards from the Morrison Formation at Fruita, Colorado. 243–248. *In* MORALES, M. (ed.). *The Continental Jurassic. Bulletin of the Museum of Northern Arizona*, **60**, 588 pp.

—— and BARBADILLO, L. J. 1997. Early Cretaceous lizards from las Hoyas, Spain. *Zoological Journal of the Linnean Society*, **119**, 23–49.

—— —— 1998. An unusual lizard from the Early Cretaceous of Las Hoyas, Spain. *Zoological Journal of the Linnean Society*, **124**, 235–265.

—— —— 1999. A short-limbed lizard from the Early Cretaceous of Spain. *Special Papers in Palaeontology*, **60**, 73–85.

—— and CHURE, D. 1998. Paramacellodid lizard skulls from the Jurassic Morrison Formation at Dinosaur National Monument, Utah. *Journal of Vertebrate Paleontology*, **18**, 99–114.

—— and MANABE, M. 1999a. Early Cretaceous lizards from the Okurodani Formation of Japan. *Geobios*, **32**, 889–899.

—— —— 1999b. A choristoderan reptile from the Lower Cretaceous of Japan. *Special Papers in Palaeontology*, **60**, 101–119.

—— —— 2000. Fossil lizards. 105–106. *In* MATSUOKA, H. (ed.). *Fossils of the Kawajima 'Kaseki-kabe' (Fossil-bluff). Scientific report on a Neocomian (Early Cretaceous) fossil assemblage from the Kuwajima Formation, Tetori Group, Shiramine, Ishikawa, Japan*. Shiramine Village Board of Education: Ishikawa Prefecture, Japan, 277 pp. [In Japanese with English abstract].

—— —— 2008. A herbivorous lizard from the Early Cretaceous of Japan. *Palaeontology*, **51**, 487–498.

—— —— COOK, E., HIRAYAMA, R., ISAJI, S., NICHOLAS, C. J., UNWIN, D. and YABUMOTO, Y. 1998. An Early Cretaceous assemblage from Gifu Prefecture, Japan. *Bulletin of the New Mexico Museum of Natural History and Science*, **14**, 183–186.

—— —— NORO, M., ISAJI, S. and YAMAGUCHI, M. 2006. A long-bodied aquatic varanoid lizard from the Early Cretaceous of Japan. *Palaeontology*, **49**, 1143–1165.

—— RAIA, P. and BARBERA, C. 2004. New lizards and sphenodontians from the Early Cretaceous of Italy. *Acta Palaeontologica Polonica*, **49**, 393–408.

—— and WANG YUAN 2005. *Dalinghosaurus*, a lizard from the Early Cretaceous Jehol Biota of northeast China. *Acta Palaeontologica Polonica*, **50**, 725–742.

—— —— 2007. A juvenile lizard from the Late Jurassic/Early Cretaceous of China. *Naturwissenschaften*, **94**, 431–439.

—— —— and LI CHUN 2005. The Early Cretaceous lizard *Yabeinosaurus* from China: resolving an enigma. *Journal of Systematic Palaeontology*, **3**, 319–335.

FEDOROV, P. V. and NESSOV, L. A. 1992. A lizard from the boundary of the Middle and Late Jurassic of north-east Fergana. *Vestnik Sankt-Petersburgskago Universiteta, Seriya 7 (Geologiya, Geografiya) [Bulletin of the St Petersburg University, Geology, Geography]* **3**, 9–14 [In Russian].

FUJITA, M. 2003. Geological age and correlation of the verte-brate-bearing horizons in the Tetori Group. *Memoir of the Fukui Prefectural Dinosaur Museum*, **2**, 3–14.

GAO KEQIN and CHENG ZHENGWU 1999. A new lizard from the Lower Cretaceous of Shandong, China. *Journal of Vertebrate Paleontology*, **19**, 456–465.

—— and HOU LIANHAI 1995. Iguanians from the Upper Cretaceous Djadochta Formation, Gobi Desert, China. *Journal of Vertebrate Paleontology*, **15**, 57–78.

—— —— 1996. Systematics and diversity of squamates from the Upper Cretaceous Djadochta Formation, Bayan Mandahu, Gobi Desert, People's Republic of China. *Canadian Journal of Earth Sciences*, **33**, 578–598.

—— and NESSOV, L. A. 1998. Early Cretaceous squamates from the Kyzylkum Desert, Uzbekistan. *Neues Jahrbüch für Geologie und Paläontologie Abhandlungen*, **207**, 289–309.

—— and NORELL, M. A. 2000. Taxonomic composition and systematics of late Cretaceous lizard assemblages from Ukhaa Tolgod and adjacent localities, Mongolian Gobi Desert. *American Museum of Natural History Bulletin*, **249**, 1–118.

GIFU-KEN DINOSAUR RESEARCH GROUP 1992. The Tetori Group in Ogamigo area, Shokawa-mura, Gifu Prefecture, Japan. *Bulletin of the Gifu Prefectural Museum*, **13**, 9–16. [In Japanese].

GILMORE, C. W. 1943. Fossil lizards of Mongolia. *Bulletin of the American Museum of Natural History*, **81**, 361–384.

HASEGAWA, Y., MANABE, M., ISAJI, S., OHKURA, M., SHIBATA, I. and YAMAGUCHI, I. 1995. Terminally re-sorbed iguanodontid teeth from the Neocomian Tetori Group, Ishikawa and Gifu Prefecture, Japan. *Bulletin of the National Science Museum, Tokyo, Series C*, **21**, 35–49.

HECHT, M. K. and HECHT, B. M. 1984. A new lizard from Jurassic deposits of Middle Asia. *Paleontologicheskii Zhurnal*, **3**, 135–138.

HIRAYAMA, R. 1996. Fossil land turtles from the Early Creta-ceous of Central Japan. *Journal of Vertebrate Paleontology*, **16** (Suppl. 3), 41A.

—— 1999. Testudinoid turtles from the Early Cretaceous (Neo-comina) of Central Japan. *Journal of Vertebrate Paleontology*, **19** (Suppl. 3), 51–52A.

—— 2000. Fossil turtles. 75–92. *In* MATSUOKA, H. (ed.). *Fossils of the Kuwajima 'Kaseki-kabe' (Fossil-bluff). Scientific report on a Neocomian (Early Cretaceous) fossil assemblage from the Kuwajima Formation, Tetori Group, Shiramine, Ishikawa, Japan.* Shiramine Village Board of Education: Ishikawa Prefecture, Japan, 277 pp. [In Japanese with English abstract].

HOFFSTETTER, R. 1967. Coup d'oeil sur les Sauriens (Lacer-tiliens) des couches de Purbeck (Jurassique supérieur d'Angle-terre). *Problemes Actuels de Paléontologie (Evolution des Vertebrés) Centre National de la Recherche Scientifique*, **163**, 349–371.

ISAJI, S., OKASAKI, H., HIRAYAMA, R., MATSUOKA, H., BARRETT, P., TSUBAMOTO, T., YAMAGUCHI, M., YAMAGUCHI, I. and SAKUMOTO, T. 2005. Deposi-tional environments and taphonomy of the bone-bearing beds of the Lower Cretaceous Kuwajima Formation, Tetori Group, Japan. *Bulletin of the Kitakyushu Museum of Natural History and Human History, Series A Natural History*, **3**, 123–133.

JI SHUAN 1998. A new long-tailed lizard from Upper Jurassic of Liaoning, China. 496–505. *In* GEOLOGY DEPART-MENT OF PEKING UNIVERSITY (ed.). *Collected works of International Symposium on Geological Science held at Peking University, Beijing, China, 1998.* Seismological Press, Beijing, 1066 pp.

—— 2005. A new Early Cretaceous lizard with well-preserved scale impressions from western Liaoning, China. *Progress in Natural Science*, **15**, 162–168.

—— and JI QIANG 2004. Postcranial anatomy of the Meso-zoic *Dalinghosaurus* (Squamata): evidence from a new speci-men of western Liaoning. *Acta Geologica Sinica*, **78**, 897–906.

—— LU LIWU and BO HAICHEN 2001. New material of *Yabeinosaurus tenuis* (Lacertilia). *[Land and Resources]*, **2001**, 41–43. [In Chinese].

—— and REN DONG 1999. First record of lizard skin fossil from China with description of a new genus (Lacertilia, Scinc-omorpha). *Acta Zootaxonomica Sinica*, **24**, 114–120.

KRAMBERGER, K. G. 1892. *Aigialosaurus*, eine neue Eidechse aus den Kreideschiefern der Insel Lesina mit Rucksicht auf die bereits beschriebenen Lacertiden von Comen und Lesina. *Glas-nik Huvatskoga Naravolosovnoga Derstva (Societas Historico-Matulis Croatica) u Zagrebu*, **7**, 74–106.

KUSUHASHI, N., MATSUOKA, H., KAMIYA, H. and SETOGUCHI, T. 2002. Stratigraphy of the late Mesozoic Tetori Group in the Hakusan Region, central Japan: an over-view. *Memoirs of the Faculty of Science, Kyoto University, Geol-ogy and Mineralogy Series*, **59**, 9–31.

LESHCHINSKIY, S. V., VORONKEVICH, A. V., FAY-NGERTZ, A. V., MASCHENKO, E. N., LOPATIN, A. V. and AVERIANOV, A. O. 2001. Early Cretaceous verte-brate locality Shestakovo, Western Siberia, Russia: a refugium for Jurassic relicts? *Journal of Vertebrate Paleontology*, **21**, 73A.

LI JINLING 1985. A new lizard from the Late Jurassic of Su-bei, Gansu. *Vertebrata PalAsiatica*, **23**, 13–18. [In Chinese with English summary].

LI PIPENG, GAO KEQIN, HOU LIANHAI and XU XING 2007. A gliding lizard from the Early Cretaceous of China. *Proceedings of the National Academy of Sciences USA*, **104**, 5507–5509.

MAEDA, S. 1961. On the geological history of the Mesozoic Tetori Group in Japan. *Journal of the College of Arts and Science, Chiba University*, **3**, 369–426. [In Japanese with English abstract].

MANABE, M. 1999. The early history of the Tyrannosauridae in Asia. *Journal of Paleontology*, **13**, 477–482.

—— and BARRETT, P. M. 2000. Dinosaurs. 93–98. *In* MATSUOKA, H. (ed.). *Fossils of the Kuwajima 'Kaseki-kabe' (Fossil-bluff). Scientific report on a Neocomian (Early Creta-ceous) fossil assemblage from the Kuwajima Formation, Tetori Group, Shiramine, Ishikawa, Japan.* Shiramine Village Board of Education, Ishikawa Prefecture, Japan, 277 pp. [In Japanese with English abstract].

—— —— and ISAJI, S. 2000a. A refugium for relicts? *Nature*, **404**, 953.

—— ROUGIER, G. W., ISAJI, S. and MATSUOKA, H. 2000b. Fossil mammals. 107–108. *In* MATSUOKA, H. (ed.). *Fossils of the Kuwajima 'Kaseki-kabe' (Fossil-bluff). Scientific report on a Neocomian (Early Cretaceous) fossil assemblage from* the Kuwajima Formation, Tetori Group, Shiramine, Ishikawa, Japan. Shiramine Village Board of Education, Ishikawa Prefecture, Japan, 277 pp. [In Japanese with English abstract].

MATEER, N. 1982. Osteology of the Jurassic lizard Ardeosaurus brevipes (Meyer). Palaeontology, 25, 461–469.

MATSUKAWA, M. and OBATA, I. 1994. Dinosaurs and sedimentary environments in the Japanese Cretaceous: a contribution to dinosaur facies in Asia based on molluscan palaeontology and stratigraphy. Cretaceous Research, 15, 101–125.

MATSUMOTO, R., EVANS, S. E. and MANABE, M. 2007. Monjurosuchus (Reptilia: Choristodera) from the Lower Cretaceous of Japan. Acta Palaeontologica Polonica, 52, 329–350.

MATSUOKA, H. 2000a. A frog fossil. 50–52. In MATSUOKA, H. (ed.). Fossils of the Kuwajima 'Kaseki-kabe' (Fossil-bluff). Scientific report on a Neocomian (Early Cretaceous) fossil assemblage from the Kuwajima Formation, Tetori Group, Shiramine, Ishikawa, Japan. Shiramine Village Board of Education: Ishikawa Prefecture, Japan, 277 pp. [In Japanese].

—— 2000b. Tritylodonts (Synapsida, Therapsida). 53–74. In MATSUOKA, H. (ed.). Fossils of the Kuwajima 'Kaseki-kabe' (Fossil-bluff). Scientific report on a Neocomian (Early Cretaceous) fossil assemblage from the Kuwajima Formation, Tetori Group, Shiramine, Ishikawa, Japan. Shiramine Village Board of Education: Ishikawa Prefecture, Japan, 277 pp. [In Japanese].

NESSOV, L. A. 1988. Late Mesozoic amphibians and lizards of Soviet Middle Asia. Acta Zoologica Cracoviensia, 31, 475–486.

—— and GAO KEQIN 1993. Cretaceous lizards from the Kyzylkum Desert, Uzbekistan. Journal of Vertebrate Paleontology, 13, 51A.

OPPEL, M. 1811. Die Ordnungen, Familien und Gattungen der Reptilien, als Prodrom einer Naturgeschichte derselben. Joseph Lindauer, Munich, 87 pp.

PROTHERO, D. R. and ESTES, R. 1980. Late Jurassic lizards from Como Bluff, Wyoming, and their palaeobiogeographic significance. Nature, 286, 484–486.

REYNOSO, V. H. 1998. Huehuecuetzpalli mixtecus gen. et sp. nov: a basal squamate (Reptilia) from the Early Cretaceous of Tepexi de Rodríguez, Central México. Philosophical Transactions of the Royal Society, Series B, 353, 477–500.

—— and CALLISON, G. 2000. A new scincomorph lizard from the early Cretaceous of Puebla, México. Zoological Journal of the Linnean Society, 130, 183–212.

RICHTER, A. 1994. Lacertilier aus der Unteren Kreide von Uña und Galve (Spanien) und Anoual (Marokko). Berliner geowissenschaftliche Abhandlungen, 13, 135–161.

ROBINSON, P. L. 1962. Gliding lizards from the Upper Keuper of Great Britain. Proceedings of the Geological Society, London, 1601, 137–146.

ROUGIER, G. W., ISAJI, S. and MANABE, M. 2007. An Early Cretaceous mammal from the Kuwajima Formation (Tetori Group), Japan, and a reassessment of triconodont phylogeny. Annals of Carnegie Museum, 76, 73–115.

RUSSELL, D. A. 1967. Systematics and morphology of American mosasaurs. Peabody Museum of Natural History, Yale, Bulletin, 23, 1–241.

SETOGUCHI, T., MATSUOKA, H. and MATSUDA, M. 1999a. New discovery of an early Cretaceous tritylodontid

(Reptilia, Therapsida) from Japan and the phylogenetic reconstruction of Tritylodontidae based on dental characters. 117–124. In WANG YUANQING and DENG TAO (eds). Proceedings of the Seventh Annual Meeting of the Chinese Society of Vertebrate Paleontology. China Ocean Press, Beijing, 274 pp.

—— TSUBAMOTO, T., HANAMURA, H. and HACHIYA, K. 1999b. An early Late Cretaceous mammal from Japan, with reconsideration of the evolution of tribosphenic molars. Paleontological Research, 3, 18–28.

SHIKAMA, T. 1969. On a Jurassic reptile from Miyama-cho, Fukui Prefecture, Japan. Science reports, Yokohama National University, 15, 25–34.

SMITH, P. E., EVENSEN, N. M., YORK, D., CHANG MEEMANN, JIN FAN, LI JINLING, CUMBAA, S. and RUSSELL, D. 1995. Dates and rates in ancient lakes: ^{40}Ar-^{39}Ar evidence for an Early Cretaceous age for the Jehol Group, north-east China. Canadian Journal of Earth Sciences, 32, 1426–1431.

SULIMSKI, A. 1975. Macrocephalosauridae and Polyglyphanodontidae (Sauria) from the Late Cretaceous of Mongolia. Palaeontologica Polonica, 33, 25–102.

SWOFFORD, D. L. 1993. PAUP* – (Phylogenetic Analysis Using Parsimony, version 3.1.1. University of Illinois, Urbana-Champaign.

—— 2001. PAUP* (Version 4.06b10 for 32-bit Microsoft Windows). Sinauer Associates Incorporated, Publishers, Sunderland, Massachusetts, USA.

TAKADA, T., MATSUOKA, H. and SETOGUCHI, T. 2001. The first multituberculate from Japan. 55–58. In DENG TAO and WANG YUAN (eds). Proceedings of the Eighth Annual Meeting of the Chinese Society of Vertebrate Paleontology, China Ocean Press, Beijing, 301 pp.

TOWNSEND, T. M., LARSON, A., LOUIS, E. and MACEY, J. R. 2004. Molecular phylogenetics of Squamata: the position of snakes, amphisbaenians, and dibamids, and the root of the squamate tree. Systematic Biology, 53, 735–757.

UNWIN, D. M. and MATSUOKA, H. 2000. Pterosaurs and birds. 51–54. In MATSUOKA, H. (ed.). Fossils of the Kuwajima 'Kaseki-kabe' (Fossil-bluff). Scientific report on a Neocomian (Early Cretaceous) fossil assemblage from the Kuwajima Formation, Tetori Group, Shiramine, Ishikawa, Japan. Shiramine Village Board of Education, Ishikawa Prefecture, Japan, 277. [In Japanese].

—— BAKHURINA, N. N., LOCKLEY, M. G., MANABE, M. and LÜ, J. 1997. Pterosaurs from Asia. Journal of the Palaeontological Society of Korea, 2, 43–65.

—— MANABE, M., SHIMIZU, K. and HASEGAWA, Y. 1996. First record of pterosaurs from the Early Cretaceous Tetori Group: a wing-phalange from the Amagodani Formation in Shokawa, Gifu Prefecture, Japan. Bulletin of the National Science Museum, Tokyo, Series C, Geology and Paleontology, 22, 37–46.

WANG XIAOLIN and ZHOU ZHONGHE 2003. Mesozoic Pompei. 19–35. In CHANG MEEMANN, CHEN PEI JI, WANG YUANQING and WANG YUAN (eds). The Jehol Biota: emergence of feathered dinosaurs and beaked birds.

Shanghai Scientific and Technical Publishers, Shanghai, China, 208 pp.

—— —— HE HUAIYU, JIN FAN, WANG YUANQING, ZHANG JIANGYONG, WANG YUAN, XU XING and ZHANG FUCHENG 2005. Stratigraphy and age of the Daohugou Bed in Ningcheng, Inner Mongolia. *Chinese Science Bulletin*, **50**, 2369–2376.

—— WANG YUANQING, ZHANG FUCHENG, ZHANG JIANGYONG, ZHOU ZHONGHE, JIN FAN, HU YAOMING, GU GANG and ZHANG HAICHUN 2000. Vertebrate biostratigraphy of the Lower Cretaceous Yixian Formation in Lingyuan, western Liaoning and its neighboring southern Nei Mongol (Inner Mongolia), China. *Vertebrata PalAsiatica*, **38**, 81–99 (in Chinese with English summary).

WANG YUAN and EVANS, S. E. 2006a. Advances in the study of fossil amphibians and squamates from China: the past fifteen years. *Vertebrata PalAsiatica*, **44**, 60–73.

—— —— 2006b. Small tetrapods of the Early Cretaceous Jehol Biota and associated faunae in China: palaeoenvironmental implications. 142–146. *In* BARRETT, P. M. and EVANS, S. E. (eds). *Ninth International Symposium on Mesozoic terrestrial Ecosystems and Biota, Abstracts and Proceedings*. Natural History Museum, London, 187 pp.

YABUMOTO, Y. 2000. Fossil fishes. 46–49. *In* MATSUOKA, H. (ed.). *Fossils of the Kuwajima 'Kaseki-kabe' (Fossil-bluff). Scientific report on a Neocomian (Early Cretaceous) fossil assemblage from the Kuwajima Formation, Tetori Group, Shiramine, Ishikawa, Japan*. Shiramine Village Board of Education, Ishikawa Prefecture, Japan, 277 pp. [In Japanese].

YOUNG, C. C. 1958. On a new locality of *Yabeinosaurus tenuis* Endo and Shikama. *Vertebrata PalAsiatica*, **2**, 151–156.

ZHOU ZHONGHE, BARRETT, P. M. and HILTON, J. 2003. An exceptionally preserved Lower Cretaceous ecosystem. *Nature*, **421**, 807–814.

APPENDIX

Data for *Sakurasaurus* as coded for the characters in Conrad (2008)

```
?000?   00101   110?0   00000   01???   ??010   10000   00000   1???0   0?10?
?0?10   02100   100-?   ??011   ??12-   00011   00?0?   10110   ???1?   ??0??
?????   00?00   00000   ???11   ?????   ?????   0?00?   00?0?   ?????   ?????
??00?   ???1?   ?000?   01???   ?001?   ?1000   02010   -000?   00?00   ???00
?0010   000?0   20000   00010   00100   A???1   2000?   ??-??   0?0??   ?????
?????   ?0???   ?????   ?????   ?????   ?????   ???1?   ?0?00   ?????   ?????
?????   ?????   ?????   ?????   ?????   ?????   ?????   ?????   ?????   ?????
?????   ?????   ????
```

[Special Papers in Palaeontology 81, 2009, pp. 61–69]

NEW CRANIAL AND DENTAL FEATURES OF *DISCOSAURISCUS AUSTRIACUS* (SEYMOURIAMORPHA, DISCOSAURISCIDAE) AND THE ONTOGENETIC CONDITIONS OF *DISCOSAURISCUS*

by JOZEF KLEMBARA

Faculty of Natural Sciences, Department of Ecology, Comenius University in Bratislava, Mlynská dolina B-1, 84215 Bratislava, Slovakia;
e-mail: klembara@fns.uniba.sk

Typescript received 1 May, 2008; accepted in revised form 22 June 2008

Abstract: Two new cranial and dental features of the Lower Permian discosauriscid seymouriamorph *Discosauriscus austriacus* are described. First, further mechanical preparation of the largest known specimen [skull length (SL) about 62 mm] has exposed the quadrate and articular, both documented for the first time. Second, the largest chemically prepared specimen (SL about 52 mm) shows well-developed dentine infolding on the basal portions of the marginal teeth and the presence of five well developed ridges (one mesial, one distal, one labial and two lingual) in the upper half of their external crown surface. It is argued here that the presence of the quadrate and articular and the morphology of the teeth, in conjunction with other anatomical information and new skeletochronological and microanatomical data, support the hypothesis that specimens of *D. austriacus* with SLs of about 45–52 mm represent postmetamorphic, late juvenile or subadult individuals and that the largest documented specimen with a SL of about 62 mm was presumably sexually mature.

Key words: *Discosauriscus austriacus,* Discosauriscidae, Seymouriamorpha, Lower Permian, Czech Republic, anatomy, skull, teeth.

ALL members of the family Discosauriscidae (Tetrapoda: Seymouriamorpha) are known exclusively from lacustrine sediments (Ivakhnenko 1987; Klembara and Meszáros 1992). Specimens with skull lengths (SLs) up to about 30–35 mm display external gills and represent the most abundant size category (Špinar 1952); no larger, fully ossified specimens were recorded. These observations led several authors to infer that discosauriscids were probably paedomorphic tetrapods (e.g. Holmes 1984; Smithson 1985; Ivakhnenko 1987). Later, Klembara (1995, 1997) described much larger specimens of *Discosauriscus austriacus* from the Lower Permian of the Boskovice Furrow in Moravia (Czech Republic) with SLs ranging from about 47 to 62 mm, and concluded that they represent postmetamorphic, juvenile stages which were presumably terrestrial, at least in part. Contrary to the smaller specimens with SLs up to about 35 mm, these larger specimens are extremely rare. Although they exhibit many features that distinguish them from smaller specimens, the degree of ossification of the braincase in the two size categories is comparable. They both display partially ossified basisphenoid (in the

region of the basipterygoid processes), basioccipital and exoccipital. Until now, no ossified quadrate and articular have been recorded, with one possible exception. Klembara (1997) reported a partially ossified quadrate in a small specimen, but in hundreds of similarly sized specimens this bone has never been documented. In early tetrapods with a biphasic lifecycle, the quadrate and articular ossify mostly after metamorphosis, and are first observed in juveniles, although in many instances, these bones ossify in later ontogenetic stages (Klembara *et al.* 2007 and references therein). Therefore, although morphological and size-related characters indicate strongly that the largest known specimens of *D. austriacus* represent postmetamorphic, juvenile stages (Klembara 1997; Klembara *et al.* 2001; Klembara *et al.* 2006, 2007), the absence of ossified quadrate and articular appeared to be at odds with the rest of the skeletal anatomy.

In the present paper, I present new information on the quadrate and articular as well as on the dentition in the largest of all known specimens of *D. austriacus*. The new data come from further mechanical and chemical

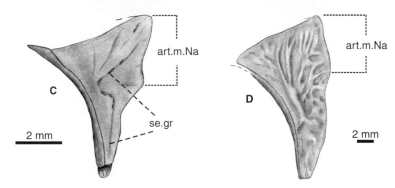

TEXT-FIG. 1. A, B, D, *Discosauriscus austriacus* (Makowsky, 1876), SNM Z 25814. A, photograph of skull in dorsal view. B, drawing of left prefrontal (the same magnification as in A). C, left prefrontal of *Discosauriscus pulcherrimus* (MHK 61803) in dorsal view. D, left prefrontal of *D. austriacus* in dorsal view.

preparation, as detailed below, and shed new light on the ontogenetic stages of these specimens.

The specimens of *D. austriacus* examined here come from two Lower Permian localities in the Boskovice Furrow and have previously been described in connexion with other aspects of their anatomy (e.g. Klembara 1995, 1997; Klembara and Bartík 2000). Specimen SNM Z 25814 from the Obora locality represents the largest individual of *D. austriacus* recorded so far with a SL of about 62 mm (Text-fig. 1A). It is carbonized, dorsoventrally flattened and not amenable to chemical preparation. However, additional mechanical preparation of several portions of the skull proved to be successful and delivered important information on the articulation area of the right lower jaw (in particular, the morphology of the quadrate and articular).

A series of chemically prepared specimens, especially SNM Z 15568, gave additional information on tooth morphology. Although the ontogeny of teeth in *D. austriacus* was briefly mentioned in the context of the recognition of the ontogenetic stages of various individuals of *Discosauriscus* (Klembara 1995), the morphology of the teeth, especially in the largest specimens, has not been described in detail.

The aims of this paper are: (1) to document the anatomy of the articulation region of the right lower jaw, such as is gleaned from observations of specimen SNM Z 25814; (2) to describe the morphology of the teeth in the largest known, chemically prepared specimens of *D. austriacus* (especially SNM Z 15568); and (3) to discuss the ontogenetic changes of *D. austriacus* on basis of all available anatomical and skeletochronological (Sanchez *et al.* 2008) data.

MATERIAL AND METHODS

The following specimens have been used for the present study (skull length (SL) measured as postparietal + parietal + frontal + nasal):

1. *Discosauriscus austriacus*, SNM Z 25814 (formerly DE OB 1, *Discosauriscus* sp.; Klembara 1995, fig. 8; SL about 62 mm); Obora locality, Boskovice Furrow, Czech Republic (Text-fig. 1A). The specimen is carbonized and dorsoventrally compressed; however, the individual bones are approximately *in situ* and the outlines of several of them are clearly traceable; in addition, the ornamentation of several bones (e.g. left prefrontal) is well preserved. The

skull of this specimen was subjected to further mechanical preparation in several places, especially the lower jaw articulation region of the right side.

2. *Discosauriscus austriacus*, SNM Z 15568 (formerly DE K 52, SL about 52 mm; Klembara 1995, fig. 7B); Kochov-Horka locality near Letovice (Klembara and Meszároš 1992). The specimen was removed from the embedding rock through chemical preparation and is three-dimensionally preserved.

Other specimens of *D. austriacus* used for comparisons include: DE KO 4 (SL – 40 mm); DE KO 16 (SL – about 21 mm); DE KO 26 (SL – 28 mm); DE KO 33 (SL – 26 mm); DE KO 41 (SL – 33 mm); DE KO 80 (SL – 47 mm); SNM Z 25744 (formerly DE K 323, SL – 49 mm; Klembara 1995, fig. 7A); DE KO 79 (SL – 51 mm); MHK 61803 (SL – 33 mm).

Taxonomical remarks. There are two currently recognized species in the genus *Discosauriscus* i.e. *D. austriacus* and *D. pulcherrimus* (for the complete list of the anatomical differences between these two species; see Klembara 1997). Specimen SNM Z 25814 (Text-fig. 1A–B) was formerly assigned to *Discosauriscus* sp. (Klembara 1995, p. 268, fig. 8). However, this specimen can be attributed to *D. austriacus* based upon the fact that the prefrontal-postfrontal suture lies in the middle of frontal length; conversely, in *D. pulcherrimus*, the posterior ramus of the prefrontal is short and the prefrontal-postfrontal suture lies at the boundary between the anterior and middle one-third of the frontal length. In *D. pulcherrimus*, the length of the prefrontal-nasal suture is of equal length or only slightly longer then the prefrontal-frontal suture (Text-fig. 1C). In *D. austriacus*, however, the posterior ramus of the prefrontal is relatively longer and the prefrontal-frontal suture is about twice as long as the prefrontal-nasal suture. Although the preservation of the frontal in SNM Z 25814 does not permit an accurate measurement of its length, the perfectly preserved left prefrontal shows that the prefrontal-frontal suture is twice as long as the prefrontal-nasal suture (Text-fig. 1D). Therefore, SNM Z 25814 is considered here to belong to *Discosauriscus austriacus* (Makowsky, 1876).

Description

Institutional abbreviations. DE: Department of Ecology, Faculty of Natural Sciences, Bratislava, Slovakia; MHK: Museum of Eastern Bohemia in Hradec Králové, Czech Republic; SNM: Slovak National Museum, Bratislava, Slovakia.

Anatomical abbreviations. Ar: articular; art.m.Na: articular margin for nasal; art.su.Pt: articular surface for quadrate ramus of pterygoid; art.su.Sq: articular surface for squamosal; den.inf: dentine infoldings; dis.ri: distal ridge; lab.ri: labial ridge; lin.ri: lingual ridge; mes.ri: mesial ridge; Par: prearticular; Pfr: prefrontal; Qu: quadrate; San: surangular; se.gr: sensory groove.

Posterior portion of check
On the right-hand side of SNM Z 25814, the region of the jaw articulation has been exposed in dorsal view after removal of the squamosal, quadratojugal and posterior portion of the jugal (Text-figs 1A, 2A). A small, stout and L-shaped skeletal element is clearly visible (Text-fig. 2A–B). The element in question is situated between the remnant of the unornamented medial margin of the squamosal otic flange, medially, and fragments of the surangular, laterally. The L-shaped element consists of a shorter posterior ramus and a considerably longer and larger main ramus. The former runs perpendicularly to the latter. The main ramus is oriented anteroposteriorly and shows a stout appearance. It is delimited medially and laterally by two walls oriented obliquely relative to its central surface. The anterior wall of the posterior ramus merges into the main ramus along a smoothly curved but abrupt inflexion. The posterior margin of the L-shaped element shows two posterior protrusions with a shallow notch between them. The outline of the entire element is clearly traceable in dorsal view. No similarly shaped element has been previously recorded in any known specimen of *Discosauriscus*, and its shape differs from that of any known skull or lower jaw bone of this tetrapod. The L-shaped element is interpreted as the quadrate. The reasons for this interpretation are as follows. First, the element is situated immediately below the posterior portions of the squamosal and quadratojugal and appears to be preserved in its original position, although it lies horizontally and is pushed slightly anterior to the position of the jaw joint (see also below). Second, the two posterior protrusions and shallow notch visible along the posterior wall of the element are reminiscent of condyles and intercondylar surface of typical quadrates, together representing the articular surface of this bone. If this interpretation is correct, then the obliquely oriented medial wall of the quadrate body (i.e. its main ramus) represents the articulation surface for the quadrate ramus of the pterygoid, whereas the obliquely oriented lateral wall represents the articulation surface for the squamosal. Chemically prepared squamosals of smaller specimens reveal that the quadrate-squamosal articulation is delimited laterally by a distinct crest running anterodorsally to-posteroventrally on the internal surface of the unornamented otic flange of the squamosal (Klembara 1997, fig. 13).

The identification of bony elements or fragments visible immediately behind the quadrate and extending to the level of the posterior end of the squamosal is problematic (Text-figs 1A, 2A, C). Remnants of three lower jaw bones, together contributing to the posterior portion of the latter, may be present. Laterally, a vertically oriented bone broken in several fragments may represent the surangular. Medially, a fragmented plate-like bone is identified as the prearticular. A third bone situated between the posteriormost portions of the putative surangular and prearticular is mediolaterally broad and stout with slightly depressed central and medial portions and a slightly raised lateral portion of its dorsal surface. This bone is identified as the articular (Text-fig. 2C). The suture between the posteriormost portion of the prearticular and the articular is clearly recognizable. The articular extends slightly

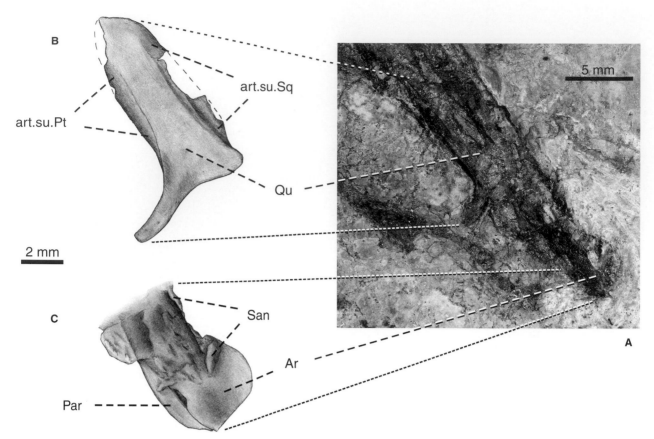

TEXT-FIG. 2. *Discosauriscus austriacus* (Makowsky, 1876), SNM Z 25814. A, photograph of enlarged area of right jaw joint region indicated by white quadrangular in Text-fig. 1A. B, drawing of quadrate. C, drawing of posteriormost portion of jaw joint.

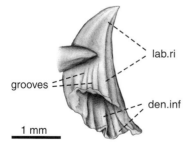

TEXT-FIG. 3. *Discosauriscus austriacus* (Makowsky, 1876), SNM Z 15568. Isolated, probably anterior tooth of left dentary in labial view showing basal portion broken off from internal wall of dentary.

more posteriorly than the prearticular. However, the anterior extension of the articular could not be determined, as it is obscured by bony fragments.

Morphology of teeth
The marginal teeth exhibit a conical shape with a pointed tip and a posterior curvature of the upper portions of their crown throughout the recorded growth stages of *D. austriacus* (Text-figs 3–4).

Specimens with skull length up to 40 mm: larval and metamorphic specimens (see below). In the early larval stage represented by DE KO 16 (SL about 21 mm), the posterior dentary teeth and a single maxillary tooth are preserved. Their external surfaces are completely smooth as are those of the slightly larger KO 33 (SL – 26 mm); in this specimen, breakages through the teeth reveal no dentine infolding inside the crowns (Text-fig. 4A–B).

In DE KO 26 (SL – 28 mm), dorsoventral grooves are present on the external surface of the base of the tooth crown, indicating the presence of dentine infolding inside the crowns. The dentine is folded inside the pulp cavity in such a way that the axis of elongation of each fold (projecting towards the centre of the pulp cavity) corresponds approximately in position to every other groove on the external surface of the tooth crown base. The length of the dentine folds is approximately one – third of the diameter of the pulp cavity. On the base of the fourth tooth in the left maxilla of DE KO 26, eight folds of dentine are present inside the pulp cavity, giving a total of sixteen grooves on the external surface of the tooth base. The grooves present on the portion of the tooth base attached to the internal surface of the maxilla or dentary are not visible; their count can be estimated on the basis of the number of dentine folds. The length of the external grooves occupies about one-fifth of the tooth length. In DE KO 41 (SL – 33 mm), the tooth morphology resembles that of DE KO 26. In the fourth teeth of the left max-

TEXT-FIG. 4. *Discosauriscus austriacus* (Makowsky, 1876). Left A, C and right E, F maxilla teeth in lingual A, C, E and labial F views and enlarged schematic drawings of their transverse sections at levels indicated by a–d based on DE KO 33 A, B, DE KO 4 C, D and SNM Z 15568 E–H.

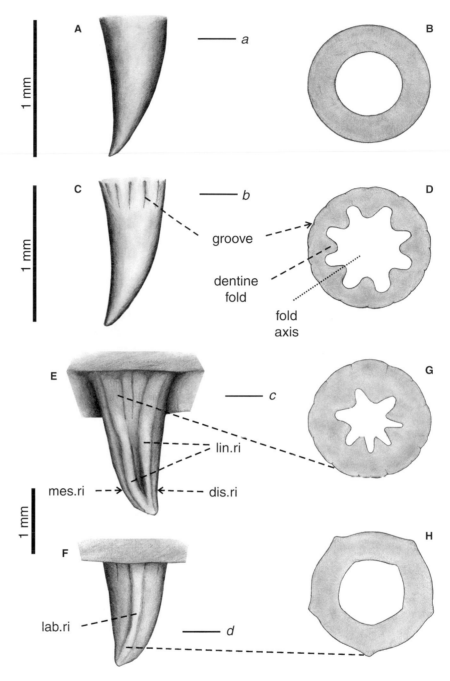

illa of DE KO 4 (SL – 40 mm), there are eight folds of dentine and thus sixteen external dorsoventral grooves (Text-fig. 4C, D). However, the number of grooves and folds may vary slightly in the marginal tooth rows of individual specimens.

Specimens with skull length 45–52 mm: juveniles or subadults (see below). Specimens belonging to these growth stages exhibit more or less distinct ridges on the external surface of the marginal teeth and palatal tusks (e.g. DE KO 80, SL about 47 mm; SNM Z 25744, SL – 49 mm; DE KO 79, SL – 51 mm). The ridges are particularly well-developed in SNM Z 15568 (SL about 52 mm) and form the basis for the following description (Text-figs 3, 4E–H).

The dorsoventral grooves on the external surface of the basal portions of the teeth extend for one-third of the tooth length. As in earlier ontogenetic stages, the number of grooves and corresponding dentine infoldings varies in the marginal dentition of individual specimens. However, in the majority of cases, eight internal folds are present, corresponding to sixteen external dorsoventral grooves (Text-fig. 4E, G). This suggests that the number of grooves and folds is already established in late larval or metamorphic specimens. The dentine folds reach approximately two-thirds of the mean length. The areas of the tooth base between adjacent dorsoventral grooves are slightly convex.

Five strong ridges are present on the remaining external surface of the teeth: one mesial, one distal, one labial and two lingual (Text-fig. 4E–H). The mesial and distal ridges are the sharpest and form cutting edges. The basal portions of all five ridges are continuous with the areas of the tooth base between adjacent dorsoventral grooves. These areas are generally more robust than those between remaining grooves. Transverse breakages in the tooth crowns show that these ridges have no distinct morphological expression on the pulp cavity surface; the latter is smooth and bears only shallow notches in the positions marked externally by the five ridges (Text-fig. 4H). Teeth are not preserved in SNM Z 25814 in which the quadrate and articular have been identified (see above).

DISCUSSION

All specimens of the two recognized species of *Discosauriscus* from the Boskovice Furrow come from lacustrine sediments. Their assignment to different ontogenetic stages has not been sufficiently resolved. Some authors have suggested that discosauriscids were probably neotenic (Holmes 1984; Smithson 1985) or facultatively neotenic (Boy and Sues 2000). The Tadzhikistan discosauriscid, *Ariekanerpeton sigalovi*, also comes from lacustrine sediments. Ivakhnenko (1987) considered the largest known specimens of *Ariekanerpeton*, with SLs of about 50 mm, to represent senile stages, although their degree of skeletal ossification is similar to that of *Discosauriscus* (Klembara 1995; Klembara and Ruta 2005a, b). Klembara (1994a, b, 1995, 1997) summarized different ontogenetic hypotheses by taking into account various aspects of the anatomy of discosauriscids and concluded that specimens of *D. austriacus* from the Boskovice Furrow encompass larval, metamorphic and juvenile stages; however, fully ossified, adult specimens of this species remain unknown.

The new skeletal and dental features described above contribute to the resolution of this intricate problem, i.e. the identification of ontogenetic stages of differently sized specimens of *D. austriacus* known so far.

Quadrate and articular

A small, partially ossified quadrate was described in one late larval specimen; however, this may represent an exceptional occurrence (Klembara 1997, fig. 17). The quadrate and articular, such as are observed in SNM Z 25814, represent the first record of these bones in *Discosauriscus*. Although the quadrate in SNM Z 25814 is not fully ossified, its general shape matches the shape of the quadrate of other stem amniotes, including *Seymouria* (White 1939; Klembara *et al.* 2005, 2006, 2007), *Proterogyrinus* (Holmes 1984) and *Pholiderpeton* (Clack 1987) and is also very similar to that of the parareptile *Procolophon* (Carroll and Lindsay 1985).

However, comparisons of morphology of the articular of *D. austriacus* with that in other early tetrapods are limited, because the preservation of the articular in the former is poor.

Teeth

The morphology of teeth, such as is elucidated by SNM Z 15568, is documented exclusively in the largest specimens of *D. austriacus*. The smallest specimen exhibiting external ridges already partially developed has a SL of about 47 mm. Unlike specimens with SLs up to about 45 mm, SNM Z 15568 is characterized by more extensive dentine infolding inside the pulp cavity (dentine folds also appear closer to the cavity centre). This condition is similar to that observed in *Seymouria baylorensis*, such as was described by Broili (1927), although the dentine appears more strongly infolded in this species, as the specimens described by Broili represent large adult stages.

The same external tooth morphology as that exhibited by SNM Z 15568 is present also in the largest specimen of *Ariekanerpeton sigalovi* (PIN 2079/262, SL about 48 mm) from the Lower Permian of Tadzhikistan. The largest teeth of this discosauriscid seymouriamorph, especially those of the premaxilla, bear the typical five keels (Klembara and Ruta 2005a, figs 1, 6C; Bulanov 2003). Analogically with the situation in *Discosauriscus austriacus*, such large specimens as PIN 2079/262 are very rare within otherwise large collection of specimens of this species (see also below).

The tooth morphology of SNM Z 15568 is also very similar to that of juvenile and possibly late juvenile or subadult specimens of the temnospondyl *Onchiodon* (Boy 1990), in which the external surface of the teeth bears ridges (cf. Text-fig. 4E–F and Boy 1990, p. 288, fig. 1). Unfortunately, Boy (1990) did not illustrate the transverse sections of the *Onchiodon* teeth, so the extension of dentine folds inside the pulp cavity remains unknown.

Ontogenetic conditions and life style

Klembara (1995, 1997) partitioned known specimens of *D. austriacus* into two groups: those with SLs up to about 45 mm are larval or metamorphic individuals tied to a lacustrine habitat (Klembara and Meszáros 1992); those with SLs ranging between 45 and 62 mm represent postmetamorphic, juvenile individuals presumably dwelling (completely or in part) on land. This conclusion is also supported by the fossil record: hundreds of specimens with SLs up to about 40 mm have been collected, but only six specimens are known in which the SL ranges from about 47 to 62 mm. In addition, specimens with SLs in the range of 40–45 mm are also very rare. The largest specimen of *Discosauriscus* in which external gills are preserved has a SL of about 32 mm (Klembara 1995). Based upon the presence of external gills, degree of development of sensory grooves and other morphological data, I concluded that specimens with SLs up to about 35 mm are larvae, those with SLs ranging from 35 to about 45 mm are metamorphic, and those with SLs ranging from about 45 to about 62 mm are postmetamorphic, juvenile stages. The occurrence of metamorphosis was inferred on the basis of morphological and proportional characters surveyed across all size classes as well as on the basis of comparisons between the ontogenetic series of *Discosauriscus* and those of some temnospondyls (e.g. Boy 1988, 1990, 1993). As reported by Klembara *et al.* (2006), the skull proportions of the largest *D. austriacus* skulls (SNM Z 15568 and SNM Z 25744) are very similar to those of adult *Karpinskiosaurus secundus* (Bulanov 2003 and pers. obs.) from the Upper Permian of Russia.

More recently, skeletochronological analyses conducted on long bones of *D. austriacus* have played a significant role in assessing the ontogenetic condition of the largest known specimens of this species; perhaps more interesting is the fact that these techniques suggest the age of individuals in different size categories. This is achieved through examination of lines of arrested growth (LAGs) (Castanet *et al.* 2003; Sanchez *et al.* 2008). The occurrence of LAGs reflects an annual periodicity pattern. Specimens with SLs of about 35 mm display 6 LAGs and were thus 6 years old when they died; specimens with SLs of about 40 mm display 8 LAGs; those with SLs of about 45 mm display 9 LAGs. The transverse section of the diaphyseal portion of the tibia of largest, chemically prepared specimen SNM Z 15568 (SL about 52 mm) exhibits 10 LAGs. In Recent amphibians and squamates, the tightening (i.e. close packing) of peripheral LAGs is associated with deceleration of growth and acquisition of sexual maturity (Castanet *et al.* 2003). The three peripheral LAGs of the tibia of SNM Z 15568 show the initial phases of this tightening and suggest that in this specimen, growth had already begun to slow down and that sexual maturity was being approached (Sanchez *et al.* 2008). These data provide independent evidence in support of the hypothesis that SNM Z 15568 was a late juvenile or subadult individual. Furthermore, the degree of compactness of the bone cortex in this specimen indicates that its limbs were already adapted to a terrestrial locomotion (Sanchez *et al.* 2008). Unlike smaller specimens in which the complete set of sensory grooves on the skull roof bones is present, the largest specimens (SNM Z 15568 and SNM Z 25744) bear only remnants of such grooves on the snout (Klem-

bara 1996, 1997), which suggests that they lived already partially or completely outside their lacustrine habitat, although probably still in its close vicinity. The perfectly preserved left prefrontal of the largest specimen SNM Z 25814 does not show the presence of sensory grooves (Text-fig. 1D). Independently of the morphological and skeletochronological investigations, the terrestrial lifestyle has also been recently inferred for the largest specimens of *Discosauriscus* on the basis of the study of the tibia microanatomy (Kriloff *et al.* 2008).

The skull of SNM Z 25814 (SL about 62 mm) is 10 mm longer than the largest skeletochronologically studied specimen, SNM Z 15568 (SL about 52 mm), which represents a late juvenile or subadult specimen. Although long bones are not preserved in SNM Z 25814, it may be speculated that it exhibited higher number of closely packed peripheral LAGs than SNM Z 15568. Thus, it might have reached sexual maturity.

Complete ontogenetic sequences are not known in any stem amniote. However, a partial growth series is documented in *Seymouria sanjuanensis* (Klembara *et al.* 2007). It includes an early and a late juvenile stage with SLs of 19 mm and 56 mm, respectively; the length of the adult skulls is up to about 95 mm. The quadrate and articular are already present in the early juvenile skull, which is also the smallest specimen of *S. sanjuanensis*. Comparisons between the smallest *S. sanjuanensis* and similarly sized skulls of *Discosauriscus* show that *S. sanjuanensis* metamorphosed at a much earlier ontogenetic stage than *Discosauriscus* (Klembara *et al.* 2007). *S. sanjuanensis* was adapted to a terrestrial environment and was able to survive highly ephemeral aquatic conditions (Berman *et al.* 2000; Klembara *et al.* 2001; Klembara *et al.* 2007). In temnospondyls, several of which are known from detailed growth series, the quadrate and articular first appear after metamorphosis, in the juvenile stages, but ossify completely quite late in ontogeny, mostly in late juvenile, subadult or adult stages (e.g. *Sclerocephalus*, Boy 1988; *Onchiodon*, Boy 1990; Werneburg and Steyer 1999; Witzmann 2005; *Micromelerpeton*, Boy 1995, fig. 8A; *Cheliderpeton*, Boy 1993; *Palatinerpeton*, Boy 1996). However, in *Acanthostomatops vorax* (Witzmann and Schoch 2006) and *Apateon gracilis* (Schoch and Fröbisch 2006), the quadrate and articular are present only in the adult stages. It is interesting to note that in adult, fully terrestrial *Acanthostomatops* specimens (SL = 75 mm), no ossified braincase elements have been reported (Witzmann and Schoch 2006). One exception is represented by the large temnospondyl *Archegosaurus decheni* in which the quadrate starts to ossify in larval stages, before the resorption of the branchial platelets (Witzmann 2006a, b). However, as concluded by Witzmann (2006a, b), *A. decheni* did not metamorphose. Data gleaned from all these comparisons, together with the presence of a quadrate and articular, suggest that the largest known specimen of *D. austriacus*, SNM Z 25814, is postmetamorphic and may represent a juvenile, subadult or adult stage. In addition, results from skeletochronological analyses tend to corroborate this ontogenetic assignment for SNM Z 25814. An early adult condition for this specimen cannot be ruled out.

The hypothesis that the largest, late juvenile or subadult specimens (SNM Z 15568 and SNM Z 25744) of *D. austriacus* were already adapted to terrestrial locomotion is also supported by anatomical data that suggest a change in diet. In larval and metamorphic specimens, several strong ridges covered in small denticles form a radiating pattern on the ventral surface of the pterygoid, whereas in large specimens, such ridges are absent and many densely arranged rows of small pointed and sharp denticles are present (Klembara 1997; Klembara *et al.* 2006, fig. 5B). Also, the strong ridges present on the marginal teeth and palatal tusks of these specimens (see above) strongly indicate adaptation to capturing larger preys on land or by lake shores. Furthermore, the postcranial skeleton of these large specimens is massively built (Klembara and Bartík 2000). Thus, the zygapophyses are proportionally larger than in smaller specimens and the neural arches are distinctly swollen and very broad in comparison with those of smaller individuals (Klembara and Bartík 2000, figs 2–4). The degree of the ossification of the humerus in *D. austriacus* is basically the same as that in the similarly sized *Seymouria sanjuanensis* representing the late juvenile ontogenetic stage (Klembara *et al.* 2001, 2006, 2007).

To summarize, both morphological and paleohistological data indicate, that (1) specimens of *D. austriacus* with SLs of about 45–52 mm represent postmetamorphic, late juvenile or subadult individuals and that the largest documented specimen with a SL of about 62 mm presumably reached already sexual maturity; and (2) the metamorphosis or the transition to the terrestrial life in *Discosauriscus* was a gradual process what is probably a plesiomorphic feature for early tetrapods (present also in many temnospondyls), in contrast to the condensed metamorphosis in *Apateon gracilis* (Schoch and Fröbisch 2006) and lissamphibians.

Acknowledgements. I would like to dedicate this paper to Dr Andrew R. Milner, who helped me in many respects throughout my career over many years. I am thankful Dr M. Ruta (University of Bristol) for critically reading the manuscript and the text correction. I thank Dr Florian Witzmann (Humbolt University, Berlin) for useful discussion regarding some aspects of the ontogeny of temnospondyls. The photographs of SNMZ 25814 in Text-figures 1A and 2A were taken by Július Kotus (Bratislava). The drawings in Text-figures 1C–D, 2B–C, 3 and 4 were made by A. Čerňanský (Comenius University in Bratislava). This project was partly supported by the Scientific Grant Agency, Ministry of Education of the Slovak Republic and the Slovak Academy of Sciences, Grant Nr. 1/0197/09.

REFERENCES

BERMAN, D. S., HENRICI, A. C., SUMIDA, S. S. and MARTENS, T. 2000. Redescription of *Seymouria sanjuanensis* (Seymouriamorpha) from the Lower Permian of Germany based on complete, mature specimens with a discussion of paleoecology of the Bromacker locality assemblage. *Journal of Vertebrate Paleontology*, **20**, 253–268.

BOY, J. A. 1988. Über einige Vertreter der Eryopoidea (Amphibia: Temnospondyli) aus dem europäischen Rotliegend (?höchstes Karbon – Perm). 1. *Sclerocephalus*. *Paläontologische Zeitschrift*, **62**, 107–132.

—— 1990. Über einige Vertreter der Eryopoidea (Amphibia: Temnospondyli) aus dem europäischen Rotliegend (?höchstes Karbon – Perm). 3. *Onchiodon*. *Paläontologische Zeitschrift*, **64**, 287–312.

—— 1993. Über einige Vertreter der Eryopoidea (Amphibia: Temnospondyli) aus dem europäischen Rotliegend (?höchstes Karbon – Perm). 4. *Cheliderpeton latirostre*. *Paläontologische Zeitschrift*, **67**, 123–143.

—— 1995. Über die Micromelerpetontidae (Amphibia: Temnospondyli). 1. Morphologie und Paläoökologie des *Micromelerpeton credneri* (Unter-Perm; SW-Deutschland. *Paläontologische Zeitschrift*, **69**, 429–457.

——1996. Ein neuer Eryopoide (Amphibia: Temnospondyli) aus dem saarpfälzischen Rotlieged (Unter-Perm; Südwest-Deutschland). *Mainzer geowissenschaftliche Mitteilungen*, **25**, 7–26.

—— and SUES, H.-D. 2000. Branchiosaurs, larvae, metamorphosis and heterochrony in temnospondyls and seymouriamorphs. 1150–1197. In HEATWOLE, H. and CARROLL, R. L. (eds). *Amphibian biology, 4, palaeontology*. Surrey Beatty & Sons, Chipping Norton, 1496 pp.

BROILI, F. 1927. Über den Zahnbau von *Seymouria*. *Anatomischer Anzeiger*, **63**, 185–188.

BULANOV, V. V. 2003. Evolution and systematics of seymouriamorph parareptiles. *Paleontological Journal*, **37** (Suppl. 1), 1–105.

CARROLL, R. L. and LINDSAY, W. 1985. Cranial anatomy of the primitive reptile *Procolophon*. *Canadian Journal of Earth Sciences*, **22**, 1571–1587.

CASTANET, J., FRANCILLON-VIEILLOT, H. and DE RICKLÈS, A. 2003. The skeletal histology of the Amphibia. 1598–1683. In HEATWOLE, H. and DAVIES, M. (eds). *Amphibian biology, 5, osteology*. Surrey Beatty & Sons, Chipping Norton, 536 pp.

CLACK, J. A. 1987. *Pholiderpeton scutigerum* Huxley, an amphibian from the Yorkshire Coal Measures. *Philosophical Transactions of the Royal Society of London, Series B*, **318**, 1–107.

HOLMES, R. B. 1984. The Carboniferous amphibian *Proterogyrinus scheelei* Romer, and the early evolution of tetrapods.

Philosophical Transactions of the Royal Society of London, Series B, **306**, 431–527.

IVAKHNENKO, M. F. 1987. Permian parareptiles of USSR. *Trudy Paleontologicheskogo Instituta Akademii Nauk*, **223**, 1–160. [In Russian].

KLEMBARA, J. 1994a. Electroreceptors in the Lower Permian tetrapod *Discosauriscus austriacus* (Makowsky 1876). *Palaeontology*, **37**, 609–626.

—— 1994b. The sutural pattern of skull-roof bones in Lower Permian *Discosauriscus austriacus* from Moravia. *Lethaia*, **27**, 85–95.

—— 1995. The external gills and ornamentation of skull roof bones of the Lower Permian tetrapod *Discosauriscus* (Kuhn 1933) with remarks of its ontogeny. *Paläontologische Zeitschrift*, **69**, 265–281.

—— 1996. The lateral line system of *Discosauriscus austriacus* (Makowsky, 1876) and the homologization of skull roof bones between tetrapods and fishes. *Palaeontographica A*, **240**, 1–27.

—— 1997. The cranial anatomy of *Discosauriscus* Kuhn, a seymouriamorph tetrapod from the Lower Permian of the Boskovice Furrow (Czech Republic). *Philosophical Transactions of the Royal Society of London*, **352**, 257–302.

—— and BARTÍK, I. 2000. The postcranial skeleton of *Discosauriscus* Kuhn, a seymouriamorph tetrapod from the Lower Permian of the Boskovice Furrow (Czech Republic). *Transactions of the Royal Society of Edinburgh: Earth Sciences*, **90**, 287–316.

—— and MESZÁROŠ, Š. 1992. New finds of *Discosauriscus austriacus* (Makowsky, 1876) from the Lower Permian of the Boskovice Furrow (Czecho-Slovakia). *Geologica carpathica*, **43**, 305–312.

—— and RUTA, M. 2005a. The seymouriamorph tetrapod *Ariekanerpeton sigalovi* from the Lower Permian of Tadzhikistan. Part I. Cranial anatomy and ontogeny. *Transactions of the Royal Society of Edinburgh, Earth Sciences*, **96**, 43–70.

—— —— 2005b. The seymouriamorph tetrapod *Ariekanerpeton sigalovi* from the Lower Permian of Tadzhikistan. Part II. Postcranial anatomy and relationships. *Transactions of the Royal Society of Edinburgh: Earth Sciences*, **96**, 71–93.

—— MARTENS, T. and BARTÍK, I. 2001. The postcranial remains of a juvenile seymouriamorph tetrapod from the Lower Permian Rotliegend of the Tambach Formation of Central Germany. *Journal of Vertebrate Paleontology*, **21**, 521–527.

—— BERMAN, D. S., HENRICI, A. and ČERŇANSKÝ, A. 2005. New structures and reconstructions of the skull of the seymouriamorph *Seymouria sanjuanensis*, Vaughn. *Annales of Carnegie Museum*, **74**, 217–224.

—— —— —— and WERNEBURG, R. 2006. Comprison of cranial anatomy and proportions of similarly sized *Seymou-*

ria sanjuanensis and *Discosauriscus austriacus*. *Annales of Carnegie Museum*, **75**, 37–49.

—— —— —— —— —— and MARTENS, T. 2007. First description of skull of Lower Permian *Seymouria sanjuanensis* (Seymouriamorpha: Seymouriidae) at an early juvenile stage. *Annals of Carnegie Museum*, **76**, 53–72.

KRILOFF, A., GERMAIN, D., CANOVILLE, A., SACHE, M. and LAURIN, M. 2008. Evolution of bone microanatomy of the tetrapod tibia and its use in palaeobiological inference. *Journal of Evolutionary Biology*, **21**, 807–826.

MAKOWSKY, A. 1876. Über einen neuen Labyrinthodonten 'Archegosaurus austriacus nov. spec'. *Sitzungsberichte der keiserischen Akademie der Wissenschaft*, **73**, 155–166.

SANCHEZ, S., KLEMBARA, J., CASTANET, J. and STEYER, S. 2008. Salamander-like development in a stem amniote revealed by paleohistology. *Biology Letters*, **4**, 411–414.

SCHOCH, R. and FRÖBISCH, N. B. 2006. Metamorphosis and neoteny: alternative pathways in an extinct amphibian clade. *Evolution*, **60**, 1467–1475.

SMITHSON, T. R. 1985. The morphology and relationships of the Carboniferous amphibian *Eoherpeton watsoni* Panchen. *Zoological Journal of the Linnean Society*, **85**, 317–410.

ŠPINAR, Z. V. 1952. Revision of some Moravian Discosauriscidae (Labyrinthodontia). *Rozpravy Ústředního Ústavu Geologického*, **15**, 1–115. [In Czech].

WERNEBURG, R. and STEYER, J. S. 1999. Redescription of the holotype of *Actinodon frossardi* GAUDRY, 1866 from the Lower Permian of France (Autun). *Geobios*, **32**, 599–607.

WHITE, T. E. 1939. Osteology of *Seymouria baylorensis* Broili. *Bulletin of the Museum of Comparative Zoology, Harvard College*, **85**, 325–409.

WITZMANN, F. 2005. Hyobranchial and postcranial ontogeny of the temnospondyl *Onchiodon labyrinthicus* (Geinitz, 1861) from Niederhäslich (Döhlen Basin, Autunian, Saxony). *Paläontologische Zeitschrift*, **79**, 479–492.

—— 2006a. Cranial morphology and ontogeny of the Permo–Carboniferous temnospondyl *Archegosaurus decheni* Goldfuss, 1847 from the Saar-Nahe Basin, Germany. *Transactions of the Royal Society of Edinburgh, Earth Sciences*, **96** (for 2005), 131–162.

—— 2006b. Developmental patterns and ossification sequence in the Permo–Carboniferous temnospondyl *Archegosaurus decheni* (Saar-Nahe Basin, Germany). *Journal of Vertebrate Paleontology*, **26**, 7–17.

—— and SCHOCH, R. 2006. Skeletal development of the temnospondyl *Acanthostomatops vorax* from the Lower Permian Döhlen Basin of Saxony. *Transactions of the Royal Society of Edinburgh, Earth Sciences*, **96** (for 2005), 365–385.

[Special Papers in Palaeontology 81, 2009, pp. 71–89]

A REVISION OF *SCINCOSAURUS* (TETRAPODA, NECTRIDEA) FROM THE MOSCOVIAN OF NÝŘANY, CZECH REPUBLIC, AND THE PHYLOGENY AND INTERRELATIONSHIPS OF NECTRIDEANS

by ANGELA C. MILNER* *and* MARCELLO RUTA†

*Department of Palaeontology, The Natural History Museum, SW7 5BD London, UK; e-mail: a.milner@nhm.ac.uk
†Department of Earth Sciences, University of Bristol, BS8 1RJ Bristol, UK; e-mail: m.ruta@bristol.ac.uk

Typescript received 1 May 2009; accepted in revised form 19 May 2009

Abstract: The morphology of the terrestrial nectridean *Scincosaurus*, from the Moscovian 'Gaskohle' near Nýřany, Czech Republic, is revised in detail. Its anatomical characters are incorporated in a new cladistic analysis to investigate the internal relationships of nectrideans and their phylogenetic position relative to other Palaeozoic tetrapod groups. We constructed a data matrix consisting of eight outgroups, 11 nectridean operational taxonomic units for the best known taxa, and 173 characters (109 cranial; 64 postcranial) using a combination of characters from previously published work, supplemented by additional first hand observations on other nectridean taxa. Our analysis retrieved nectrideans as a clade, in contrast to some other recent studies. Two monophyletic sister groups are represented within the clade. The first is urocordylids, with the two subfamilies, sauropleurines and urocordylines, paired as sister taxa. The second group is represented by the terrestrial *Scincosaurus* as sister taxon to the diplocaulids, which are all aquatic. Nectrideans as a group received unexpectedly low bootstrap support, as did the nodes subtending urocordylids and the scincosaurid-diplocaulid clade. Despite the large number of putative shared derived characters of the group, the analysis produced low bootstrap percentage and decay index for nectrideans as a whole. This might result from the fact that nectridean synapomorphies are mostly confined to their postcranial skeleton; indeed, the majority of those pertain to the construction of vertebrae. Therefore, it is possible that the signal carried by those synapomorphies is swamped by the distribution of other, particularly cranial, features.

Key words: Anatomy, Nectridea, phylogeny, *Scincosaurus*.

NECTRIDEANS are a clade of small, mostly aquatic tetrapods recorded in the Pennsylvanian and Permian of Europe, North America, and Africa. They are among the earliest discovered groups of Palaeozoic tetrapods (Brownrigg 1866). They were grouped together in a single order by Miall (1875) on the basis of characteristic vertebral morphology, although they exhibit two very different skull morphologies. Synoptic reviews of nectrideans have been provided by A. C. Milner (1980), based on Milner (1978), Milner (1993), and Bossy and Milner (1998), based on Bossy (1976) and Milner (1978).

Nectrideans include the three families Urocordylidae, Diplocaulidae, and Scincosauridae. The long-bodied urocordylids possessed long deep swimming tails. The diplocaulids, most readily recognisable by the boomerang-headed Permian genus *Diplocaulus*, were short-bodied forms with horned skulls. In Carboniferous diplocaulids, a connection between the skull horns and the shoulder girdle presumably facilitated a damping mechanism that restricted head oscillation during swimming. Both families possessed membrane bone vertebrae with a single ossification per segment (Gardiner 1983), and with expanded and ornamented fan-shaped neural and haemal spines. Accessory articulations were present above the zygapophyses in the dorsal vertebrae and variably between haemal arches in the caudal vertebrae.

The third family, the scincosaurids, is exemplified by *Scincosaurus*. This genus shares the characteristic hallmark vertebral structure of all nectrideans, but exhibits a remarkably different overall body shape, being a heavily ossified terrestrial form. The size distribution of *Scincosaurus* specimens, all recorded from the single geographically restricted locality at Nýřany, was interpreted by A. R. Milner (1980) as representing a single age class of terrestrial or semi-terrestrial adults that had returned to a shallow coal-swamp breeding pool and suffered a mass mortality. The opportunity is taken here to revise Fritsch (1881) original description of this highly unusual tetrapod in the context of investigating the interrelationships of nectrideans.

© The Palaeontological Association doi: 10.1111/j.1475-4983.2009.00883.x

MATERIAL AND METHODS

A binocular microscope and drawing tube were used to make observations and drawings either from specimens preserving original bone, or from silicone or latex peels. The latter were prepared from natural moulds of specimens etched with dilute hydrochloric acid to remove decayed and fragmented bone.

A data matrix of cranial and postcranial characters was assembled for the phylogenetic section of the present work. The data matrix collated information mostly from two recent cladistic analyses of early tetrapods, namely Anderson (2007) and Ruta and Coates (2007), and adopted protocols similar to those highlighted by Ruta and Clack (2006) and Clack and Finney (2005). Specifically, firstly, characters common to both analyses were identified that could be coded for the taxonomic sample in the present work. All those characters that were not duplicated in the two analyses were then added. Throughout this procedure, coding of taxa was checked for all characters. A small number of additional characters were taken from Ruta and Bolt (2006) and Ruta and Clack (2006); see also Clack and Finney 2005).

Institutional abbreviations. BSM, Bayerische Staatssammlung für Geologie und Paläontologie, Munich; MB, Museum für Naturkunde, Berlin; NHM, The Natural History Museum, London; NMP, Narodní Museum, Prague; NMW, Naturhistorisches Museum, Vienna; SMF, Senckenberg Museum, Frankfurt; UMZC, University Museum of Zoology, Cambridge; ZP, Zapadočeske Museum, Plzeň.

SYSTEMATIC PALAEONTOLOGY

TETRAPODA Haworth, 1825 [*sensu* Goodrich, 1930]
NECTRIDEA Miall, 1875
Family SCINCOSAURIDAE Jaekel, 1909

Genus SCINCOSAURUS Fritsch, 1876

Type species. *Scincosaurus crassus* Fritsch, 1876.

Included species. *Scincosaurus crassus* Fritsch, 1876; *S. spinosus* Civet, 1982.

Scincosaurus crassus Fritsch, 1876
Text-figures 1–9

v*1876 *Scincosaurus crassus* Fritsch, p. 72.
1879 *Scincosaurus crassus* Fritsch 1879, p. 28.
v*1881 *Keraterpeton crassum* (Fritsch); Fritsch, p. l36, figs 82–87, pls 27–30.

1890 *Ceraterpetum crassum* (Fritsch); Lydekker 1890, p. 198.
1895 *Scincosaurus crassum* Fritsch; Andrews 1895, p. 84.
1897 *Scincosaurus crassus* Fritsch; Woodward 1897, p. 298.
1903 *Scincosaurus crassus* Fritsch; Jaekel 1903, p. 116.
1938 *Scincosaurus crassus* Fritsch; Steen, p. 213.

Lectotype. NMP M 460 (Fritsch Cat. No. 91, orig. 2, ČGH 2099) designated by Bossy and Milner (1998), counterparts of an imperfect skull and almost complete skeleton preserved as far as the 42nd caudal with girdles, limbs, and scales, figured by Fritsch 1881, pl. 27, fig. 1 (whole skeleton); pl. 28, fig. 11 (tarsus); pl. 29, fig. 4 (pectoral girdle); pl. 30, fig. 3 (forelimb), fig. 7 (pectoral girdle).

Emended diagnosis. Nectridean amphibian with proportionately small, short-snouted skulls, ratio of skull length to total body length *c.* 1:16. Orbits small and widely separated, sclerotic ring present. Supratemporal and postparietal absent, postfrontal and tabular enlarged, tabular contacts postfrontal and postorbital, no parietal-squamosal contact. No tabular horn. Dermal ornament of densely packed deep circular pits. Short marginal tooth row with spatulate, sharply carinate, slightly incurved tooth crowns. Vertebral column consisting of 24 presacrals, one sacral, and up to 48 caudals. Trunk vertebrae with laterally expanded zygapophyses, accessory articulations poorly developed. Neural arch inserted along full length of centrum in caudal vertebrae with elongate zygaphoysyes. Haemal spines rectangular with a straight edge, haemal arches inserting along full length of centrum. Transverse processes retained in distal caudals. Appendicular skeleton heavily ossified, including carpals and tarsals. Cleithrum with unexpanded rounded head. Interclavicle broadly T-shaped, developing into a massive equilaterally triangular shape and bearing large circular dermal pits with a less pitted medial ridge. Humerus showing 90° torsion, and proximal and distal expansions; entepicondylar foramen present. Olecranon process on ulna. Four-digit manus and five-digit pes. Ventral dermal scales ornamented, with a single row of deep circular pits along scalloped posterior edge, and arranged en chevron.

Referred material. More than 70 specimens, held principally in NMP, ZP, NMW, MB; also BSM, NHM, SMF, UMZC. Figured specimens only are listed here. NMP M 461 (F.C. No. 92, orig. 70), trunk and caudal vertebrae; figured Fritsch 1881, pl. 27, fig. 2 (pelvic and proximal caudal region); pl. 30, fig.2 (pelvic region and hind limb). NMP M 462 (F.C. No. 93, orig. 71), tail section; figured Fritsch 1881, pl. 27, fig. 3 (mid-caudal vertebrae). NMP M 463 (F.C. No. 94, orig. 73), disarticulated trunk in counterpart including well-preserved interclavicle; figured Fritsch 1881, pl. 27, fig. 4 (interclavicle); pl. 29, fig. 7 (interclavicle); pl. 30, fig. 6 (?hyoid elements, although these were misidentifications of a clavicle and a phalanx). NMP M 465 (F.C. No. 96, orig. 200, ČGH 3004), two skulls in dorsal aspect and pectoral girdle elements; figured Fritsch 1881, pl. 28, fig. 1 (skull roof), fig. 2 (marginal teeth), fig. 3 (sclerotic ring), fig. 4 (cleithrum), fig. 7 (isolated frontal and parietal); pl. 29, figs 2, 3 (marginal dentition). NMP M 466 (F.C. orig. 97 (misprinted as 198), ČGH 3013), partially disarticulated skeleton in counterpart; figured Fritsch 1881, pl. 28, fig. 5 (rib), fig. 8 (skull roof), fig. 10 (manus). NMP M 467 (F.C. No. 98, orig. 241), complete pelvis; figured Fritsch 1881, pl. 28, fig. 9. NMP M 469 (F.C. No. 99, orig. 159), almost complete skeleton; figured Fritsch 1881, pl. 29, fig. 1 (skull roof); pl. 30, fig. 1 (pelvic region), fig. 9 (rib). NMP M 982 (ČGH 3002/3011), almost complete skeleton in counterpart; figured Steen 1938, pl. 2, fig. 1 (dorsal aspect). NMP M 1059 (F.C. orig. 199), pl. 30, fig. 5 (ventral scales in external view). NMP M 617 (F.C. orig. 244), Fritsch 1881, p. 142, text-fig. 85 (two posterior dorsal vertebrae), p. 86, text-fig. 86, (two proximal caudal vertebrae). NMP M 466 (F.C. Orig. 198), Fritsch 1881, p. 142, text-fig. 84 (two anterior dorsal vertebrae). ZP 705 (M.P.393 III D.29, č 705), skeleton in ventral aspect; figured Steen 1938, text-figs 6b, 7 (palate). ZP 711 (M.P. III D.35, č 711), skeleton in dorsal aspect; figured Steen 1938, p. 214, text- fig. 5a (skull roof); p. 216, text-fig. 7 (skull roof reconstruction); p. 271, text-fig. 43d (restoration of skeleton). ZP 406, partial skull and skeleton in dorsal aspect (currently missing from the Zapadočeske Museum, Plzeň); figured Steen 1938, p. 211, text-fig. 3a (dorsal vertebrae); p. 214, text-fig. 5a (skull roof). ZP 381, partial trunk, figured Steen 1938, p. 212, text-fig. 4d (dorsal and caudal vertebrae); p. 218, text-fig. 8b (scapula).

Locality, horizon, and age.   Nýřany 'Gaskohle' from Nýřany Colliery near Nýřany, south-west of Plzeň, Czech Republic; apex of Lower Grey Beds of the Plzeň-Manětín Basin, Asturian (Upper Westphalian D), Moscovian, Late Carboniferous.

Description

General features.   The species is represented by more than 70 specimens from Nýřany. The majority are crushed, disarticulated and/or fragmentary. The most complete skeletons range from 125mm (ZP 711) to c.150 mm (MB Am. 27) in total head to tail length, although smaller skeletons are known, notably NMW 1898 X 40, in which the skull plus trunk length of 28 mm indicates a complete individual less than 80 mm long. The overall skeletal proportions are quite different from all other nectrideans; the skull is very small, around 0.16 of the trunk length compared to 0.30–0.40 in other taxa; the tail is very long, so that the proportion of head to total skeleton length is approximately 1:20 (Text-fig. 1A, B). The postcranial skeleton is unusually heavily ossified for such a small animal; the limbs are well developed and include carpals and tarsals which never ossify in other nectrideans. Two of the most characteristic features are the massive ornamented interclavicle and tightly interlocked heavy ventral dermal scalation.

Skull

The cranial material is frequently extensively crushed, suggesting that the skull was quite delicate and relatively deep. The skull is broad and short-snouted, its breadth being approximately equal to its length, and with an interorbital width of approximately one-third of the skull length. The orbits are small and dorsolaterally positioned; a ring of nine or 10 sclerotic plates is present. The posterior border of the skull is straight. The skull table lacks postparietal, intertemporal, and supratemporal bones, and there is no trace of an otic notch. The characteristic ornament of the dermal bones of the skull roof consists of densely distributed deep circular pits. Slight ridges occur on some bones in the areas furthest from the centres of ossification. They are radially aligned as in many basal tetrapods. A pit-line of the supraorbital lateral line sulcus runs along the nasal and frontal, visible most clearly in ZP 711 (Text-fig. 2A); Steen (1938, p. 214, text-fig. 5A) noted a postfrontal groove and a few pits on the postorbital in ZP 406. No other lateral line sulci are visible on any other specimens and their presence in only two individuals, and on one side of the skull roof only, might best be interpreted as a relict of a more amphibious ancestry.

Skull roof.   The skull roof is preserved most completely in ZP 711 (Text-figs 2A, 3A), although the premaxilla is missing. A complete premaxilla in NMP M 465 shows a narrow tooth-bearing ramus with a pronounced square-ended posterior median process contacting the nasal. It bears a row of five marginal teeth (Fritsch 1881, pl. 28, fig. 1) (see 'Dentition' below). The maxilla is slender and does not enter the lateral orbit margin, *contra* Steen (1938, p. 214, fig. 5b). The maxillary tooth count is not determinable in any specimen, although, assuming equal spacing, there could be up to eight teeth. The lacrimal is triangular and forms the posterolateral segment of the narial margin, anteriorly, and the dorsolateral segment of the orbit margin, posteriorly. The nasal is broadly expanded anteriorly and borders the medial and anterior margins of the naris. It sutures with the premaxilla well anterior to the naris; among other nectrideans, a similar configuration is found only in *Ptyonius* (Bossy and Milner 1998, p. 77, text-fig. 53G). The frontal is approximately twice the mid-line length of the nasal, although Steen (1938, p. 214, text-fig. 5b) depicted it as a much shorter bone equal in length to the nasal.

The parietal is the largest bone in the skull roof and is twice its anterior width along the posterior margin, which delimits the skull table. The posterolateral margin is relatively straight. A small pineal foramen is situated on the medial margin a short distance from the frontal suture. The dermal ornament is replaced by smooth, slightly fluted bone along the posterior margin. The absence of postparietals is unique among nectrideans; among other lepospondyl groups, postparietals are also lacking in brachystelechid microsaurs – which is attributed to the small skull size of these tetrapods (Carroll 1998) – and in phlegethontiid aïstopods (Anderson 2002).

The prefrontal is long and slender and contacts the narial margin anteriorly, as in most diplocaulids. The postfrontal is large with a posterior process contacting the tabular, a unique condition in nectrideans, but which occurs commonly in microsaurs (Carroll 1998). The postorbital forms the posterior orbit margin and contacts the postfrontal, jugal, and squamosal, as in other nectrideans; in addition, it contacts the tabular posteromedially, a condition typical of microsaurs. The jugal has a short

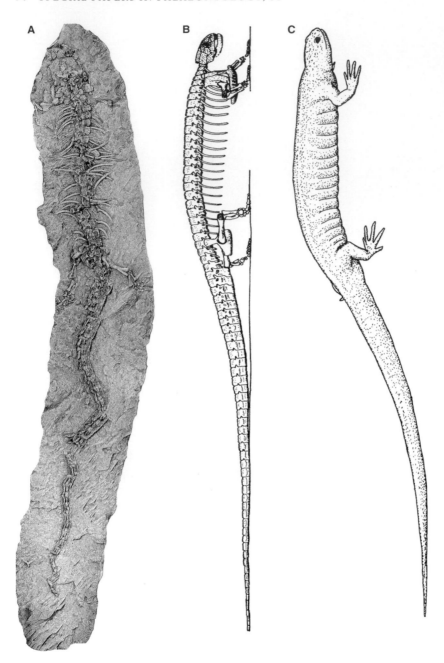

TEXT-FIG. 1. *Scincosaurus crassus* Fritsch. A, NMP M 982, ammonium chloride coated latex peel of acid-etched skeleton in dorsal view. B, composite skeleton reconstruction (modified from Milner 1980a, b). C, whole animal restoration (modified after Milner 1980a, b). All views 1.5x.

anterior orbital arm that forms the posterolateral orbit border and contacts the maxilla anteriorly, although the length of the jugal-maxilla suture is not determinable. The expanded posterior region of the jugal contacts the postorbital, squamosal, and presumably the quadratojugal. The quadratojugal is not preserved in any specimen: the element depicted by Steen (1938, p. 214, text-fig. 5b) as a quadratojugal is in fact a squamosal. The squamosal occupies the posterolateral region of the skull and contacts the jugal, postorbital, and presumably the quadratojugal. It lacks a parietal contact because of the intervening postfrontal and tabular suture, in contrast to diplocaulids. The retention of supratemporals and tabulars in the skull table of urocordylids precludes a parietal-squamosal contact. The squamosal has a

wide exposure on the palate, in common with diplocaulids (Bossy and Milner 1998) (Text-fig. 1B).

The tabular is long and narrow and extends anteriorly between the parietal and squamosal to contact the postfrontal and postorbital. This pattern is unique among nectrideans. In diplocaulids, a broad parietal-squamosal contact separates the tabular from the postfrontal and postorbital, whereas in urocordylids, the tabular is restricted to the posterior skull roof and is separated from the circumorbital elements by a supratemporal (Bossy and Milner 1998). The tabular occupies a similar area of the skull roof and contacts the same bones as in tuditanomorph microsaurs (Carroll 1998). The posterior tabular margin is convex and extends beyond the parietal margin. The dermal

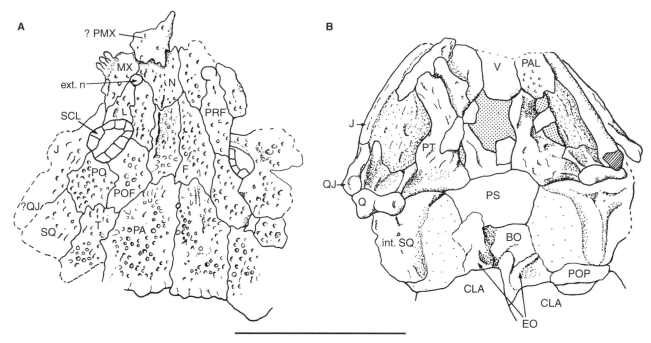

TEXT-FIG. 2. *Scincosaurus crassus* Fritsch. A, ZP 711, skull roof in dorsal view. B, ZP 705, palate in ventral view. Interpretive drawings from silicone peels of acid-etched specimens. Scale bar represents 5 mm. BO, basioccipital; CLA, clavicle; EO, exoccipital; ext.n, external naris; F, frontal; int. SQ, internal flange of squamosal; J, jugal; L, lacrimal; MX, maxilla; N, nasal; PA, parietal; PAL, palatine; PMX, premaxilla; PS, parasphenoid; PT, pterygoid; POF, postfrontal; POP, Paroccpital process of opisthotic; PRF, prefrontal; Q, quadrate; QJ, quadratojugal; SCL, sclerotic plate; SQ, squamosal; V, vomer.

TEXT-FIG. 3. *Scincosaurus crassus* Fritsch. A, composite reconstruction of skull roof. B, reconstruction of the palate based principally on ZP 705. Scale bar represents 5 mm. Both modified from Bossy and Milner, 1998. Abbreviations as in Text-figure 2.

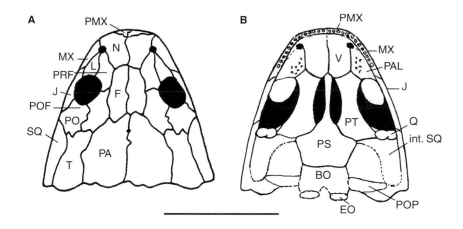

ornament is replaced by smooth fluted bone, as in the parietal. In contrast to diplocaulids, the tabular bears no horn, although Fritsch (1881, pl. 28, fig. 1) depicted, erroneously, a tabular with a slender unornamented horn. The 'horn' is in fact a cleithrum lying between a clavicle and a dermal roof fragment (A. C. Milner 1980).

Palate. The palate is preserved relatively completely in only one specimen, ZP 705 (Text-fig. 2B). The palate is slightly open with small interpterygoid vacuities; the quadrate condyles are positioned approximately two-thirds of the way along the skull length, and are considerably anterior to the occipital condyles. The rectangular vomer bears neither teeth nor denticles, in contrast to diplocaulids (Bossy and Milner 1998). There is a posterior median notch between the vomers for reception of the cultriform process of the parasphenoid. The squarish to rectangular palatine bears a small patch of denticles on its posterior region. Steen (1938, p. 216, text-fig. 6b) depicted the choana as being bordered medially by the palatine although this feature cannot be determined from the casts examined in this study. The pterygoid has a vertical palatine ramus contacting the vomer and palatine anteriorly and suturing with the basal plate of the parasphenoid posteriorly. The medial margin of the palatine ramus is slightly concave; small interpterygoid vacuities are present, which suggests that the pterygoids do not meet anteriorly (Text-fig. 3B). The quadrate ramus is directed laterally to the quadrate. The curvature of the lateral margin of the pterygoid suggests that a wide subtemporal fossa was present (Text-fig. 3B).

The quadrate bears a well developed trochlear condyle and has a dorsally directed flange abutting against the dorsolateral margin of the pterygoid ramus. The quadrate is supported posteriorly by a transverse ridge which also underlies the quadrate ramus of the pterygoid. This ridge is continued posteriorly as a tongue-like process merging into the underside of the skull roof. Steen (1938, p, 216, text-fig. 6b) interpreted this process as a 'paraotic plate' of the pterygoid. Although this area of the palate is difficult to interpret, there seems to be no continuation of the pterygoid posterior to the parasphenoid and quadrate. This process is interpreted herein as an internal development of the squamosal, the anterior end of which supports the quadrate and pterygoid. Thus, the posterior process braces the cheek region (Text-fig. 2B). This feature is shared with diplocaulids: the contact between the squamosal and the pterygoid is accentuated by the anterior position of the jaw articulation relative to the occiput. The parasphenoid consists of a large rectangular basal plate and an anteriorly directed cultriform process. The anterior end of the cultriform process is not preserved, but is apparently accommodated by the median notch between the vomers.

Occiput. The exoccipitals have separate condyles, as in diplocaulids. Steen (1938, p. 215, text-fig. 6a) figured the condyles as convex ovals in an unreferenced specimen, which accords with the structure in diplocaulids. No median suture and no lateral sutures are visible to demarcate the extent of the exoccipitals. The basioccipital is represented by a raised medial area which is sutured anteriorly to the parasphenoid basal plate. A structure between the left exoccipital and the internal squamosal ridge may represent the paroccipital process of the opisthotic. A fenestra ovalis was figured by Steen (1938, p. 215, text-fig. 6a) in an unreferenced specimen.

Mandible. Partial mandibles preserved in ZP 705 and ZP 1317 indicate that the jaw is slightly convex in plan view, as might be expected from the shape of the skull (Text-fig. 3). The inner aspect of a left dentary (ZP 712; Text-fig. 4B) suggests a relatively shallow jaw with a short tooth row. Eleven spatulate teeth are present, nine on the dentary margin and two set medial to them.

Dentition. The teeth of *Scincosaurus* are very unusual among Palaeozoic tetrapods; they possess long waisted crowns with spatulate tips in both upper (Text-fig. 4A) and lower jaws (Text-fig. 4B). This morphology, described as 'pedicellate' by A. C. Milner (1980), is not comparable to the pedicellate condition in some dissorophoid temnospondyls, such as *Doleserpeton* (Bolt 1977), and in several lissamphibians, in which there is a dividing zone of weakly mineralised tissue towards the base of the tooth crown (Davit-Beal *et al.* 2007). In lingual view, the tips of the crowns are slightly incurved with two distinct sharp carinae inset from the anterior and posterior tooth edges (Text-fig. 4B). This morphology is similar to the crown tips of some early Cretaceous scincomorph lizards, which show well-defined anterior and posterior stria dominans on the lingual surface (Richter 1995; Evans and Searle 2002).

Postcranial skeleton

Axial skeleton. No specimen shows a completely articulated and fully exposed presacral column in dorsal view; for this reason, estimates of the vertebral count are difficult. Steen (1938, text-fig. 43d) restored 23 presacral vertebrae followed by one sacral and two postsacral vertebrae bearing short curved sacral ribs, based on ZP 711, although the presacral count is fairly certainly 24 in NMP M982 (Text-fig. 1A, B). The first seven presacral vertebrae in ZP 406 gradually increase in height and length so that the seventh is approximately 1.5 times the linear dimensions of the first, as figured by Steen (1938, p. 211, text-fig. 3c). The crushed atlas was depicted as bearing a pair of ribs in the same figure, a condition that does not occur in any other nectridean. The height of each vertebra is about 1.5 times its centrum length. The neural arches are fused to the full length of the centra throughout the vertebral column, and bear laterally expanded flat zygapophyses with wide horizontal articulating surfaces (Text-fig. 5). Zygantrum-zygosphene articulations are present above the zygapophyses, as in urocordylids and diplocaulids (Bossy and Milner 1998), but they are confined to the vertebrae in the posterior region of the trunk and tail. Expanded neural spines are continuous with the full length of the neural arches and have a pronounced striated dorsal ridge. The sacral vertebra bears a comparatively large transverse process situated towards the middle of the centrum. The first three caudal vertebrae lack haemal arches. Two pairs of short sacral ribs were figured by Fritsch (1881, pl. 30, fig. 1) in NMP M 469, although they are not preserved in other specimens. In ZP 714, the third caudal bears a long, slightly posteriorly curved square-ended process; there is no visible suture, so it is not possible to determine whether the process in question is a long transverse process or a short one with a fused sacral rib (Text-fig. 5). The remaining caudal vertebrae bear haemal arches, with expanded spines, inserted along the full length of the centrum. The pre and post-zygapophyses are connected by a thin, laterally expanded bony shelf. The transverse processes are narrow and laterally expanded, and lie parallel to the zygopophysial shelves. Both the transverse processes and the bony shelves are retained throughout the length of the tail, which suggests that the latter would have been rounded in profile (Text-fig. 1C). The vertebrae abut throughout their full depth and there are accessory apophyses between adjacent neural arches and adjacent haemal arches. Accessory haemal apophyses are also present in urocordylids, and haemal spines abut or overlap in most nectrideans (Bossy and Milner 1998), although none have haemal spines tightly appressed throughout their depth. The ventral ends of the haemal spines in the proximal half of the tail are flattened and expanded transversely, thus resulting in a slightly T-shaped cross-section. They bear heavily pitted ornament and bestow a slightly flattened skid-like base to the tail. The caudal vertebrae gradually reduce in height and length distally as the tail tapers. ZP 711 and MB Am. 47 preserve 46 and 48 caudals respectively, the most distal ones with the neural and haemal spines almost completely reduced. The tail was apparently relatively rigid; some specimens are preserved with a degree of curvature in the tail, as in NMP M 460 (Fritsch 1881, pl. 27, fig. 1) and NMP M 982 (Text-fig. 1A), although the tail is

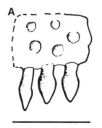

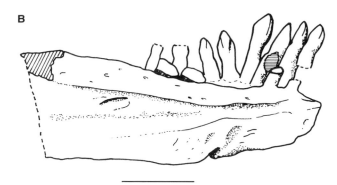

TEXT-FIG. 4. *Scincosaurus crassus* Fritsch. A, ZP 711, tooth-bearing fragment of premaxilla. B, ZP 712, left dentary in lingual view. Interpretive drawings from silicone peels of acid-etched specimens. Scale bars represent 0.5 mm.

more frequently preserved broken into straight articulated sections (Text-fig. 1A).

Dorsal ribs. The dorsal ribs are slender, curved and double-headed. The rib heads are oblique, the tuberculum articulates with the corresponding transverse process (Text-fig. 5). The long curved capitulum apparently articulates either intercentrally or with the centrum of the preceding vertebra, as in diplocaulids. The ribs are progressively reduced in length in the posterior part of the trunk and the last five pairs are very short. The sacral rib is short, straight and robust; Fritsch figured it as having a slender capitulum (1881, pl. 30, fig. 1) in NMP M 460.

Appendicular skeleton

Pectoral girdle and forelimbs. The dermal elements of the pectoral girdle are heavily ossified. The clavicles are small and cap the anterior end of the interclavicle (Text-fig. 6A). The ventral plates bearcircular pitted ornament; the dorsal process, which arises from the posterolateral margin of the plate, is unornamented. The interclavicle is very large and is equilaterally triangular in shape with rounded edges in large individuals. The ventral surface is massively ornamented with large deep circular pits interrupted by a prominent, less heavily ornamented, strut-like median ridge (Text-fig. 6A). In smaller individuals, the interclavicle is much less expanded laterally, and there is only a small area of ornamented bone on each side of the median ridge (Fritsch 1881, pl. 29, figs 5, 6, 7). The cleithrum consists of a smooth vertical shaft that tapers ventrally and a lightly ornamented rounded head, *contra* Steen (1938, p. 217) and figure (1938, p. 218, text-fig. 8c) of a T-shaped head. Internally, the shaft bears a strongly developed mid-line ridge (Text-fig. 6B) to brace the scapulocoracoid. A cleithrum preserved between a dermal roof bone fragment and a clavicle in NMP M463 was identified as an epiotic horn by Fritsch (1881, pl. 28, fig. 1). A partly preserved scapulocoracoid in ZP 712 shows an expanded coracoid portion and a broken dorsal scapular portion (Text-fig. 6A). A scapulocoracoid figured by Steen (1938, p. 218, text-fig. 8B) from an unreferenced specimen depicts a glenoid with marked supra and infraglenoid buttresses and articular surfaces which would restrict dorsoventral movement of the forelimb.

The humerus shows torsion through 90 degrees so that its expanded ends are set at right angles to each other. (Text-fig. 7A). The head and proximal articular surface are not well preserved. Well-developed deltoid and pectoral crests are present together with a pronounced capitellum for articulation with the radius and a trochanter for reception of the articular surface of the ulna. An entepicondylar foramen is present in a slit-like groove. The humerus structure is typical of terrestrial tetrapods and is very similar to that of the microsaur *Pantylus* (Carroll 1998). The ulna is slender, longer than the radius, and with little distal expansion (Text-fig. 7B). The shaft is slightly twisted and curved distally so that the proximal and distal articular surfaces are at an angle. A distinct olecranon process and well-developed sigmoid notch are present. The radius has a stout cylindrical shaft with a broad convex distal expansion for articulation with the carpus. Seven ossified carpals are preserved, shown in close articulation in ZP 1317 (Text-fig. 7B). Three elements in the proximal row are interpreted topologically as a squarish radiale, a broad intermedium with two distal articular facets, and a large rectangular ulnare. A distal row of four elements lies near the bases of the metacarpals. Other carpals primitively present in the tetrapod wrist are presumed to have been lost or fused to other elements, as suggested by Carroll (1998) to account for the reduced number of carpals in terrestrial microsaurs such as *Pantylus*. Four stout metacarpals are present; the size of the first three increases in sequence, while the fourth is smaller than the first. A complete set of phalanges is preserved in MB Am. 31 (Text-fig. 7C), with a formula of 2, 3, 3, 2. The phalanges are long and slender with slender, bluntly pointed unguals. The manus is unambiguously four-toed *contra* Bossy and Milner (1998, text-fig. 66g) who figured a pes in error showing a partial fifth digit.

Pelvic girdle and hind limbs. The pelvic girdle is preserved as a co-ossified structure (Text-fig. 5). The ilium bears a dorsal process that is broadly expanded distally. The acetabulum is positioned centrally and restricted to the ilium. A pronounced lipped ridge marking the junction with the puboischiadic plate and the ventral border of the acetabulum is preserved in MB Am. 29. A suture between the pubis and ischium was figured by Steen (1938, p. 218, text-fig. 8a) in an unreferenced specimen but has not been observed in any specimens by the authors. The femur is quite robust with a slender shaft. Its

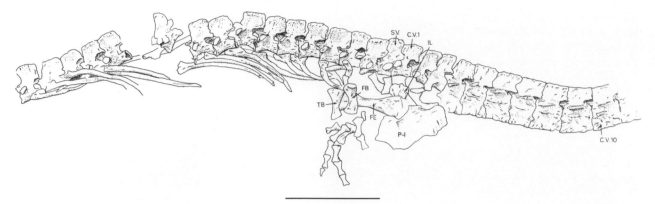

TEXT-FIG. 5. *Scincosaurus crassus* Fritsch. ZP 714, posterior trunk, pelvic region and proximal caudal vertebrae in lateral view. Interpretive drawing from a silicone peel of an acid-etched specimen. Scale bar represents 10 mm. C.V., caudal vertebra; FB, fibula; FE, femur; IL, ilium; P-I, pubischiadic plate; S.V., sacral vertebra; TB, tibia.

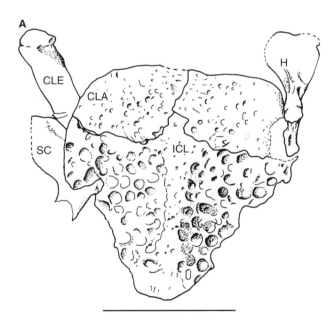

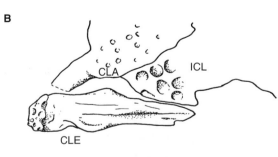

TEXT-FIG. 6. *Scincosaurus crassus* Fritsch. A, ZP 712, pectoral girdle in ventral view (modified from Bossy and Milner 1998). B, ZP 1196, cleithrum in medial view. Interpretive drawings from silicone peels of acid-etched specimens. Scale bars represent 5 mm. CLA, clavicle; CLE, cleithrum; H, humerus; ICL, interclavicle; SC, scapulocoracoid.

head is anteroposteriorly elongate and terminal (Text-fig. 5; Fritsch 1881, pl. 30, fig. 2). A shallow intertrochanteric fossa and an internal trochanter continued as an adductor ridge are preserved in ZP 712 and MB Am. 29. The distal condyles are separated by a deep intercondylar groove and a well-developed adductor crest. The tibia exhibits a typical terrestrial tetrapod pattern with an expanded proximal articular surface, and is almost equally expanded distally with a slight demarcation of two facets for articulation with the tarsus. The fibula is slightly longer than the tibia and has a convex head; the shaft is slender and broadens distally with two distinct tarsal facets, a short straight lateral one and a slightly concave one. A distinct flattened lateral crest is present on the medial edge between the proximal and distal expansions. The tarsus is preserved in several individuals, most completely in MB Am. 28 (Text-fig. 8) where it is articulated and therefore probably represents the complete complement of seven elements, fewer than in other terrestrial Palaeozoic tetrapod groups. Three proximal elements are present, the central and the largest of which is identified as an intermedium. This is also the largest element in other early tetrapods, including most terrestrial microsaurs (Carroll 1998), the anthracosaur *Proterogyrinus* (Holmes 1984), the basal temonspondyl *Balanerpeton* (Milner and Sequeira 1994); in contrast, in the stem-group amniote (*sensu* Ruta and Coates 2007) *Seymouria* the fibiale is the largest tarsal element (Berman *et al.* 2000). The intermedium is flanked by a quadrangular fibulare and a small rounded tibiale. There are four closely articulating rectangular tarsals in a distal row. If these represent distal tarsals then centrale elements are lacking. The tarsus is therefore unusually reduced, even more than in some microsaurs, notably *Tuditanus* in which only one centrale is present (Carroll and Baird 1968). The tarsus also appears to lack a fifth element in the distal row so that metacarpal V articulated with the fibulare. The pes is 5- toed. Metatarsals I to IV are robust, II to IV are subequal in length, and III is broader and slightly longer that II and IV. The phalangeal formula is clearly 2, 3, 4, 3, 2 (Text-fig. 8). The unguals are bluntly pointed with faint longitudinal striated markings.

TEXT-FIG. 7. *Scincosaurus crassus* Fritsch. A, ZP 705, right humerus in ventral view (from Bossy and Milner 1998). Scale bar represents 5 mm. B, ZP1317, left epipodials and carpus in anterior view (modified from Bossy and Milner 1998). Scale bar represents 3 mm. C, MB Am. 31, left forelimb in anterior view (underlying carpals omitted for clarity). Scale bar represents 5 mm. Interpretive drawings from silicone peels of acid-etched specimens. delt, deltopectoral crest; ent. f., entepicondylar foramen; H, humerus; IN, intermedium; MC, metacarpal; o.pr., olecranon process; RA, radius; RC, radial condyle; RD, radiale, U, ulna; UL, ulnare.

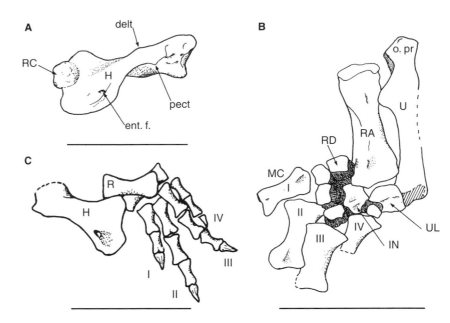

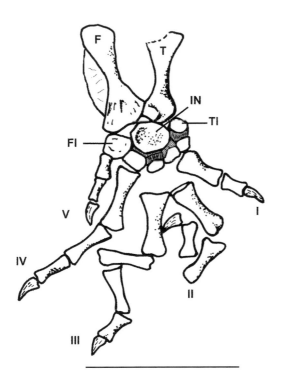

TEXT-FIG. 8. *Scincosaurus crassus* Fritsch. MB Am. 28, right tarsus and pes in anterior view. Interpretive drawing from a silicone peel of an acid-etched specimen. Scale bar represents 5 mm. F, femur; FI, fibulare; IN, intermedium; T, tibia; TI, tibiale.

Ventral dermal scales

Thick ventral scales are arranged in overlapping V-shaped chevrons between the pectoral and pelvic girdles. The individual scales are rectangular, becoming more elongate in the lateral rows (Text-fig. 9A, B). In ventral aspect, the posterior edge of each scale is crenulated and bears a single row of ornament pits (Text-fig. 9A). The dorsal surfaces are essentially smooth with fine striate markings generally radiating out from the mid-line. The anterior edge of each scale is indented by a wide deep U-shaped pocket for reception of the posterior edge of the preceding scale so that vertical rows are locked firmly together (Text-fig. 9C). Dorsal scales have not been observed on any specimen.

PHYLOGENETIC ANALYSIS

Data matrix

The data matrix consists of eight outgroups, 11 nectridean OTUs, and 173 characters (109 cranial; 64 postcranial) based primarily on those of Ruta *et al.* (2003), Ruta and Coates (2007), and Anderson (2007) (see Appendix). The nectridean sample encompassed over 73 per cent of the valid genera recognised by Bossy and Milner (1998). The sauropleurines *Montcellia* and *Lepterpeton*, the scincosaurid *Sauravus*, and the *incertae sedis* nectridean *Arizonerpeton* were excluded because of their poor preservation. As the present study focuses on the interrelationships of the best known genera, a few adjustments were made to the taxonomic exemplar. Specifically, *Sauropleura* was treated as a composite taxon, with morphological information gleaned mostly from two of the three recognised species, i.e. *S. pectinata* and *S. scalaris*, using data from Bossy and Milner (1998). In the case of *Keraterpeton*, we considered only the type species, *K. galvani*. Data for *K. galvani*, *Batrachiderpeton*, and *Diceratosaurus* were taken from Milner (1978) and Bossy and Milner (1998),

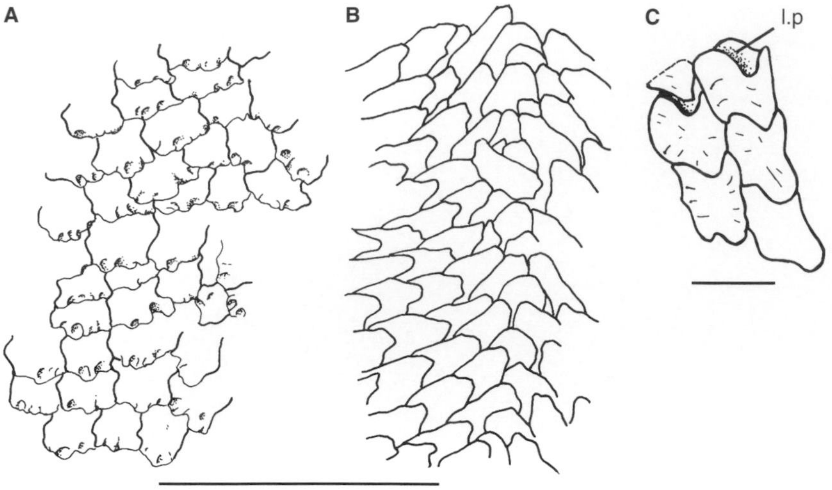

A   B   C   l.p

TEXT-FIG. 9. *Scincosaurus crassus* Fritsch. A, ZP 704, ventral dermal scales in ventral view (modified from Bossy and Milner 1998). B, NMP M 367, ventral dermal scales in dorsal view (modified from Bossy and Milner 1998). Interpretive drawings from silicone peels of acid-etched specimens. Scale bars represent 5 mm. C, MB Am. 29, ventral dermal scales in ventral view. Scale bar represents 3 mm. Anterior to top in all views. l.p., locking pocket.

whereas data for *Diploceraspsis* were taken from Beerbower (1963). *Diplocaulus* was coded using primarily data from *D. magnicornis* (see also Anderson 2007 for a more inclusive species-level sample of nectrideans). *Scincosaurus* was coded from the present work.

The data matrix was subjected to a maximum parsimony analysis heuristic search; 1000 random stepwise addition replicates; tree bisection-reconnection branch-swapping algorithm saving a single tree in memory at end of each replicate; subsequent swapping on all trees at end of all replicates and with multiple tree saving option in effect; e.g. see Quicke *et al.* (2001).

The amount of statistical support for tree nodes was evaluated via bootstrap (e.g. Felsenstein 1985) and decay index values (e.g. Bremer 1994). Bootstrap percentages – obtained through 10000 replicates using the fast stepwise addition sequence option in PAUP* – are reported in Text-figure 10, and refer to a 50 per cent majority-rule consensus in which groupings with less than 50 pe rcent bootstrap support are also reported. Decay index values (Text-fig. 10) show the number of additional steps required to collapse a node, and were obtained by performing parsimony analyses in which all trees having length equal to or less than a specified number of steps were saved. Increasing numbers of steps were then selected until all nodes were collapsed in the strict consensus of all trees saved from the relevant parsimony run.

Results

Three equally most parsimonious trees differing exclusively in the mutual relationships of the three aïstopods were obtained. When all characters were reweighted based upon the maximum value (i.e. best fit) of their rescaled

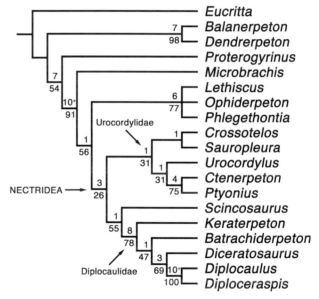

TEXT-FIG. 10. Strict consensus of three equally parsimonious trees retrieved from the cladistic analysis (see text for details). Numbers above the tree branches are decay index values; number below the branches are bootstrap percentages. The 10+ notation means that nodes are not collapsed at 10 extra steps, but the precise number of steps was not found due to excessive time required by calculations.

consistency indexes, a single tree was found in which aïstopods are resolved as follows: [*Lethiscus* (*Ophiderpeton Phlegethontia*)]. The three shortest trees from the original unweighted analysis have a length of 392 steps, with C.I. = 0.5306, R.I. = 0.6441, and R.C. = 0.3418, and are rooted along the internode that separates *Eucritta* from remaining taxa. The pattern of nectridean relationships conforms largely to that proposed by A. C. Milner (1980), but departs somewhat from those found by

Anderson (2001, 2007), Ruta *et al.* (2003), and Ruta and Coates (2007) (see also remarks below).

DISCUSSION

Nectrideans emerge as a clade consisting of two mono-phyletic sister groups. The first group is represented by the urocordylids, with the two subfamilies sauropleurines (*Crossotelos Sauropleura*) and urocordylines [*Urocordylus* (*Ctenerpeton Ptyonius*)] paired as sister taxa. The second group is represented by the scincosaurids (*Scincosaurus*) as sister taxon to the diplocaulids. The latter family is structured as follows: (*Keraterpeton, Batrachiderpeton, Diceratosaurus, Diplocaulus Diploceraspis*). The highest bootstrap percentage support is assigned to the (*Diplocaulus Diploceraspis*) node (100), to the diplocaulids as a whole (78), and to the (*Ctenerpeton Ptyonius*) node (75). Most remarkable is that nectrideans as a group receive very low support (26), as do the nodes subtending the urocordylids (31) and the scincosaurid-keraterpetontid clade (55).

Inspection of decay index values shows that most nodes in the nectridean phylogeny collapse at very few additional steps relative to the length of the shortest trees. Nodes with low decay index values also show low bootstrap percentage support (Text-fig. 10). The highest decay index values are given to the (*Ctenerpeton Ptyonius*) node (3), the [*Diceratosaurus* (*Diplocaulus Diploceraspis*)] node (3), the node subtending diplocaulids (8), and the (*Diplocaulus Diploceraspis*) node (10+). The 10+ notation for the latter implies that more than 10 steps are required to collapse this node. A precise number of steps, however, was not calculated due to computation time.

The low bootstrap percentage and decay index for nectrideans as a whole are surprising, given both the relatively small taxonomic exemplar in the present analysis (which would tend to decrease the general levels of homoplasy; Sanderson and Donoghue 1989) and the large number of putative shared derived characters of this group. However, we point out that large-scale cladistic analyses of early tetrapods – e.g. Anderson (2001), Ruta *et al.* (2003), and Ruta and Coates (2007) – have not lent support to nectridean monophyly. This might result from the fact that synapomorphies for this group are mostly confined to their postcranial skeleton (indeed, the majority of those pertain to the construction of vertebrae). Therefore, it is possible that the signal carried by those synapomorphies is swamped by the distribution of other features (in particular cranial) in those analyses (discussion in Ruta *et al.* 2003). However, Anderson (2007) has reinstated nectridean monophyly, albeit with a low bootstrap support (MR, unpublished results based upon

TABLE 1. Multiple linear regression for branch lengths, decay index values, and bootstrap percentage values.

Variable	*b*	*t*	P-perm	P-param
Intercept	8.87534	1.4342	0.08966	
1	0.2125	0.2167	0.408	0.4162
2	0.10401	0.75817	0.252	0.23214

	EDO	ERY	DVI	DIS
EDO	0	0.0007	0	0
ERY	0.0105	0	0.0002	0
DVI	0	0.003	0	0
DIS	0	0	0	0

$R^2 = 0.17702$; $F = 1.18306$; p-perm = 0.327; p-param = 0.34248. Branch lengths are used as the response variables, whereas decay index and bootstrap percentage values are used as predictor variables.

The first column (*b*) lists the values for the intercept of the multiple linear regression equation and the values (slopes 1 and 2) of the partial regression coefficients.

The second column (*t*) lists the associated *t*-statistics for those parameters.

The third column lists the permutational probabilities (999 replicates) for the slope parameters.

The fourth column lists the parametric probabilities of all three parameters.

The bottom row reports the coefficient of multiple determination, the *F*-statistic, and the permutational and parametric probabilities for the multiple regression.

Anderson 2007 data matrix) and with a novel hypothesis of interrelationships for the scincosaurid-diplocaulid clade (see below).

Significant in this context, is the observation that none of the seven character-state changes supporting nectridean monophyly in the three shortest trees are uniquely derived, their consistency index values varying between 0.333 and 0.75. We further explored briefly correlations among the lengths of the tree branches, their bootstrap percentage support, and their decay indexes were further explored briefly. For this purpose, a multiple linear regression analysis was employed on those three sets of values (the software is available at: http://www.bio.umontreal.ca/casgrain/en/labo/regression.html; see Legendre 2002). The results (Table 1) show that the correlation among the three sets of values is very weak. Statistical tests for the regression parameters yield nonsignificant results.

Experiments with topological constraints

Recent conflicting hypotheses of the internal relationships of nectrideans (e.g. Anderson 2007; Ruta and Coates

2007) were evaluated in the light of the data matrix presented here. Specifically, we wanted to test whether such alternative hypotheses are a considerably worse fit for our data matrix than the three most parsimonious trees. For this purpose, Templeton, Kishino-Hasegawa, and Winning-sites tests were carried out on tree length differences between the three shortest trees and the trees obtained by enforcing topological constraints that reflect alternative tree shapes.

We began with a consideration of Anderson's (2007) tree topology, and discussion was restricted to the nectridean exemplar in that study. Anderson's (2007) analysis retrieved nectrideans as a clade consisting of urocordylids, on the one hand, and a scincosaurid-diplocaulid group, on the other. The element of novelty in this analysis was the arrangement of taxa within the latter group. Thus, the genus *Scincosaurus* was nested within keraterpetontids as sister taxon to (*Diploceraspis Diplocaulus*) with, in increasingly less apical position, *Diceratosaurus* and a (*Batrachiderpeton Keraterpeton*) clade. Except for the absence of *Crossotelos* and the inclusion of all three *Sauropleura* species, the arrangement of urocordylids in Anderson (2007) study is identical to that in the present paper. A topological constraint was built in MacClade matching Anderson (2007) preferred interrelationships for nectrideans and maximum parsimony analysis was run in PAUP* to find the shortest trees that are compatible with such a constraint. *Crossotelos* was placed in a polytomy with other nectrideans, and outgroups are left unresolved near the root of the tree. PAUP* yielded three trees compatible with the selected constraint (length = 402 steps; C.I. = 0.5174; R.I. = 0.6248; R.C. = 0.3233). Although these trees are 10 steps longer than the most parsimonious trees, they are not a significantly worse fit for the data than the latter (p > 0.05 for Templeton, Kishino-Hasegawa, and Winning-sites tests).

We then built a topological constraint that matched Ruta and Coates (2007) preferred pattern of nectridean internal relationships. In this constraint, *Scincosaurus* appears as the sister taxon to all remaining nectrideans. In addition, aïstopods are placed as sister group to urocordylids. *Crossotelos* and *Ctenerpeton* (not included in Ruta and Coates 2007) are positioned in a basal polytomy with the urocordylids and the aïstopods, to avoid imposing further topological constraints. PAUP* retrieves five minimal trees compatible with the selected constraint (length = 405 steps; C.I. = 0.5136; R.I. = 0.619; R.C. = 0.3179). Both Templeton and Kishino-Hasegawa tests show that the constraint-based trees are significantly worse than the most parsimonious trees (p < 0.05), whereas the Winning-sites test reveals that differences between the former and the latter are only marginally nonsignificant (p values only slightly larger than 0.05).

Partitioned data sets

To assess the influence of character partitions on the tree topology, we decided to run a parsimony analysis using exclusively the 109 cranial characters. The results were dismaying, but perhaps not totally unexpected. Thus, PAUP* found 66 trees at 238 steps (C.I. = 0.5168; R.I. = 0.6577; R.C. = 0.3399), the strict consensus of which was completely unresolved. However, an agreement subtree (a pruned tree showing only taxa for which all most parsimonious trees agree upon relationships) showed that the only unstable taxa were *Ctenerpeton* and *Phlegethontia*. The basic geometry of the 66 shortest trees was one in which, with the exception of *Ctenerpeton*, the pattern of nectridean relationships was identical to that of the original analysis, but with one important difference: the microsaur *Microbrachis* appeared nested between the urocordylids and the scincosaurid-diplocaulid clade. Among the outgroups, *Proterogyrinus* occurred between *Eucritta* and the *Balanerpeton-Dendrerpeton* clade. These results are similar, in very broad terms, to those found by Anderson (2001) and Ruta and Coates (2007); see also Ruta *et al.* 2003), who also retrieved nectrideans as a paraphyletic group (see also preceding section on topological constraints).

Cranial characters are thus unable to provide support for nectridean monophyly. The biological implications of this observation require scrutiny. If the remarkable differences in the skull structure of urocordylids and scincosaurids-diplocaulids reflect phylogenetic distinctiveness, then it could be postulated that the membrane bone vertebra, which consists of a single fused structure in all nectrideans, arose at least twice independently. Furthermore, diplocaulids would be a secondarily aquatic clade, their ancestry being linked to a *Scincosaurus*-like grade of tetrapods.

However, if the features of nectridean vertebrae are synapomorphies for all species in this group, then the variety of skull shapes and proportions attests to a significant variety of morphofunctional adaptations, even within the same family. One difficulty with this scenario is the fact that the skull of scincosaurids does not appear to represent a suitable ancestral condition for the group as a whole (as per Ruta and Coates 2007 tree topology) or even for diplocaulids (as per the tree topology in this paper). Interestingly in this context, Anderson's (2007) tree places scincosaurids at the apical end of a paraphyletic diplocaulid assemblage, which suggests that scincosaurids were secondarily terrestrial. In this evolutionary scenario, scincosaurids appear as miniaturised, heavily ossified hornless relatives of the highly derived boomerang-headed diplocaulids. This assemblage is placed as the sister group of the aquatic genus *Diceratosaurus* from the

Middle Pennsylvanian of Linton, Ohio. *Diceratosaurus* retains postparietals, horned tabulars and a short trunk with a presacral count of 14 (Bossy and Milner 1998). The evolutionary implications of this arrangement include the loss of postparietals and tabular horns and an increase in the number of presacral vertebrae from 14 to 24 in the branch leading to scincosaurids.

One interesting change in the results of the analysis is observed when modifications to one particular character (64, Lateral line canals: absent (0); present (1)) are introduced. Observations of lateral line canals are difficult and are confined to two specimens. We cannot rule out the possibility that this character is variable, and may even change during ontogeny. Most specimens (where observations are possible) fail to reveal the presence of canals, and, where present, these are confined to very short tracts on one side of the skull roof (ontogenetic changes in the lateral line system have been documented in other early tetrapods). In our analysis, we have coded *Scincosaurus* as lacking lateral lines (despite occurrences in two specimens). If, however, the coding is changed, then the results of the cladistic analysis reveal some differences from those discussed here. Thus, a PAUP* analysis retrieves six equally parsimonious trees. In three of these trees, nectrideans are monophyletic, and their intrinsic relationships are identical to those found above. In the remaining trees, however, aistopods and urocordylids form sister groups, and this broader clade joins diplocaulids, with *Scincosaurus* lying outside the aistopod-urocordylid-diplocaulid clade. In all these trees, *Crossotelos* and *Sauropleura* form a paraphyletic array relative to urocordylines, whereas *Batrachiderpeton* and *Keraterpeton* are sister taxa.

Large-scale analyses of early tetrapods show contrasting results, and the amount of statistical support for each of the multiple nectridean topologies appears to be invariably low. An additional difficulty is represented by the impact of different taxon and character sampling in various studies. At present, it is unclear whether addition of characters from ever-expanding tetrapod phylogenies (e.g. Ruta and Clack 2006; Clack and Finney 2005; Anderson 2007; Ruta and Coates 2007) is likely to increase systematic bias towards one hypothesis or another, but a synthesis of available data sets is certainly urgently needed. Comparisons between Anderson (2001, 2007) analyses, and between Ruta *et al.* (2003) and Ruta and Coates (2007) analyses, show that refinement and increase of data sets may entail different results for some groups. Provisionally, we assume, in agreement with traditional views and the results from this study, that nectrideans are monophyletic, but note that such results may change dramatically in future. It is obviously apparent that nectrideans represent the derived end points of an earlier radiation of small bodied forms for which there is currently no fossil record. Real progress on further understanding their interrelationships requires the discovery of small tetrapods in Viséan and Tournaisian time.

Acknowledgements. The help of The Natural History Museum Images Resources in producing Text-figures 1–9 is gratefully acknowledged. ACM thanks staff at Bayerische Staatssammlung für Geologie und Paläontologie, Munich; University Museum of Zoology, Cambridge; Museum für Naturkunde, Berlin; Narodní Museum, Prague; Naturhistorisches Museum, Vienna; Senckenberg Museum, Frankfurt; and Zapadočeske Museum, Plzeň, Czech Republic, for access to specimens in their care. MR's research is funded by NERC (ARF NE/F014872/1). Comments from Jennifer Clack and Florian Witzmann have greatly improved the text and we are especially grateful for their very prompt responses. We dedicate this paper to Andrew R. Milner in recognition of his outstanding work on early tetrapods.

REFERENCES

ANDERSON, J. S. 2001. The phylogenetic trunk: maximal inclusion of taxa with missing data in an analysis of the Lepospondyli (Vertebrata, Tetrapoda). *Systematic Biology*, **50**, 170–193.

—— 2002. Revision of the aïstopod genus *Phlegethontia* (Tetrapoda: Lepospondyli). *Journal of Paleontology*, **76**, 1029–1046.

—— 2007. Incorporating ontogeny into the matrix: a phylogenetic evaluation of developmental evidence for the origin of modern amphibians. 182–227. *In* ANDERSON, J. S. and SUES, H.-D. (eds). *Major transitions in vertebrate evolution.* Indiana University Press, Bloomington, 417 pp.

ANDREWS, C. W. 1895. Notes on a specimen of *Keraterpeton galvani* Huxley, from Staffordshire. *Geological Magazine*, **2**, 81–84.

BEERBOWER, J. R. 1963. Morphology, paleoecology and phylogeny of the Permo-Pennsylvanian amphibian *Diploceraspis*. *Bulletin of the Museum of Comparative Zoology, Harvard*, **130**, 31–108.

BERMAN, D. S., HENRICI, A. C., SUMIDA, S. S. and MARTENS, T. 2000. Redescription of *Seymouria sanjuanensis* (Seymouriamorpha) from the Lower Permian of Germany based on complete, mature specimens with a discussion of paleoecology of the Bromacker locality assemblage. *Journal of Vertebrate Paleontology*, **20**, 253–268.

BOLT, J. R. 1977. Dissorophoid relationships and ontogeny, and the origin of the Lissamphibia. *Journal of Paleontology*, **51**, 235–249.

BOSSY, K. V. K. 1976. Morphology, paleoecology and evolutionary relationships of the Pennsylvanian urocordylid nectrideans (Subclass Lepospondyli, Class Amphibia). *Dissertation Abstracts, (B)*, **37**, 2731.

—— and MILNER, A. C. 1998. Order Nectridea Miall, 1875. 73–131. *In* WELLNHOFER, P. (ed.). *Encyclopedia of Paleoherpetology. Part 1. Lepospondyli.* Verlag Dr. Friedrich Pfeil, Munich, 216 pp.

BREMER, K. 1994. Branch support and tree stability. *Cladistics*, **10**, 295–304.

BROWNRIGG, W. B. 1866. Note on part of the Leinster Coalfield, with a record of some fossils found therein. *Journal of the Royal Geological Society of Ireland*, **1**, 145–146.

CARROLL, R. L. 1998. Order Microsauria Dawson 1863. 1–71. *In* WELLNHOFER, P. (ed.). *Encyclopedia of Paleoherpetology. Part 1. Lepospondyli.* Verlag Dr. Friedrich Pfeil, Munich, 216 pp.

—— and BAIRD, D. 1968. The Carboniferous amphibian *Tuditanus* and the distinction between microsaurs and reptiles. *American Museum Novitates*, **2337**, 1–50.

CIVET, C. 1982. Etude d'un nouvel Amphibien Fossile du Bassin Houiller de Montceau-les-Mines *Scincosaurus spinosus* nov. sp. *"La Physiophile": Société d'études des sciences Naturelles et Historiques de Montceau-les-Mines*, **96**, 73–79.

CLACK, J. A. and FINNEY, S. M. 2005. *Pederpes finneyae*, an articulated tetrapod from the Tournaisian of Western Scotland. *Journal of Systematic Palaeontology*, **2**, 311–346.

DAVIT-BEAL, T., CHISAKA, H., DELGADO, S. and SIRE, J.-Y. 2007. Amphibian teeth: current knowledge, unanswered questions, and some directions for future research. *Biological Reviews of the Cambridge Philosophical Society*, **82**, 49–81.

EVANS, S. E. and SEARLE, B. 2002. Lepidosaurian reptiles from the Purbeck Limestone Group of Dorset, southern England. *Special Papers in Palaeontology*, **68**, 145–159.

FELSENSTEIN, J. 1985. Confidence limits on phylogenies: an approach using the bootstrap. *Evolution*, **39**, 783–791.

FRITSCH, A. 1876. Über die Fauna der Gaskohle des Pilsner und Rakonitzer Beckens. *Sitzungsberichte der Königlichen Böhmischen Gesellschaft der Wissenschaften Prague*, **1875**, 70–78.

——1879. *Fauna der Gaskohle und der Kalketeine der Permformation Böhmens.* Band I, Heft I, Selbstverlag, Prague, 1–92.

——1881. *Fauna der Gaskohle und der Kalketeine der Permformation Böhmens.* Band I, Heft III, Selbstverlag, Prague, 93–126.

GARDINER, B. G. 1983. Gnathostome vertebrae and the classification of the Amphibia. *Zoological Journal of the Linnean Society*, **79**, 1–59.

GOODRICH, E. S. 1930. *Studies on the structure and development of vertebrates.* Macmillan and Co, London, 837 pp.

HAWORTH, A. H. 1825. A binary arrangement of the class Amphibia. *Philosophical Magazine and Journal*, **65**, 372–373.

HOLMES, R. 1984. The carboniferous amphibian *Proterogyrinus scheelei* Romer, and the early evolution of tetrapods. *Philosophical Transactions of the Royal Society of London, Series B*, **306**, 431–524.

JAEKEL, O. 1903. Über *Ceraterpeton, Diceratosaurus* und *Diplocaulus. Neues Jahrbuch für Mineralogie, Geologie und Paläontologie*, **1**, 109–134.

——1909. Über die Klassen der Tetrapoden. *Zoologischer Anzeiger*, **34**, 193–212.

LEGENDRE, P. 2002. *Program for multiple linear regression (ordinary or through the origin) with permutation test – User's notes.* Département de Sciences Biologiques, Université de Montréal. 11 pp.

LYDEKKER, R. 1890. *Catalogue of the fossil reptiles and amphibians in the British museum (natural history), Part IV.* British Museum (Natural History), London, 205 pp.

MIALL, L. C. 1875. Report of the committee consisting of Professor Huxley LL.D., F.R.S., Professor Harkness, F.R.S., Henry Woodward, F.R.S., James Thompson, John Brigg, and L. C. Miall, on the structure and classification of the labyrinthodonts. *Report of the British Association for the Advancement of Science*, **1874**, 149–192.

MILNER, A. C. 1978. Carboniferous Keraterpetontidae and Scinosauridae (Nectridea, Amphibia) – a review. Unpublished PhD thesis. University of Newcastle upon Tyne, 192 pp.

——1980. A review of the Nectridea (Amphibia). 377–405. *In* PANCHEN, A. L. (ed.). *The terrestrial environment and the origin of land vertebrates.* Academic Press, London, 633 pp.

MILNER, A. R. 1980. The tetrapod assemblage from Nýřany, Czechoslovakia. 439–498. *In* PANCHEN, A. L. (ed.). *The terrestrial environment and the origin of land vertebrates.* Academic Press, London, 633 pp.

——1993. The Paleozoic relatives of lissamphibians. *Herpetological Monographs*, **7**, 8–27.

—— and SEQUEIRA, S. E. K. 1994. The temnospondyl amphibians from the Viséan of East Kirkton, West Lothan, Scotland. *Transactions of the Royal Society of Edinburgh: Earth Sciences*, **84**, 331–361.

QUICKE, D. L. J., TAYLOR, J. and PURVIS, A. 2001. Changing the landscape: a new strategy for estimating large phylogenies. *Systematic Biology*, **50**, 60–66.

RICHTER, A. 1995. Lacertier aus der Unterer Kreide von Uña und Galve (Spanien) und Anoual (Marokko). *Berliner Geowissenschaftliche Abhandlungen*, **14**, 1–174.

RUTA, M. and BOLT, J. R. 2006. A reassessment of the temnospondyl amphibian *Perryella olsoni* from the Lower Permian of Oklahoma. *Transactions of the Royal Society of Edinburgh: Earth Sciences*, **97**, 113–165.

—— and CLACK, J. A. 2006. A review of *Silvanerpeton miripedes*, a stem amniote from the Lower Carboniferous of East Kirkton, West Lothian, Scotland. *Transactions of the Royal Society of Edinburgh: Earth Sciences*, **97**, 31–63.

—— and COATES, M. I. 2007. Dates, nodes and character conflict: addressing the lissamphibian origin problem. *Journal of Systematic Palaeontology*, **5**, 69–122.

—————— and QUICKE, D. L. J. 2003. Early tetrapod relationships revisited. *Biological Reviews of the Cambridge Philosophical Society*, **78**, 251–345.

SANDERSON, M. J. and DONOGHUE, M. J. 1989. Patterns of variation in levels of homoplasy. *Evolution*, **43**, 1781–1795.

STEEN, M. C. 1938. On the fossil Amphibia from the Gaskohle of Nýřany and other deposits in Czechoslovakia. *Proceedings of the Zoological Society of London (B)*, **108**, 205–283.

WOODWARD, A. S. 1897. On a new specimen of the stegocephalan [sic] *Ceraterpeton galvani* Huxley from the Coal Measures of Castlecomer, Kilkenny, Ireland. *Geological Magazine*, **4**, 293–298.

APPENDIX

Character list. Character selection reflects variation in morphological conditions for the taxonomic exemplar included in this study; therefore, it encompasses fewer characters than those in some of the most recent analyses of Palaeozoic tetrapods. Characters are taken mostly from Anderson (2007) and Ruta and Coates (2007) data matrices, and reference to those is symbolised by letters and numbers which identify, respectively, author and character position in the original matrices (A = Anderson; RC = Ruta and Coates). A few additional characters are taken from Ruta and Bolt (2006; RB) and Ruta and Clack (2006; RCl).

1. Premaxilla alary process: absent (0); present (1). A19, RC1
2. Premaxillae more (0) or less than (1) two-thirds as wide as skull table RC5
3. Paired nasals: absent (0); present (1). RC10
4. Nasals more (0) or less than (1) one-third as long as frontals. RC11
5. Parietal/nasal length ratio less than (0) or greater than 1.45 (1). RC13
6. Prefrontal less than (0) or more than (1) three times longer than wide. RC15
7. Prefrontal entering nostril margin: no (0); yes (1). A21 (*partim*), RC19
8. Prefrontal/maxilla suture: absent (0); present (1). RC20
9. Lacrimal without (0) or with (1) dorsomesial digitiform process. RC24
10. Portion of lacrimal lying anteroventral to orbit abbreviated: absent (0); present (1). RC26
11. Posterior extremity of maxilla extending behind level of orbit posterior margin (0) or lying anterior to this margin (1). RC27
12. Maxilla contributing to orbit margin: no (0); yes (1). A49, RC28
13. Posterior extremity of maxilla not lying (0) or lying (1) at level of posterior extremity of vomers. RC32
14. Maxilla longer (0) or shorter (1) than palatine. A52
15. Frontal unpaired (0) or paired (1). A16, RC33
16. Frontals shorter than (0), longer than (1), or approximately equal in length (2) to parietals. RC34
17. Frontal excluded from (0) or contributing to (1) orbit margin. A17, RC35
18. Frontal anterior margin deeply wedged between nasal posterolateral margins: absent (0); present (1). RC37
19. Parietal/tabular suture: absent (0); present (1). A36, RC38
20. Parietal/postorbital suture: absent (0); present (1). A34, RC39
21. Parietals more (0) or less (1) than two and a half times as long as wide. RC42
22. Parietal/squamosal suture extending in part onto dorsal surface of skull table: no (0); yes (1). A35, RC43
23. Parietal/frontal suture strongly interdigitating: no (0); yes (1). RC44
24. Parietal/postparietal suture strongly interdigitating: no (0); yes (1). RC45
25. Separately ossified postparietal: present (0); absent (1). A37 (*partim*), RC46
26. Postparietal less than (0) or more than (1) four times wider than long. RC48
27. Postparietals without (0) or with (1) median lappets. RC49
28. Postparietal/exoccipital suture: absent (0); present (1). RC50
29. Postparietals without (0) or with (1) broad, concave posterior emargination. A25 (*partim*), RC54
30. Nasals not smaller (0) or smaller (1) than postparietals. RC55
31. Postfrontal posterior margin lying flush with jugal posterior margin: no (0); yes (1). RC58
32. Separately ossified intertemporal: present (0); absent (1). A1, RC59
33. Intertemporal shaped like a small, subquadrangular bone, less than half as broad as the supratemporal: absent (0); present (1). RC62
34. Separately ossified supratemporal: present (0); absent (1). A2, RC63
35. Narrow, strap-like supratemporal, at least three times as long as wide: absent (0); present (1). RC65
36. Supratemporal/squamosal suture: smooth (0); interdigitating (1). RC66
37. Tabular/squamosal suture extending onto skull table dorsal surface: present (0); absent (1). A6 (*partim*), RC70
38. Tabular (including its ornamented surface) elongate posterolaterally or posteriorly in the form of massive, horn-like process, conferring boomerang-like shape to skull outline in dorsal or ventral view: absent (0); present (1). A43, RC72
39. Parietal-parietal width smaller than (0) or greater than (1) distance between skull table posterior margin and orbit posterior margin, measured along skull mid-line RC73
40. Tabular horns directed posteriorly (0) or diverge posterolaterally in adults (1). A44
41. Postorbital without (0) or with (1) ventrolateral digitiform process fitting into deep vertical groove along jugal lateral surface. A32, RC78
42. Postorbital contributing to (0) or excluded from (1) orbit margin. A33, RC79
43. Postorbital irregularly polygonal (0) or broadly crescentic, narrowing to a posterior point (1). A192, RC80
44. Postorbital not wider (0) or wider (1) than orbit. RC82
45. Postorbital at least one-fourth of the width of the skull roof at the same transverse level: absent (0); present (1). RC83
46. Anterior part of squamosal lying posterior to (0) or anterior to (1) parietal mid-length. RC85
47. Squamosal without (0) or with (1) broad, concave embayment. A29, RC87
48. Squamosal without (0) or with (1) internal shelf bracing quadrate from behind. A28, RC89
49. Squamosal bordering posterior skull margin (0) or excluded by enlarged tabular (1). NEW
50. Squamosal triradiate: present (1); absent (0). NEW
51. Jugal not contributing (0) or contributing (1) to skull table ventral margin. RC91
52. Jugal not extending (0) or extending (1) anterior to orbit anterior margin. RC96
53. Snout blunt (0) or pointed (1). A26

54. Interorbital distance greater than (0), smaller than (1), or subequal to (2) half of skull table width. RC106

55. Interorbital distance greater than (0), smaller than (1), or subequal to (2) maximum orbit diameter. RC107

56. Orbit anteroposterior diameter shorter (0), longer (1), or subequal to (2) distance between orbit posterior margin and suspensorium anterodorsal margin. RC110

57. Orbit centre closer to anterior extremity of premaxillae than to posterior margin of skull roof (0), occupying approximately mid-length between anterior extremity of premaxillae and posterior margin of skull roof (1), or closer to posterior margin of skull roof than to anterior extremity of premaxillae (2). RB83

58. Orbit centre closer to anterior extremity of premaxillae than to posterodorsal margin of squamosal (0), occupying mid-length between anterior extremity of premaxillae and posterodorsal margin of squamosal (1), or closer to posterodorsal margin of squamosal than to anterior extremity of premaxillae (2). RB84

59. Minimum interorbital distance smaller (0), subequal to (1), or greater (2) than distance between posterior margins of orbits and mid-point of posterior margin of skull table. RB85

60. Pineal foramen occurring posterior to (0), at the level of (1), or anterior to (2) interparietal suture mid-length, or absent (3). A38 (*partim*), RC111

61. Postorbital region of skull table abbreviated, at least one-third wider than long: absent (0); present (1). RC114

62. Broad opening in skull postorbital region (aïstopod pattern): absent (0); present (1). RC116

63. Anteroposteriorly narrow, bar-like squamosal: absent (0); present (1). RC119

64. Lateral line canals: absent (0); present (1). A46, RC120, 121

65. Dermal cranial sculpture consisting mostly of a pit and ridge radiating pattern (0), circular pitting (1), mostly formed of striations and sulci (2), or relatively smooth (3). A47 (*partim*)

66. Ventral, exposed surface of vomers (i.e. excluding areas of overlap with surrounding bones) narrow, elongate, and strip-like, without extensions anterolateral and posterolateral to the choana, and two and a half to three times longer than wide: absent (0); present (1). RC122

67. Vomer with (0) or without (1) row-like clump of teeth (3+). RB105 (*partim*)

68. Vomer with (0) or without (1) fangs comparable in size to or larger than marginal teeth (premaxillary or maxillary). A72 (*partim*), 80 (*partim*), RC123

69. Vomer without (0) or with (1) small teeth (denticles) forming continuous shagreen or discrete patches, and the basal diameter and/or height of which is less than 30 per cent of that of adjacent marginal teeth (premaxillary or maxillary) and remaining vomer teeth (if present). A77, RC124

70. Vomer excluded from (0) or contributing to (1) interpterygoid vacuities. RC125

71. Vomer contact with pterygoid palatal ramus: present (0); absent (1). RC129

72. Vomers with distinctive tooth row close to medial margin of choana: absent (0); present (1). NEW

73. Palatine with (0) or without (1) fangs comparable in size to or larger than marginal teeth (premaxillary or maxillary). A72 (*partim*), 80 (*partim*), RC133

74. Palatine without (0) or with (1) small teeth (denticles) forming continuous shagreen or discrete patches, and the basal diameter and/or height of which is less than 30 per cent of that of adjacent marginal teeth (maxillary) and remaining vomer teeth (if present). A78, RC134

75. Palatine with (0) or without (1) row of teeth (3+) comparable in size to, or greater than marginal teeth (maxillary) and parallel to these. RC135

76. Separately ossified ectopterygoid: present (0); absent (1). A92, RC140

77. Ectopterygoid with (0) or without (1) fangs comparable in size to or larger than marginal teeth (premaxillary or maxillary) and remaining ectopterygoid teeth (if present). A72 (*partim*), 80 (*partim*), RC141

78. Ectopterygoid longer than/as long as (0) or shorter than (1) palatine. RC143

79. Ectopterygoid with (0) or without (1) row of teeth (3+) comparable in size to, or greater than marginal teeth (maxillary) and parallel to these. RC144

80. Pterygoid without (0) or with (1) posterolateral flange. RC149

81. Pterygoids sutured with each other: no (0); yes (1). A90, RC150

82. Pterygoid denticles: present (0); absent (1). A75, RB127

83. Pterygoid -basisphenoid articulation: mobile (0); not mobile (1). A82

84. Interpterygoid vacuities: present (0); absent (1). A89 (*partim*), RC160

85. Interpterygoid vacuities occupying at least half of palatal width: absent (0); present (1). RC161

86. Interpterygoid vacuities concave along their whole margins: absent (0); present (1). RC162

87. Interpterygoid vacuities together broader than long: absent (0); present (1). RC163

88. Occipital profile of skull low and wide (0), high and wide (1), or high and narrow (2). A64

89. Exoccipitals enlarged to form flattened, widely spaced occipital condyles: absent (0); present (1). A67 (*partim*), RC168

90. Parasphenoid without (0) or with (1) elongate, strut-like cultriform process. RC176

91. Parasphenoid without (0) or with (1) a pair of posterolaterally oriented, ventral thickenings (ridges ending in basal tubera). RC178

92. Parasphenoid without elongate, broad posterolateral processes (0), or with processes that are less than (1), or at least half as wide as (2) parasphenoid plate. RC179

93. Parasphenoid without (0) or with (1) single median depression. RC180

94. Jaw articulation lying posterior to (0), level with (1), or anterior to (2) occiput. A70 (*partim*), RC187

95. Well-developed retroarticular process: absent (0); present (1). A102, RB160

96. Jaw articulation position relative to marginal dentary tooth row: above (0); same level (1); below (2). NEW

97. Mandible slender and elongate (0) or stout and abbreviated (1). NEW

98. Dentary with (0) or without (1) anterior fangs generally comparable in size with or greater than other dentary teeth, lying close to symphysial region, and usually mesial to marginal dentary teeth. RC193

99. Posteriormost extension of splenial mesial lamina closer to anterior margin of adductor fossa than to anterior extremity of jaw, when the lower jaw ramus is observed in mesial aspect and in anatomical connection (i.e. symphysial region oriented towards observer): absent (0); present (1). RC198

100. Separately ossified postsplenial: present (0); absent (1). A98 (*partim*), RC201

101. Angular not reaching (0) or reaching (1) lower jaw posterior end. RC207

102. Separately ossified anterior coronoid: present (0); absent (1). A106 (*partim*), RC212

103. Separately ossified middle coronoid: present (0); absent (1). A106 (*partim*), RC216

104. Posterior coronoid with (0) or without (1) small teeth (denticles) forming continuous shagreen or discrete patches, and the basal diameter and/or height of which is less than 30 per cent than that of adjacent marginal dentary teeth. RC222

105. Posterior coronoid without (0) or with (1) posterodorsal process. RC224

106. Number of maxillary teeth greater than 30 (0), between 20 and 29 (1), or smaller than 20 (2). A57

107. Number of premaxillary teeth greater than 10 (0), between 5 and 9 (1), or smaller than 9 (2). A56

108. Labyrinthine infolding: present (0); absent (1). A63

109. Dentary tooth count greater than (0) or smaller than 20 (1). NEW

110. Cleithrum dorsal end: smoothly broadening to spatulate dorsal end (0); distal expansion marked from narrow stem by notch or process or decrease in thickness (1); tapering (2); with T-shaped dorsal expansion (3). RC236 (*partim*), A155 (*partim*)

111. Cleithrum lacking ornament (0) or ornamented (1). NEW

112. Clavicles meet anteriorly: yes abutting (0); no (1); yes interdigitating (2). A156, RC239 (*partim*)

113. Interclavicle rhomboidal with posterior part longer (0) or shorter (1) than anterior part. A 151 (*partim*), 152 (*partim*), RC242

114. Posterior border of interclavicle pointed (0), rounded or convex (1), or straight (2). NEW

115. Interclavicle fimbriated anterior margin: present (0); absent (1). A153

116. Distinct supinator process of humerus projecting anteriorly: absent (0); present (1). A163, RC250

117. Entepicondyle foramen: present (0); absent (1). A160, RC253

118. Humerus without (0) or with (1) waisted shaft. RC258

119. Position of radial condyle: terminal (0); ventral (1). RC259

120. Width of entepicondyle greater (0) or smaller (1) than half humerus length. RC263

121. Portion of humerus shaft length proximal to entepicondyle smaller (0) or greater (1) than humerus head width. RC264

122. Humerus length greater (0) or smaller (1) than combined length of two and a half mid-trunk vertebrae. A164 (*partim*), RC266

123. Radius longer (0) or shorter (1) than humerus. A165 (*partim*), RC268

124. Olecranon process of ulna: absent (0); present (1). A166, RC271

125. Dorsal iliac process: absent (0); present (1). A170 (*partim*), RC272

126. Transverse pelvic ridge: absent (0); present (1). RC275

127. Posterodorsal process of ilium less (1) or more (0) than three times as long as wide. NEW

128. Ischium at least marginally longer than tall (0) or about as tall as long (1) in lateral view. NEW

129. Ischium dorsal margin without (0) or with (1) upward curvature in lateral view. NEW

130. Number of pubic obturator foramina: multiple (0); single (1); absent (2). RC279

131. Puboischiadic plate spans six (0), five (1), four (2), or three (3) vertebrae. A171.

132. Pubic plate anteriorly acute and narrowing or squared off (0) or with polygonal outline (1). NEW

133. Internal trochanter separated from general surface of femur shaft by distinct, trough-like space: absent (0); present (1). RC281

134. Fourth trochanter of femur with distinct rugose area: no (0); yes (1). RC282

135. Cervical ribs with (0) or without (1) flattened distal ends. A142, RC296

136. Ribs mostly straight (0) or ventrally curved (1) in at least part of the trunk. RC297

137. Number of caudal ribs: five or more (0); four (1); three (2); two or fewer (3). A147

138. Trunk ribs needle-like: yes (1); no (0). NEW

139. Sacral rib distinguishable by size: shorter than trunk ribs and longer than presacrals (1); same length as presacrals (0). RCl 306

140. Sacral rib distinguishable by shape: broader than immediate presacrals but not broader than mid-trunk proximal shafts (0); broader than mid-trunk proximal shafts (1). RCl 307

141. Axis arch not fused (0) or fused (1) to axis (pleuro)centrum. RC303

142. Odontoid process, or tuberculum interglenoideum, on anterior surface of atlas body: absent (0); present (1). RC304

143. Extra articulations above zygapophyses in at least some trunk and caudal vertebrae: absent (0); present one (1); present two or more (2). A 112, RC305 (*partim*)

144. Neural spines rectangular to fan-shaped in lateral view: no (0); yes (1). A121, RC306 (*partim*)

145. Neural and haemal spines aligned dorsoventrally: absent (0); present (1). A124 (*partim*), RC307

146. Haemal spines not fused (0) or fused (1) to caudal centra. A124 (*partim*), RC308

147. Extra articulations on haemal spines: absent (0); present (1). A126 (*partim*), RC309

148. Trunk pleurocentra fused midventrally: no (0); yes (1). RC311
149. Trunk pleurocentra fused middorsally: no (0); yes (1). RC312
150. Neural arches of trunk vertebrae fused to centra: no (0); yes (1). A117 (*partim*), RC314
151. Bicipital rib bearers on trunk centra: absent (0); present (1). RC315
152. Trunk intercentra: present (0); absent (1). A115, RC316
153. Striated ornament on vertebral centra: absent (0); present (1). RC320
154. Height of ossified portion of neural arch in mid-trunk vertebrae greater (0) or smaller (1) than distance between pre- and postzygapophyses. RC329
155. Crenulations or fimbriate sculpture along dorsal margin of ossified portion of neural spines: absent (0); present (1). A122, RC330
156. Intravertebral foramina for spinal nerves in at least some trunk vertebrae: absent (0); present (1). A130, RC331
157. Dorsal edge of neural spine narrow and unornamented (0) or flat-topped and ornamented (1). NEW
158. Proximal haemal spines abutting (0), overlapping (1), articulating (2), or separated (3). NEW
159. Transverse processes: on arch pedicle (0); on centrum (1). A132
160. Second cervical arch proportions relative to more posterior presacral arches: expanded (0); subequal (1); shorter (2). A140
161. Haemal arches to neural spine length: longer (0); shorter (1). A125

162. Haemal arches: triangular (0); rectangular (1). A127, RC306 (*partim*)
163. Base of neural to haemal spine length: equal to or greater (0); smaller (1). A118
164. Manus with five or more (0), four (1), or three (2) digits. A168, RC333 (*partim*), 334 (*partim*), 335 (*partim*), 336 (*partim*)
165. Length of longest metacarpal less than twice its maximum width (0), at least twice as wide as long (1), or more than twice as long (2). NEW
166. Length of longest metacarpal no more than half length of radius (0) or between half and total length of radius (1) NEW
167. Pes with five or more (0) or four (1) digits. A179
168. Length of longest metatarsal reaching up to maximum oblique width of distal extremity of tibia (0), up to 33 per cent longer (1), or up to at least twice as long (2). NEW
169. Length of longest metatarsal less than twice its maximum width (0), at least twice as wide as long (1), or more than twice as long (2). NEW
170. Presacral vertebral count: 25–35 (0); less than 25 (1); more than 35 (2). A180
171. Skull/trunk length ratio greater than 0.45 (0), between 0.30 and 0.45 (1), between 0.20 and 0.29 (2), or less than 0.20 (3). A182
172. Presence (0) or absence (1) of limbs. NEW
173. Tail posterior extremity tapering (0) or deep with sudden end (1). A128

Data matrix. Polymorphic characters (states delimited by an ampersand) and uncertain characters (states delimited by a forward slash) are symbolised by capital letters, as follows: 0&1 = A; 0/1 = B; 1&2 = C; 1/2 = D; 0/1/2 = E; 0&2 = F.

Balanerpeton woodi
```
1010000000   0000110000   10AA000?00   001001100?   0010011000   0001120002
1000001011   0001100011   0000111001   0000000000   0000?00001   0101000100
011100000?   001?000010   0000000000   0000000000   0001100111   000
```
Dendrerpeton acadianum
```
1010000000   0000100000   1011010?00   000001100?   0010011000   0A020000?2
0000001011   0?01100011   0000110001   0000000??0   000??000?1   0101000100
0111010001   101?0000??   ??00000000   0000000000   0101100111   00?
```
Eucritta melanolimnetes
```
0010000000   000?120000   10000?0?00   001001100?   0010000000   ?0011100?0
000?F??010   0?01100?0?   10?1???0?0   1010000???   ?????0???1   0100000000
0?111??00?   ????0000??   ??????????   ??????????   ???????10?   00?
```
Lethiscus stocki
```
0011100000   01??1????1   0011000?01   0??010?00?   0001100?00   ?0011020?2
01103?????   ??????????   ??????????   ????1B01??   ?????D??1?   ??????????
??????????   ?????0?1??   1?0??????1   01?00??A?    ?????????2   31?
```
Microbrachis pelikani
```
0111100001   1000100011   1?10010100   01?1??001?   0000010000   1002002002
1001211110   0011201111   ?000000?01   0101000010   1000111102   0100000011
1010101001   3000012011   1100000110   0101000001   0002100122   200
```
Ophiderpeton brownriggi + O. kirktonense
```
??10100000   01??12000?   0011001?00   01?1???00?   0110000001   100000001E
01103?????   ??????????   ??????????   ??????????   1????2?11?   ??????????
??????????   ????10?1??   ??1????11?   ???100????   ?????????2   ?1?
```

Phlegethontia linearis + P. longissima

001100001??	11??0?????	??1?1?????	11?1????0?	??????0001	1011100012
01103?????	??????????	??000?020?	???2000101	?11??22102	0?????????
??????????	????1001??	1?1?????111	010A110?01	?????????2	310

Proterogyrinus scheelei

0110010000	0000110010	1000001000	000000100?	0010010000	1?01120012
000121???0	0000100?00	1000000200	1010000010	0000001001	0102100010
0111110000	0001010011	0000000100	0000000000	0100000010	10?

Batrachiderpeton reticulatum

0010100001	1101100011	1110000111	01?1??0111	0001110110	1000002012
0001000100	011001???0	1011???110	0?02111111	1110021113	1211??????
??????00?	??????1???	?011120111	0100100202	100??????1	?00

Crossotelos annulatus

??????????	01??????11	0000000?0?	01?010?00?	101??10?00	?0???????2
000?0?????	??????????	??????????	???0?B0???	1????2?1??	??????????
??????????	?????130??	??11120111	010011030?	000??????0	???

Ctenerpeton remex

??????????	??????????	??????????	??????????	??????????	??????????
??????????	??????????	??????????	??????????	??????????	??????????
???100011?	30??12000	??21111111	010010030?	0111DD1DD0	?01

Diceratosaurus brevirostris

0010111001	11??120011	1110000?11	?1?1??0100	0001110110	1002002012
00010?????	????2????0	?1101110?0	???21211??	???????1?3	1212?01??0
000100100?	?1???13001	1011130111	0110101001	0001D10DD1	100

Diplocaulus magnicornis

010??00101	1011011?11	110000011?	11?1??0111	0101110110	1101?02111
1001100101	111001???0	1110111010	0002021???	?110021113	12121010?0
0010???00?	????003001	1011110111	1110001112	0001D10DD1	100

Diploceraspis burkei

010???1???	1011011?11	110000011?	01?1??0111	0100010110	1101202111
1001100101	10100??1?0	1110111010	0002121111	111012110?	?2121?????
??????????	??????????	1121120111	1110001?12	000???????	?0?

Keraterpeton galvani

0010101001	10??100011	1111000?01	01?1??0100	0000110110	1002002012
00010?????	?1??01??0	?01?0000??	???11111??	1????22113	1211?010?0
0001001???	?????13001	1011120111	0100100201	1001D10DD1	200

Ptyonius marshii

0010001011	0000110111	0011001000	01?010100?	1001110000	1001100002
0000000110	0010001000	1100100101	01010001??	1111021100	0111001011
1011000112	31??12000	1121111111	0100100301	0111211221	10?

Sauropleura pectinata + S. scalaris

001AA11011	0100010111	0011000?00	01?010000?	1001110000	0011100002
000001010?	001001???0	?000000C00	0000000111	1111021A00	0110011011
1011000002	30??11001	1011131111	01001A1101	0001210220	100

Scincosaurus crassus

0010101001	1010100010	101?1?????	01?1??001?	0001010100	1102002002
1000010011??	101121???0	0110000?10	0001??1???	?????21110	0201?10111
1011001002	2000?12001	1011111111	0100100101	1110210221	300

Urocordylus wandesfordii

001???1?11	00????????	????????0?	???0??00??	10011?0000	000??0100?
00000?????	?01?01????	?000000???	???10001??	111102?101	0111?00010
0011001002	3101?13011	1021131111	010010010?	1A1?210221	201

[Special Papers in Palaeontology 81, 2009, pp. 91–120]

PATTERNS OF MORPHOLOGICAL EVOLUTION IN MAJOR GROUPS OF PALAEOZOIC TEMNOSPONDYLI (AMPHIBIA: TETRAPODA)

by MARCELLO RUTA

Department of Earth Sciences, University of Bristol, Bristol BS8 1RJ, UK; e-mail: M.Ruta@bristol.ac.uk

Typescript received 7 May 2008; accepted in revised form 21 August 2008

Abstract: A maximum parsimony analysis of a data matrix including 42 temnospondyl species and 14 outgroup species coded for 246 characters results in eight shortest trees. Temnospondyls emerge as two distinct radiations: one includes edopoids and eryopoids-basal archegosauriforms; the other includes dissorophoids and dvinosaurs. Both receive low statistical support. The branching pattern of dissorophoids remains elusive and is obscured by recurrent homoplasies. The two *Micropholis* morphs appear as sister taxa near the base of the dissorophoid clade. *Perryella* is either placed as sister taxon to dvinosaurs or nested within them. Temnospondyls such as the genera *Balanerpeton* and *Dendrerpeton* occur close to the node that subtends the dvinosaur–dissorophoid separation. Disparity analysis using Principal Coordinate Analysis of both Manhattan and Euclidean intertaxon distance matrices reveals that all major temnospondyl groups are widely separated in morphospace. Levels of disparity are comparable in the various clades examined, but the dvinosaurs tend to be morphologically more diverse than other groups when variance-derived metrics are used. For range-derived metrics, dissorophoids are more disparate than other groups. Overall disparity is similar in edopopids and eryopoids-basal archegosauriforms, as is in dissorophoids and dvinosaurs. However, the latter two groups are invariably slightly more disparate than the former two.

Key words: Carboniferous, characters, clades, disparity, Permian, phylogeny, Temnospondyli.

THE study of biological form has deep historical roots (Thompson 1942) and provides an important framework for theoretical and empirical investigations into patterns of evolutionary transformations and the origin and modifications of organism complexity. Quantitative analyses of shape have been employed to explore the emergence and radiation of major systematic groups and thereby inform our understanding of key events in the history of life (e.g. Fortey *et al.* 1996; Valentine 2004; Shen *et al.* 2008). Erwin (2007) summarized recent advances in this field and discussed its wide range of applications from both neontological and palaeontological standpoints.

Morphological diversity, hereafter referred to as disparity (see Runnegar 1987; Gould 1989; Foote 1993), elucidates tempo and mode of evolution (*sensu* Simpson 1944); indeed, an important step change in palaeobiological research has been the use of various disparity metrics as indicators of evolutionary trends within a phylogenetic context (e.g. Briggs *et al.* 1992; Smith 1994; Wills *et al.* 1994; Wagner 1996, 1997; Wills 1998*a*, *b*, 2001*a*, *b*; Eble 2000*a*, *b*; Stockmeyer Lofgren *et al.* 2003; Webster 2007). Detailed case studies can be found in Adrain *et al.* (2001) and MacLeod and Forey (2002).

The literature on disparity is vast (e.g. Foote 1996, 1997; Wills 2001*a*, *b*; Hammer and Harper 2006), and its chief taxonomic targets, in particular in palaeontology, have included (mostly) abundant and well-represented groups of marine benthic invertebrates. Fossil vertebrates have been subjected to very few quantitative studies, and this is particularly evident in the case of Palaeozoic tetrapods. However, several recent papers have begun to redress the balance (e.g. Laurin 2004; Ruta *et al.* 2006, 2007; Stayton and Ruta 2006; Wagner *et al.* 2006; Marjanovic and Laurin 2007; Kriloff *et al.* 2008). Importantly, these contributions have emphasized the fundamental role of phylogeny (see comments in Smith 1994), a cardinal component of palaeobiological analyses that is largely underrepresented in the invertebrate literature (but see Wills *et al.* 1994; Wagner 1996, 1997; Wills 1998*a*, *b*; Stockmeyer Lofgren *et al.* 2003; McGowan and Smith 2007; Webster 2007).

This work forms part of ongoing exploration of phylogeny-based evolutionary patterns in early tetrapods as a means to quantify and test scenarios of diversification of major groups, to understand the basis and modifications of morphological complexity and to untangle the

© The Palaeontological Association

doi: 10.1111/j.1475-4983.2009.00857.x

deep-time ancestry of a key component of animal diversity. Here, I address morphological diversification among (mostly) Palaeozoic temnospondyls. Temnospondyls are the most speciose, most diverse, and one of the best characterized among all groups of early tetrapods (e.g. Milner 1988, 1990; Holmes 2000; Laurin and Steyer 2000; Steyer and Laurin 2000). However, the large amount of comparative information now available for this group has not been followed by a uniform spread of phylogenetic scrutiny. Whilst the most diverse temnospondyl clade, i.e. the largely Mesozoic stereospondyls (see Schoch and Milner 2000; Steyer and Laurin 2000; Warren 2000), has been subjected to numerous cladistic studies, the interrelationships of major Palaeozoic clades, based upon large species-level exemplars and computer-assisted analyses, have been mostly neglected.

Milner's (1990) widely cited account of the family level phylogeny of the group (reproduced in a slightly modified version in Holmes 2000) has never been subjected to scrutiny (certainly given the enormous diversity of the group). However, several computer-generated cladograms have lent support to large portions of Milner's (1990) phylogeny. Yates and Warren's (2000) study targeted mostly stereospondyls and their closest relatives (i.e. basal archegosauriforms; Schoch and Milner 2000), although they also included representatives of other temnospondyl clades. In their analysis, Yates and Warren (2000) found a close relationship between the archegosauriform–stereospondyl clade and the dvinosaur clade, together forming their new taxon Limnarchia. The latter was placed as sister group to their new taxon Euskelia (dissorophoid–eryopoid clade). Laurin and Steyer (2000) (see also Steyer and Laurin 2000) proposed a slightly modified scheme of relationships for several families and superfamilies of temnospondyls and described it as a consensus phylogeny based upon information from Milner and Sequeira (1994, 1998), Yates and Warren (2000) and Steyer (2000). According to Laurin and Steyer (2000) and Steyer and Laurin (2000), edopoids, dissorophoids and eryopoids form successive sister groups, in that order, to limnarchians. Several other studies have tackled Palaeozoic species (e.g. see citations in Damiani *et al.* 2006; Laurin and Soler-Gijón 2006; Ruta and Bolt 2006), but none of them have produced a detailed species-based phylogeny for all major groups. Rather, the focus has been on the resolution of the affinities of one or few key species using limited exemplars outside the species or groups of interest (comments *in* Ruta and Bolt 2006).

Ruta and Bolt (2006) attempted to rectify this situation by undertaking a cladistic analysis of the major clades of Palaeozoic temnospondyls. I emphasize the fact that, although their data matrix is far from exhaustive in its taxonomic coverage, it represents the most complete treatment of Permian and Carboniferous temnospondyls

to date. A slightly revised, updated and enlarged version of Ruta and Bolt's (2006) matrix is employed here to quantify and compare disparity in nonstereospondyl temnospondyls. Five main questions are tackled:

1. Do various Palaeozoic temnospondyls show remarkably different levels of disparity?
2. How are such levels affected by the use of different disparity metrics and taxon sampling sizes?
3. What is the overall pattern of morphological character-state space occupation for the group?
4. Do different clades exhibit a significant amount of separation or overlap in morphospace?
5. What does disparity tell us about temnospondyl evolution during the Palaeozoic?

To some extent, the nature of the present paper is exploratory. Complementary approaches to the study of disparity (e.g. landmark-based geometric morphometrics) offer alternative ways in which morphological changes can be examined and quantified (e.g. Stayton 2005, 2006; Stayton and Ruta 2006; Angielczyk and Sheets 2007; Botha and Angielczyk 2007; Pierce *et al.* 2008).

DISPARITY AS AN INDICATOR OF EVOLUTIONARY DIVERSIFICATION

The key issue here is the exploration of both the potential and the underlying rationale of analyses of morphological diversification conducted within a phylogenetic framework. The reconstruction of a phylogeny provides the most important preliminary step for the study of evolutionary patterns (Smith 1994). Although analyses of morphological disparity are still relatively 'young' compared with other areas of macroevolutionary inquiry, they offer exciting and promising venues (e.g. see online column at: http://www.palass.org/modules.php?name=palaeo_math). In general terms, calculations of disparity can be performed in two ways: (1) using geometric landmarks or physical measurements (such as lengths, widths, ratios, angles and curvatures) (e.g. Elewa 2004; Zelditch *et al.* 2004) and (2) using a data matrix of morphological characters (e.g. Wills *et al.* 1994; Wills 2001*a, b*). Wills *et al.* (1994) dealt with the advantages and limitations associated with each of these two categories of methods, though note that this dichotomy represents a considerable simplification of the treatment provided by Wills *et al.* (1994, table 1) who identified and discussed fourteen disparity and distance metrics. A comprehensive analysis of the performance of various metrics and an evaluation of their sensitivity to various parameters (e.g. taxon sample size; character number; percentage of missing data) can be found in Ciampaglio *et al.* (2001).

The analyses undertaken here, and most of the disparity metrics that are derived from them, are based upon a multivariate treatment of a data matrix of morphological characters. In this respect, these metrics have 'phenetic connotations', because they are employed to evaluate '…character-state variability [as well as] amount of morphological attribute space occupied' by different clades. Furthermore, such phenetic methods '…operate on data coded as discrete homologous character states (this facility is also a requirement of cladistics), [and these] are …more appropriate…for comparing disparity in markedly dissimilar forms' (Wills *et al.* 1994, p. 93).

Wills *et al.*'s (1994) statement reflects the fact that their own analysis addressed disparity in very different arthropod taxa from a variety of time periods, and for which a geometric morphometric study of morphological diversity using landmarks offered limited potential (not least because of the difficulty in choosing landmarks across the spectrum of species). Wills *et al.*'s (1994) methodology, however, finds application in all those cases in which a taxon-character data matrix is available, including cases in which morphological features are directly comparable (as with the temnospondyl study presented herein; see also Stayton and Ruta 2006).

In this work, I employ a variety of disparity indexes as each index captures different aspects of morphological diversity. As a result, their combined use permits a more informed assessment of evolutionary patterns (e.g. see Foote 1991, 1996; Wills *et al.* 1994; Wills 2001*a*, *b*; Navarro 2003; McGowan 2004; Villier and Eble 2004; Villier and Korn 2004; Lefebvre *et al.* 2006; Erwin 2007; Moyne and Neige 2007; Adamowicz *et al.* 2008).

Null disparity models

An often neglected but focal aspect of disparity analyses is the selection of appropriate null, or even minimal (*sensu* Patterson 1994), models against which observed patterns are evaluated. Exploratory approaches to the study of morphological variation require either minimal or no initial hypothesis to test, and their results provide a direct way of quantifying patterns. This is the guiding principle in the present study. However, simulations of taxon diversification have shown that, in several cases, the geometry of morphospace occupation might not require a specific causal explanation. Thus, Pie and Weitz (2005) noted the presence of heterogeneous clusters of taxa (i.e. clouds of data points) when random walk branching models of diversification were subjected to different origination/extinction rates (including mass extinctions) or to developmental canalization (*sensu* Arthur 1997).

Mathematical treatments aside, Pie and Weitz's (2005) results are best summarized by their own observations that irregular '…occupation of morphospace reflect[s] the diffusion of character traits that share a common phylo-genetic history. [Thus, heterogeneous] occupation arises because related lineages undergo correlated random walks. [A]lthough…character traits [used in] simulation[s] are random, their relatedness is inevitable. However…discrete clusters…[do not characterize all branching random walks], [and are] present only in cases of logistic lineage diversification[s] and in the presence of developmental entrenchment' (Pie and Weitz 2005, pp. 7–8; see also Erwin 2007).

While phylogenetic relatedness may exert an important causal role in the retrieval of 'clouds' of data points, heterogeneous occupation of morphospace may not necessarily mirror patterns of taxon groupings in a tree. Empirical evidence gleaned from a variety of analyses indicates that, although a cluster of taxa in morphospace (either a geometric morphometric space or a character-state space) is usually observed, the concordance between phylogenetic and phenetic clustering does not represent the norm. Clusters of phylogenetically related taxa may, in fact, occasionally overlap one another. If an overlap is observed, then one can rule out the possibility that discrete fields of points are merely due to stochastic walks of related taxa (but see also McGowan 2007 for further remarks). This same conclusion also applies to those situations in which phylogenetically separate groups appear close to each other in morphospace (see examples *in* Stayton and Ruta 2006). A more detailed discussion of this issue can be found in Wills *et al.* (1994) and Stayton (2008).

Temnospondyls as a case study

The diversity of temnospondyls during the Palaeozoic can be briefly described as a succession of at least five main radiations, i.e. edopoids, dvinosaurs, basal archegosauriforms (plus stereospondyls), eryopoids and dissorophoids (see Milner 1990). The largest clade is represented by stereospondyls, the ancestry of which is rooted in a diverse group termed the basal archegosauriforms (e.g. Schoch and Milner 2000; Warren 2000; Yates and Warren 2000). For general introductions to the anatomy and diversity of temnospondys, see also Milner (1990), Holmes (2000), Schoch and Milner (2000), Steyer and Laurin (2000) and Warren (2000).

A few caveats apply to taxon sampling in the present study. The exploratory nature of this paper, and the fact that comprehensive phylogenies for Palaeozoic temnospondyls are not available (see also comments above), imply that results from this work ought to be regarded as empirical (*sensu* McGhee 1999; McGowan 2007). In other words, the construction of a character-state space and the results deriving from the disparity analyses best summarize the available phylogenetic evidence, in terms of both taxon and character selection. However, additional studies will be required to evaluate morphological diversification for increasingly larger taxon samples. This is not a limitation for the general purposes of this paper, because the focus is the quantification of disparity in major clades of Palaeozoic temnospondyls immediately following their phylogenetic separation. In this respect, the taxonomic sample is adequate, as it covers most of the basal as well as some of the more derived representatives of Permian and Carboniferous clades, such as have been diagnosed and defined in the recent literature.

Edopoids were reviewed recently by Milner and Sequeira (1998), Sequeira (2004, 2009 this volume) and Steyer *et al.* (2006). Seven of the nine taxa that are currently known have been included in this work. Two recently described taxa, *Nigerpeton* and *Saharastega* (Sidor *et al.* 2005), have not been considered here. *Nigerpeton* is nested within edopoids, as a member of the family Cochleosauridae (Steyer *et al.* 2006), but the publication of its full description appeared shortly after the completion of the analyses presented here. *Saharastega* has been placed as sister taxon to edopoids as a whole (Damiani *et al.* 2006). However, I express reservations about the edopoid affinities of *Saharastega*, and suspect that its proximity to edopoids results from the paucity of taxa in the cladistic analyses that accompanied its description. Both *Nigerpeton* and *Saharastega* will be considered at a later date in an enlarged and revised data matrix of Palaeozoic temnospondyls.

The second group is one of the most difficult to characterize and delimit. Based upon the results of the phylogenetic analysis discussed below, the group includes part of the eryopoid radiation (e.g. Milner 1990; Milner and Sequeira 1994, 1998; Sequeira 2004; Ruta and Bolt 2006) and part of the basal archegosauriforms–stereospondyls (e.g. Schoch and Milner 2000). An exemplar of the best-known eryopids, zatracheids and basal archegosauriforms encompasses seven species in total in this study. This group is termed the eryopoids-basal archegosauriforms hereafter (but see Milner 1996 and Schoch and Milner 2000 for a discussion of *Capetus*). One of the most abundant and best preserved basal archegosauriforms, *Archegosaurus decheni*, has been redescribed in considerable detail (e.g. Witzmann 2006*a*, *b*; Witzmann and Schoch 2006; Witzmann and Scholz 2007), but will be considered in a phylogenetic analysis at a later stage. *Archegosaurus* may occupy a less basal position among archegosauriforms than previously surmised (Schoch and Milner 2000; Witzmann and Schoch 2006; Schoch and Witzmann 2009, this volume; MR, unpublished data).

The third clade, the dvinosaurs, is represented by approximately 22 species, some based upon fragmentary or poorly diagnosable remains. For a recent revision of the group, see Englehorn *et al.* (2008). Ten of the best-known dvinosaur taxa have been included in this paper (see also Yates and Warren 2000; Milner and Sequeira 2004; Ruta and Bolt 2006).

Finally, dissorophoids are represented by 15 species in this paper, as follows: the two morphs of *Micropholis* (as described by Schoch and Rubidge 2005); one dissorophid; four trematopids; one micromelerpetontid; three branchiosaurids; and four amphibamids. Recent discoveries of dissorophoid material, together with new information on previously known species, have resulted in a concerted effort to readdress long-standing issues of dissorophoid phylogeny. Recently, a flurry of papers has begun to tease apart patterns of character acquisition and transformation within amphibamids (e.g. Schoch and Rubidge 2005; Huttenlocker *et al.* 2007; Anderson *et al.* 2008*a*, *b*), and comparable investigations are being undertaken for each of the remaining dissorophoid families (e.g. Schoch and Milner 2008). New amphibamid taxa reported by Huttenlocker *et al.* (2007), Anderson *et al.* (2008*a*, *b*) and Fröbisch and Reisz (2008) were published after the analytical part of the present paper was completed and could not be included in the cladistic analysis. The influence of these and other species on the overall geometry of the temnospondyl tree will be assessed after assembly of a new data set (MR, unpublished data).

MATERIAL AND METHODS

Phylogenetic analysis

The modified version of Ruta and Bolt's (2006) matrix includes 56 taxa (42 temnospondyls; 14 outgroups) and 246 binary and multistate characters (198 cranial; 48 postcranial). MACCLADE v. 4.08 (Maddison and Maddison 2000) and PAUP* v. 4.0b10 (Swofford 2002) were used to assemble the matrix and to process it under maximum parsimony respectively. Characters were left unweighted and unordered. To increase the chance of retrieving optimal islands of trees, a heuristic search was carried out with 5000 random stepwise addition replicates and tree bisection-reconnection branch-swapping algorithm (TBR). Only a single tree was kept at the end of each replicate in the first part of the tree search; in the second part of the search, i.e. upon completion of the 5000 replicates, TBR was performed on all trees in memory, but with the option of saving multiple trees. After additional swapping on the trees retrieved at the end of the second part of the search, neither shorter nor novel trees were found (see also Quicke *et al.* 2001). Bibliographic references for all the species included in this paper are listed in Ruta and Bolt (2006).

The Appendix (see below) includes the coding of a new character added to Ruta and Bolt (2006), as well as the scores attributed to each of the two *Micropholis* morphs for all characters. I point out that the new character now occupies the tenth column in the modified version of the Ruta and Bolt (2006) data set; all characters following it are thus shifted by one column to the right relative to the original matrix, but their description is the same as in Ruta and Bolt (2006).

Although the cladistic analysis addresses primarily Palaeozoic temnospondyls, a large number of outgroups have been used. Simultaneous unconstrained analysis (*sensu* Clark and Curran 1986; see also Maddison *et al.* 1984; Nixon and Carpenter 1993) of ingroup and outgroup taxa is a necessary prerequisite for assessments of character polarity and for accurate reconstructions of character-state conditions near the base of the group of interest. Furthermore, a key aspect of disparity studies is that they have empirical connotations (McGhee 1999), i.e. they provide the best fit for intertaxon distance data based upon the exemplar that is selected for a particular study. Although the disparity analyses that derive from the cladistic data set may not require an extensive sample of taxa outside temnospondyls, the chosen outgroups are generally regarded as being phylogenetically proximal either to temnospondyls or to the base of the crown tetrapod radiation (e.g. Clack 2002; Ruta *et al.* 2003; Ruta and Coates 2007; Ahlberg *et al.* 2008). Therefore, inclusion of these outgroups is useful to visualize patterns of divergence of temnospondyl groups (e.g. species dispersal; morphospace occupation) from assorted stem tetrapod clades.

A separate issue concerns character inclusion, and in particular the influence of autapomorphies on disparity calculations. So far as possible, chosen characters (Ruta and Bolt 2006 and this paper) aim to cover the entire range of morphological conditions for the whole skeleton in all taxa included in the data matrix. As advocated elsewhere (Ruta *et al.* 2003; Ruta and Coates 2007), a comprehensive treatment of characters is a necessary prerequisite for stringent tests of taxon interrelationships (as is also taxon inclusion). Traditionally, autapomorphic features are not considered in phylogenetic analyses. However, this is not a problem as far as disparity calculations in this work are concerned, for two main reasons. First, unique conditions have been linked, so far as applicable, to alternative conditions in multistate characters, though selection of genuine autapomorphies is often problematic. Where applicable, I chose the latter option in the construction of the data set (see also Ruta and Bolt 2006). Therefore, unique traits are represented (so far as possible) in the data matrix and thus are accounted for in the calculations of intertaxon distances. Second, it is generally difficult to identify an apomorphic condition for a given structure in a taxon simply because each taxon is specialized in its own right. Furthermore, whilst it is

always possible to spot differences between any two taxa, such differences may not necessarily be easily scored in a data matrix. In this respect, morphometric techniques offer suitable and effective solutions, as landmarks and/or measurements may capture subtle phenetic differences more easily than cladistic characters do (e.g. Richtsmeier *et al.* 2002).

Disparity analysis: distance matrix output

The protocol for undertaking disparity analyses is summarized in numerous papers as well as in the documentation accompanying available statistical software (see Foote 1991; Wills *et al.* 1994; Wills 2001*a*, *b*; Stockmeyer Lofgren *et al.* 2003; Hammer and Harper 2006; Foote and Miller 2007). The details may differ, depending upon the techniques used but also upon the issues addressed. To avoid confusion, the following sections provide a succinct step-by-step summary of the procedure adopted in this work.

The first step consisted of deriving an intertaxon distance matrix from the cladistic data matrix. Two methods were employed, each giving different types of intertaxon distances. The first of these was the construction of Manhattan distances. These were obtained from PAUP* using either the mean character difference between any two taxa (difference in character-state scorings for each taxon pair divided by the total number of characters) or the total character-state difference (same as above, but without dividing the differences by the total number of characters) (Wills *et al.* 1994). Disparity analyses using the mean or total character differences delivered very similar results. For this reason, I will discuss only results based upon the mean differences. The second method was the construction of Euclidean distances, obtained by processing the original MacClade data set with the software MATRIX v. 1.0 (Wills 1998*a*) (note that MATRIX also outputs Manhattan distances).

Wills *et al.* (1994) dealt with different properties associated with both Manhattan and Euclidean distances and advocated the use of the latter in disparity analyses, for the following main reasons. Pairwise distance measures between species (or other taxonomic units) can be subjected to various kinds of multivariate analysis, e.g. for the purpose of representing data points (taxa) in morphospace (in this case, a space built from character-state variables) and for calculations of disparity metrics. A widely employed category of multivariate methods builds a series of 'synthetic' variables that best summarize information contained in the original set of pairwise distances. One of the commonly used multivariate methods is Principal Coordinate Analysis (PCoA). PCoA subjects all distances to an ordination along 'synthetic axes'. Such an ordination involves the decomposition of the original distance matrix into a set of eigenvalues and eigenvectors. When an ordination is carried out using the Manhattan distances, PCoA frequently delivers '…dimensions…with negative eigenvalues [which] explain a negative proportion of the variance' (Wills *et al.* 1994, p. 104). This negative proportion is not easily interpretable. However, if ordination is based upon the Euclidean distances, then the impact of negative eigenvalues on the analysis is less dramatic, resulting in the production of many more positive eigenvalues, which indicate a larger number of 'real' axes (as opposed to the 'imaginary' axes associated with negative eigenvalues; see also Legendre and Legendre 1998).

The software MATRIX provides several options for the Euclidean matrix output (all details of these options are provided in the accompanying software manual). All binary characters were treated with a 'b' code. This option emphasizes the role of derived similarity associated with discrete states. The multistate characters were given a 'm' code, which again stresses the relevance of derived states in determining similarity, but is used for multiple unordered states.

MATRIX produces three kinds of Euclidean matrices; these differ in the ways in which weighted and ordered characters are treated. However, all outputs are identical in this case, as characters were left unweighted and unordered. Finally, some modifications of the cladistic data set were introduced before this could be processed in MATRIX. Specifically, all polymorphisms and uncertainties were rescored to include the smallest of all states represented in the original codings. Unknown and inapplicable states, typified by the use of a '?' in the original data matrix, were rescored as '9'.

These modifications enabled MATRIX to handle missing, polymorphic and uncertain character conditions. Results of the disparity analyses do not differ in any significant way if polymorphisms and uncertainties are rescored using alternative character-states. As far as the missing entries are concerned, these are replaced by a 'mean weighted value' that takes into account characters that can be compared. To achieve this, MATRIX first computes a 'weighted mean fractional similarity' (fs), described (as per information in the manual accompanying the software) as 'the total distance for all scorable comparisons divided by the total potential distance for all scorable comparisons'. For the calculation of Euclidean distances in the presence of missing entries, the latter '…are assigned a character distance equal to [weighted mean fractional similarity] times the maximum distance for that character alone. All character distances are subsequently squared and added. This sum is rooted to derive a Euclidean distance for that taxon/taxon comparison.' Uninformative characters are removed from calculations in MATRIX.

Disparity analysis: ordination procedure

The second step in the disparity calculation was the multivariate analysis of intertaxon distances. Euclidean and Manhattan matrices were subjected to PCoA using the program GINKGO v. 1.5.5, a multi-platform Java application (http://biodiver.bio.ub.es/vegana; see also Cáceres *et al.* 2007). For each of the two matrices, imported in GINKGO in ASCII format, PCoA was carried out on all 56 species (both ingroup and outgroups); in addition, their centroid (that is, the point with coordinates averaged over the PCo coordinates of all species) was chosen as the origin of the coordinate axes. Corrections for negative eigenvalues were based upon the methods of Lingoes (1971) and Cailliez (1983). As the two methods yield very similar results, I will focus exclusively on PCoA analyses (and other sets of calculations) deriving from the Cailliez correction, both for the Manhattan matrix and for the Euclidean matrix (see Legendre and Anderson 1998; Legendre and Legendre 1998).

Among the analytical facilities provided by GINKGO is a scree plot, depicting eigenvalue scores associated with each eigenvector (i.e. each PCo axis). Another useful, complementary facility is the tabulation of eigenvectors (PCo axes) and their related eigenvalues with the percentage of the total variance that each eigenvalue entails. In GINKGO, one can produce lists of cumulative percentages of variance. These have a practical and immediate utility, because they indicate how many axes can be selected that, together, account for a chosen threshold percentage value of total variance: 90 per cent was chosen as the threshold value. Thus, of all 56 PCo axes (this number equals the number of taxa in the present work), the first n axes were considered that, altogether, summarize at least 90 per cent of the total variance (PCo axes are listed in decreasing order of their respective contributions to total variance): n = 45 for the Euclidean matrix; n = 47 for the Manhattan matrix (see also below for brief remarks on the choice of axes).

GINKGO was further employed to visualize the plot of taxa in a character-state space defined by the first three PCo axes. For both Manhattan and Euclidean PCoA-processed matrices, three screen captures were produced that illustrate the relative positions of taxa in two dimensions (i.e. using the following PCo axes pairs: 1 and 2; 1 and 3; 2 and 3). In addition, three-dimensional screen captures were produced. The three-dimensional plots were rotated manually several times in GINKGO until a satisfactory configuration of data points was found that best illustrated all ingroup and outgroup taxa with minimum overlap among them. The two- and three-dimensional screen captures were first saved in GINKGO and then opened and processed in Adobe Photoshop v. 7.0. The results of PCoA performed on Manhattan and

Euclidean distance matrices were further checked with the free statistical software DISTPCOA (Legendre and Anderson 1998), available at: http://www.bio.umontreal.ca/Casgrain/en/labo/index.html.

Disparity analysis: indexes of morphological diversity

Principal coordinate analysis scores (a set of coordinates for each taxon on all 56 PCo axes) were extracted from GINKGO in tabulated format, and used as input for calculating disparity indexes. The following four indexes have been extensively used in disparity studies (e.g. Foote 1993; Wills *et al.* 1994; Ciampaglio *et al.* 2001): sum of ranges; product of ranges; sum of variances; and product of variances (see Van Valen 1974; Foote 1992). These indexes were produced using the morphological rarefaction program RARE v. 1.1 (Wills 1998*a*). Each of the two PCoA score sets (one from the Euclidean matrix and one from the Manhattan matrix) was subjected to the same search settings in RARE, as follows. First, only axes that explain at least 90 per cent of the total variance were considered (45 axes for Euclidean data and 47 axes for Manhattan data).

Second, a complete rarefaction profile was reconstructed for each of the four indexes computed by RARE and for each of the temnospondyl clades. The rarefaction profile was built by extracting all subsamples of taxa comprised between 2 and the total number included in each clade (that is: 7 edopoids; 7 eryopoids-archegosauriforms; 10 dvinosaurs; 15 dissorophoids), and using a bootstrap routine of 1000 replicates (sampling with replacement).

Third, 90 per cent confidence intervals were placed around each disparity value in each of the rarefaction profiles. RARE was also run without the rarefaction option, which allows this software to calculate the total disparity values for each group (i.e. all taxa are included at once).

Rarefaction is useful for contrasting disparity levels that have been recalibrated to take into account different sizes of various clades. In short, it permits standardized comparison of values when different numbers of taxa are considered. For instance, given two groups, we may want to see how the disparity value for the more diverse (speciose) group is affected when different subsamples of its constituent taxa are employed (e.g. when a subsample identical in size to the entire sample of the less speciose group is considered). For a discussion of rarefaction and its applications, see Raup (1975), Gotelli and Colwell (2001) and Hammer and Harper (2006).

As regards the disparity indexes, the range and variance describe slightly different (and, to some extent, complementary) aspects of morphological diversity. The sum of ranges (that is, the sum of maximum Euclidean distances between any two taxa; Ciampaglio *et al.* 2001) measures the amount of morphospace occupied by a sample of taxa (Vil-

lier and Korn 2004). Conversely, the sum of variances (each character possessing its own value of univariate variance) quantifies the 'spread' (i.e. the dispersal) of taxa in that sample, and can be likened to a measure of degree of dissimilarity among them (Villier and Korn 2004). The product of ranges and product of variances can be intuitively likened to measures of 'hypervolume' occupied by different groups (Wills *et al.* 1994). As the values of the two product indexes tend to become rapidly inflated with the addition of PCo axes, I took the nth root of both as a measure of the hypervolume, where n = 45 and 47 (see above) for data based upon the Euclidean and Manhattan matrices respectively. Through this procedure, the two products are scaled to a single dimension (Wills 2001*b*; Wills *et al.* 1994; Harper and Gallagher 2001) so that their values for different groups can be readily compared and easily plotted.

The calculations for the root-products of ranges and variances and for the sum of ranges require further clarification. These calculations were carried out by subdividing the tabulated data of PCoA scores for all species into different subsets, each corresponding to a specific clade. However, Wills *et al.* (1994) pointed out that this procedure is only valid if one considers the PCo axes derived from the all-inclusive analysis to be representative of the entire spectrum of morphologies. In short, a set of axes from a global ordination procedure '...will be sensitive to [those] regions of the original character space that are actually occupied by taxa [in the various] groups, and their use as reference axes therefore minimizes the amount of nonviable [i.e. unrepresented] morphospace [reflected by] the hypervolume estimate' (Wills *et al.* 1994, p. 107). However, it is important to bear in mind that the PCo axis of greatest variance in an all-encompassing PCoA analysis may not correspond exactly to the axis of greatest variance associated with a subset of taxa that are analysed independently. The sum of variances is not affected by the orientation of axes (Wills *et al.* 1994).

The various groups of temnospondyls were also analysed separately and each was subjected to PCoA runs on relevant subsets of the Manhattan and Euclidean distance matrices. This procedure was employed in calculations of the PCo volume, one of the various disparity indexes discussed by Ciampaglio *et al.* (2001). Following Ciampaglio *et al.*'s (2001) definition, the PCo volume for a group consisting of N taxa is the product of the two largest eigenvalues deriving from the distance matrix decomposition divided by the square of N (PCo volume = $\lambda_1 \lambda_2 / N^2$). It is easy to see why PCo volume calculations are derived from subsets of the Manhattan and Euclidean matrices. The denominator of the formula tends to increase rapidly with N. Given the same $\lambda_1 \lambda_2$ product for a set of calculations that uses the entire Manhattan and Euclidean matrices, the PCo volume becomes rapidly very small when N increases. Yet, groups may actually appear fairly disperse in a character-state morphospace. Calculations of PCo volume based upon portions of Manhattan and Euclidean matrices use different values of the $\lambda_1 \lambda_2$ product (one for each group) and recalibrate such products using the N^2 value, where N may differ in each group.

Finally, I compared the pairwise Manhattan and Euclidean intertaxon distances for the four main groups to test for differences in the distribution of their distance values (see Wills *et al.* 1994 for the use of additional metrics).

RESULTS

Temnospondyl interrelationships

PAUP* finds eight equally parsimonious optimal trees 936 steps long (Consistency Index = 0.3015; Retention Index = 0.6254; Rescaled Consistency Index = 0.1904). The strict consensus is shown in Text-figure 1. If the data

matrix is reanalysed with all characters reweighted according to the best fit of their rescaled consistency indexes (that is, the maximum value of the index that each character displays over all minimal trees), then PAUP* yields a single tree (Text-fig. 2A). In this, the relative position of outgroups, edopoids and eryopoids-archegosauriforms are identical to those observed in the trees from the original, unweighted analysis (therefore, they are not shown in Text-figure 2A).

Obvious similarities exist between the results of this study and those presented by Ruta and Bolt (2006). However, some important differences between the two studies require additional discussion. First, edopoids emerge as the sister group to eryopoids-basal archegosauriforms. This relationship is a clear departure from other published trees, in which edopoids form the most basal temnospondyls (e.g. Milner 1988; Holmes 2000; Ruta and Coates 2007). Boy (1981, fig. 9) illustrated a simplified temnospondyl cladogram with edopoids and eryopoids as sister taxa. It is also interesting to note that in at least

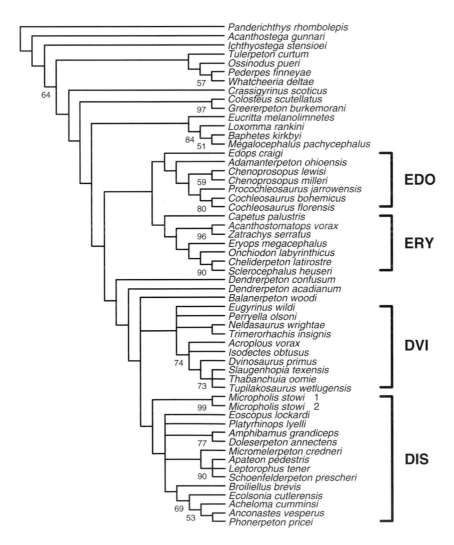

TEXT-FIG. 1. Strict consensus of eight most parsimonious trees deriving from the cladistic analysis (see text for details). Numbers on branches represent bootstrap percentages. Abbreviations of clade names (in this and in other figures) are as follows: DIS, dissorophoids; DVI, dvinosaurs; EDO, edopoids; ERY, eryopoids-basal archegosauriforms; *Micropholis stowi* 1 and 2 represent, respectively, the long- and short-skulled morphs of Schoch and Rubidge (2005). Note that the 50 per cent majority-rule consensus from all bootstrap runs shows two further clades, namely: *Acroplous* plus *Isodectes* with a 72 per cent bootstrap support; *Thabanchuia* plus *Tupilakosaurus* with a 70 per cent bootstrap support.

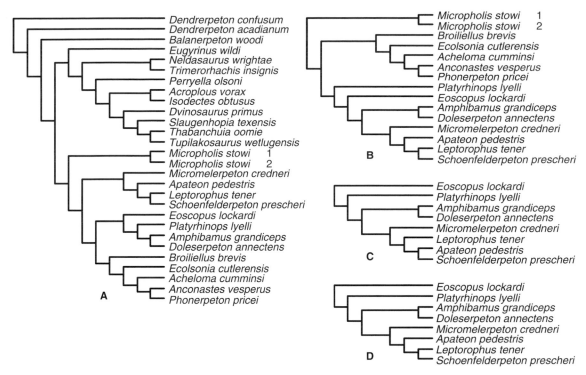

TEXT-FIG. 2. A, the dvinosaur–dissorophoid part of the single shortest tree that derives from the reweighting of characters by the best fit of their rescaled consistency index (see text for details). The mutual relationships of dissorophoids are identical to those seen in five of the eight shortest trees from the unweighted parsimony analysis, specifically trees 3, 4, 5, 6 and 8. B–D, the mutual relationships of dissorophoids in trees 1, 2, and 7, respectively, from the unweighted parsimony analysis; trees 2 and 7 differ from tree 1 both in the positions of *Eoscopus* and *Platyrhinops*, and in the positions of *Apateon* and *Leptorophus*.

one recent analysis (Ruta and Bolt 2006, fig. 25), the phylogenetic position of *Edops* is destabilized, although this genus still appears close to remaining edopoids (see Witzmann *et al.* 2007 for a discussion of eryopids and their proximity to basal archegosauriforms).

Second, the two *Dendrerpeton* species branch from adjacent nodes, and thus form a paraphyletic array relative to the dvinosaur–dissorophoid clade, whereas in Ruta and Bolt (2006), *Dendrerpeton confusum* is sister taxon to dissorophoids and *Dendrerpeton acadianum* is sister taxon to (*Balanerpeton* + dvinosaurs). Four of the shortest trees show *Balanerpeton* as the immediate sister taxon to the dissorophoid–dvinosaur clade whereas in the other four trees, *Balanerpeton* is sister taxon to dissorophoids only.

Third, partial loss of resolution characterizes dvinosaurs. In five of the trees, the sequence of taxa near the base of the dvinosaur part of the cladogram includes *Eugyrinus*, a clade of trimerorhachids (*Neldasaurus Trimerorhachis*) and *Perryella* (in order of increasing proximity to other dvinosaurs). In the remaining trees, *Perryella* is the most basal dvinosaur, whilst *Eugyrinus* and trimerorhachids are the second and third most basal dvinosaur taxa respectively. The Appendix reports character-state changes supporting the dvinosaur–dissorophoid clade under the accelerated transformation of state

changes in PAUP* (ACCTRAN: character-state changes are placed as close to the tree root as possible). This clade has not been found in previous temnospondyl phylogenies (except for Ruta and Bolt 2006). All lists of state changes (one for each minimal tree) reveal highly homoplastic characters, with consistency index values ranging from 0.077 to 0.5. This is unsurprising, given the variable placement of taxa such as *Balanerpeton* and *Dendrerpeton* near the node subtending dissorophoids and dvinosaurs as well as the recurrent features associated with neoteny or paedomorphism in these two groups.

Fourth, the interrelationships of the dissorophoid families are less resolved than in Ruta and Bolt (2006). However, there are numerous interesting implications for the phylogeny of dissorophoids as a group. The agreement subtree (a pruned tree which includes only those taxa for which all minimal trees agree upon relationships) shows that the branching pattern of dissorophoids that is common to all most parsimonious solutions is one in which the two *Micropholis* morphs, together, branch from the base of the dissorophoid clade. All other dissorophoids in this study consist of an amphibamid clade thus structured: [*Eoscopus* (*Amphibamus Doleserpeton*)], and a dissorophid–trematopid clade thus structured: [*Broiliellus* (*Ecolsonia* (*Acheloma* (*Anconastes Phonerpeton*)))]. Thus, it appears that the only unstable dissorophoids are represented by a clade encompassing micromelerpetontids (*Micromelerpeton*) plus unresolved branchiosaurids (*Apateon Leptorophus Schoenfelderpeton*). The micromelerpetontid–branchiosaurid clade represents the second basalmost dissorophoid group, after *Micropholis*, in five of the eight trees. In the remaining three trees, however, this clade branches in an apical position among dissorophoids with amphibamids forming a paraphyletic array relative to it. The amphibamid *Platyrhinops* is the only unstable amphibamid (it is pruned from the agreement subtree).

The intrinsic relationships of dissorophoids are in a state of flux. Many new discoveries are now beginning to clarify the sequence of branching events and the status of various families. However, the most recent analyses that have included a sufficiently large number of dissorophoids have also revealed a disconcerting lack of agreement. Largely irreconcilable phylogenies display almost every possible permutation of the five main families of dissorophoids.

Ruta *et al.* (2003) could not provide sufficient resolution in the apical portion of the dissorophoid component of their early tetrapod phylogeny. In their analysis, trematopids and dissorophids formed a paraphyletic array, in that order, relative to a group consisting of a resolved micromelerpetontid–branchiosaurid clade branching from a polytomy that included amphibamids (the latter were all collapsed). Using a revised, corrected and expanded version of Ruta *et al.*'s (2003) data, Ruta and Coates (2007) found partial resolution for amphibamids. These emerged either as a paraphyletic group (relative to crown lissamphibians) or as a polyphyletic array with the micromelerpetontid–branchiosaurid clade being more crownward than *Eoscopus* and *Platyrhinops* but less crownward than *Amphibamus* and *Doleserpeton*.

Schoch and Rubidge (2005) found that micromelerpetontids, trematopids and dissorophids are arranged in order of increasing proximity to amphibamids. *Micropholis* emerged as sister taxon to remaining amphibamids. A modified version of Schoch and Rubidge's (2005) data was presented by Huttenlocker *et al.* (2007), who found branchiosaurids and micromelerpetontids to be successive sister groups to a clade comprising amphibamids and dissorophids plus trematopids. In their tree, amphibamids form two discrete groups, with *Micropholis* and their new taxon *Plemmyradytes* in an apical position within a group that included *Eoscopus* and a paraphyletic *Tersomius*, and with other amphibamids (*Platyrhinops*; *Amphibamus*; *Doleserpeton*) clustered in a separate group.

Anderson *et al.* (2008a) placed a dissorophid–trematopid clade (their new group Olsoniformes) at the base of the dissorophoids, with micromelerpetontids as sister taxon to amphibamids. Within the latter, they retrieved the *Micropholis–Tersomius* clade as sister taxon to other amphibamids, and found their new taxon *Georgenthalia* in three different and equally parsimonious positions relative to *Eoscopus*, *Platyrhinops*, *Amphibamus* and *Doleserpeton* (see also Anderson 2007; Anderson *et al.* 2008b).

The most significant addition to the amphibamid literature is Anderson *et al.*'s (2008b) description of a new genus, *Gerobatrachus*, placed as the most crownward stem group member of the batrachians (salamanders–frogs clade). In their phylogeny, Anderson *et al.*'s (2008b) placed olsoniforms at the base of dissorophoids, followed by a micromelerpetontid–branchiosaurid clade and by amphibamids as a paraphyletic set of taxa relative to batrachians. *Tersomius* and *Micropholis* appear as sister taxa and *Eoscopus*, *Platyrhinops*, *Amphibamus*, *Doleserpeton* and *Gerobatrachus* branch in crownward succession, in that order, relative to batrachians (see also Fröbisch and Reisz 2008 for a further take on amphibamid interrelationships).

Schoch and Milner's (2008) analysis of the internal relationships of branchiosaurids finds the latter as a monophyletic group consisting of a basal taxon, *Branchiosaurus*, placed as sister taxon to two discrete radiations, one including different species of *Apateon*, the other consisting of species of the genus *Melanerpeton* forming a paraphyletic array relative to *Schoenfelderpeton* plus *Leptorophus*. Furthermore, micromelerpetontids (*Micromelerpeton*), olsoniforms (*Ecolsonia*) and paraphyletic amphibamids (*Micropholis* plus an *Amphibamus–Platyrhinops* clade) constitute successive outgroups (as listed, in basal to apical succession) to branchiosaurids.

The results presented in this paper offer yet another solution. Specifically, two different patterns of relationships are summarized by the eight minimal trees. The simpler pattern shows *Micropholis* and the micromelerpetontid–branchiosaurid clade as successive sister taxa to a group encompassing monophyletic amphibamids and olsoniforms (Text-fig. 2A). The more elaborate pattern retrieves *Micropholis* and olsoniforms as successive sister taxa to a group in which amphibamids (Text-fig. 2B–D) are paraphyletic relative to the micromelerpetontid–branchiosaurid clade.

Intriguingly, *Micropholis* does not cluster with amphibamids (but see Schoch and Rubidge 2005) and emerges instead as a basal dissorophoid in the present work. The permutations of some species in the most parsimonious trees are interesting in this respect, and call for a few comments on the affinities of *Perryella*. Thus, it was noted above that *Perryella* is sister taxon to all other dvinosaurs in some of the most parsimonious trees. Although no firm conclusions can be drawn from this result (pending the assembly of a more comprehensive data matrix), I suggest the possibility that the separation between dvinosaurs and dissorophoids may be sought from among a *Micropholis*- and/or *Perryella*-grade array of Permo-Triassic temnospondyls. The overall similarities between *Perryella* and *Micropholis* are striking. Therefore, to assess the impact of these two temnospondyls on the phylogeny, I undertook a set of experiments with topological constraints, devised to uncover their reciprocal influence on the relationships recovered.

When *Perryella* is constrained to group with dissorophoids, PAUP* yields 69 minimal trees that are compatible with this constraint. These trees are two steps longer than the shortest trees from the original run (Consistency Index = 0.3009; Retention Index = 0.6243; Rescaled Consistency Index = 0.1897). When *Perryella* is constrained to group with amphibamids, PAUP* finds ten trees 12 steps longer than the shortest trees (Consistency Index = 0.2977; Retention Index = 0.6185; Rescaled Consistency Index = 0.1859).

For brevity, trees based upon constraints will not be discussed in detail or illustrated here, but are available upon request from the author. However, the effects of these constraints on dissorophoids are remarkable. Thus, when the *Perryella*–dissorophoids constraint is enforced, a significant loss of resolution affects dissorophoids: these are placed in a heptachotomy of three species and four small clades. In one of these clades, *Micropholis* and *Perryella* appear as sister taxa. However, when a *Perryella*-amphibamid constraint is enforced, dissorophoids are completely resolved, except for a trichotomy subtending branchiosaurids. In addition, trematopids and dissorophids form successive sister taxa, in that order, to remaining dissorophoids. The latter consist of a micromelerpetontid–branchiosaurid clade placed as sister group to a clade in which the two morphs of *Micropholis* and *Perryella* are successive sister taxa to amphibamids. Thus, it is clear that even minimal constraints impact to a considerable extent the relationships of dissorophoids and the position of *Micropholis*, at least when such constraints involve different placements for *Perryella* (but see below). The trees that derive from the constraints are not a significantly worse fit for the data than the shortest trees (p > 0.05 for Templeton, Kishino–Hasegawa and Winning-sites tests).

When one enforces a constraint in which *Micropholis* joins amphibamids (e.g. see Schoch and Rubidge 2005; Huttenlocker *et al.* 2007; Anderson *et al.* 2008*a*, *b*), the phylogenetic placement of *Perryella* among dvinosaurs is not affected. There are 48 trees at 939 steps that are compatible with such a constraint (Consistency Index = 0.3005; Retention Index = 0.6237; Rescaled Consistency Index = 0.1893).

Disparity

Range- and variance-based metrics. The four metrics output by RARE are based upon the PCoA-processed Manhattan (Text-fig. 3) and Euclidean (Text-fig. 4) distances. The top four graphs (Text-figs 3A–D, 4A–D) report the values of the metrics for all species included in each of the four major temnospondyl groups. The bottom four graphs (Text-figs 3E–H, 4E–H) report the same metrics, but recalibrated according to the number of taxa in the smallest groups (i.e. edopoids and eryopoids-basal archegosauriforms, each with seven included species). A complete rarefaction curve for all disparity indexes is shown in Text-figure 5 (again, the top four and bottom four graphs are based upon Manhattan and Euclidean distances respectively).

The indexes in Text-figure 3 are not appreciably different from those in Text-figure 4 in terms of their relative values in the four groups (although the absolute values of the indexes may differ). For example, the value of the sum of ranges for dissorophoids is always higher than the values of the same index for other groups. Note that he graphs illustrate mean values (and their associated error bars) for the four indexes. For each major group, I also report the value of each index calculated without rarefaction (i.e. for the full set of taxa; Table 1).

Both edopoids and eryopoids-basal archegosauriforms have similar values for the sum of ranges (Text-figs 3A, 4A), although this index is consistently slightly greater in the latter group (especially for the Euclidean distances). The values are distinctly greater for dvinosaurs and dissorophoids. The recalibrated sum of ranges values (Text-figs 3E, 4E) are similar for dvinosaurs and dissorophoids, lower than the original (i.e. uncalibrated) values for these two groups, but still higher than those for edopoids and eryopoids-basal archegosauriforms.

The distribution of values for the root-product of ranges in the four groups mirrors the profiles of the sum of ranges (Text-figs 3B, 4B), including patterns of recalibrated values (Text-figs 3F, 4F).

The sum of variances (Text-figs 3C, G, 4C, G) is slightly higher in dvinosaurs than in each of the other groups. The profile of this index is conserved in the recalibrated graphs.

Finally, the profile for the root-product of variances (Text-figs 3D, H, 4D, H) resembles that for the recalibrated sum of ranges. As in the sum of variances, dvinosaurs display a consistently higher value than other groups, although only slightly so relative to dissorophoids.

When the analyses in RARE are performed without random resampling of taxa, the four metrics can be taken as a measure of global disparity for the various groups. The unrarefied disparity values based upon Manhattan and Euclidean distances are shown in Table 1. Edopoids and eryopoids-basal archegosauriforms are invariably less disparate than either dvinosaurs or dissorophoids. The only exception is represented by the Manhattan-based sum of variances for dissorophoids,

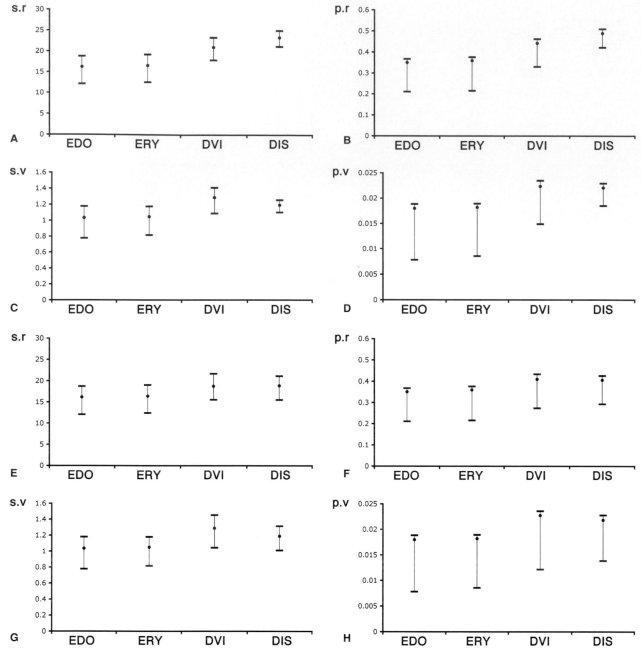

TEXT-FIG. 3. Disparity values for the four major temnospondyl groups (for abbreviations, see Text-fig. 1) based upon PCoA-processed intertaxon Manhattan distances; the error bars are bootstrap-derived 90 per cent confidence intervals (see text for details). A–D, unrarefied disparity. E–H, recalibrated disparity based upon rarefaction. The disparity indexes are: sum of ranges (A and E), root-product of ranges (B and F), sum of variances (C and G) and root-product of variances (D and H) (index abbreviations: s.r; p.r; s.v; p.v).

the value of which is greater than that for edopoids, but lower than those for dvinosaurs and for eryopoids-basal archegosauriforms. In addition, regardless of whether disparity is calculated using Manhattan or Euclidean distances, range-based indexes for dissorophoids are greater than those for dvinosaurs, whereas the reverse is true for the variance-based indexes.

Rarefaction. Text-figure 5 shows rarefaction curves for range and variance indexes. There are clear similarities between the curves based upon Manhattan distances (top four graphs) and those derived from Euclidean distances (bottom four graphs). For each graph, the disparity values are recorded on the vertical axis whilst the numbers of taxa appear on the horizontal axis. Thus, for any given

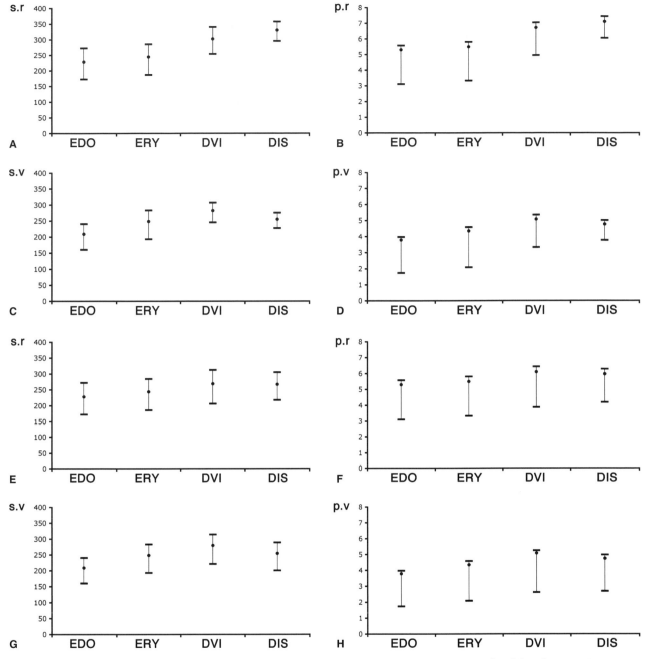

TEXT-FIG. 4. Disparity values for the four major temnospondyl groups (for abbreviations, see Text-fig. 1) based upon PCoA-processed intertaxon Euclidean distances; the error bars are bootstrap-derived 90 per cent confidence intervals (see text for details). A–D, unrarefied disparity. E–H, recalibrated disparity based upon rarefaction. The disparity indexes are: sum of ranges (A and E), root-product of ranges (B and F), sum of variances (C and G) and root-product of variances (D and H) (index abbreviations: s.r; p.r; s.v; p.v).

taxon sample size there is a corresponding value of each disparity index, and these values are plotted for all major groups. In each graph, the circles, squares, triangles and rhombs represent, respectively, edopoids, eryopoids-basal archegosauriforms, dvinosaurs and dissorophoids. The curves for the sum and the root-product of ranges based upon the Manhattan distance matrix (Text-fig. 5A, B)

run in two tightly 'appressed' pairs (i.e. the two curves in each pair nearly overlap one another). One of the two pairs consists of edopoids and eryopoids-basal archegosauriforms whilst the other consists of dvinosaurs and dissorophoids. In the case of Euclidean distances (Text-fig. 5E, F), the two curves for the sum and the root-product of ranges are slightly more widely spaced,

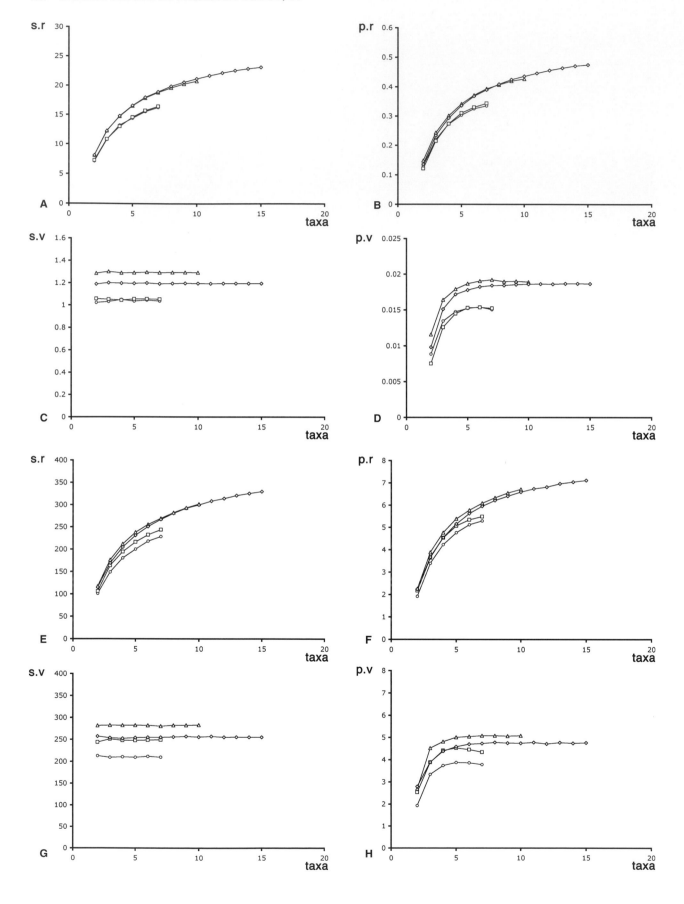

TABLE 1. Disparity values for the major temnospondyl groups.

	EDO	ERY	DVI	DIS
s.r	19.412525	19.867835	24.517203	26.31734
	281.036108	293.327383	356.960623	375.245749
p.r	0.382485	0.393254	0.495731	0.546164
	5.79067	5.9981	7.45259	7.83792
s.v	1.210332	1.226219	1.435684	1.276658
	243.944782	289.563297	312.670441	272.979026
p.v	0.0197025	0.0200964	0.0247456	0.0242628
	4.10088	4.73759	5.55689	5.12142
PCo volume	7.147×10^{-6}	1.452×10^{-5}	1.93×10^{-4}	1.137×10^{-4}
	5.485	17.201	28.864	16.546

Abbreviations: see Text-figure 1. For each index, the upper and lower values are derived from PCoA of Manhattan and Euclidean intertaxon distances respectively (index abbreviations: s.r; p.r; s.v; p.v).

although some overlap characterizes the terminal portions of the dvinosaur and dissorophoid curves for the sum of ranges, and the initial portions of the dissorophoid and eryopoid-basal archegosauriform curves for the root-product of ranges.

Curves for the variance-based metrics tend to be slightly more widely spaced than curves for the range-based metrics. I note the proximity of the edopoid and eryopoid-basal archegosauriform curves for the Manhattan-derived sum of variances (Text-fig. 5C). A similar pattern is observed for the dissorophoid and eryopoid-basal archegosauriform curves (Euclidean-derived sum of variances; Text-fig. 5G). Finally, both the Manhattan- and the Euclidean-derived curves for the root-product of variances (Text-fig. 5D, H) show overlap patterns similar to those of the sum of variances. Edopoid and eryopoid-basal archegosauriform curves for the Manhattan-derived root-product of variances show a terminal overlap. Eryopoid-basal archegosauriform and dissorophoid curves for Euclidean-derived root-product of variances overlap initially but diverge markedly terminally.

The error bars associated with each rarefaction curve have also been calculated, but these are not reported in the graphs for clarity of illustrations and to prevent crowding of lines (e.g. Miller and Foote 1996).

Most of the rarefaction curves for the four disparity indexes show a clear separation of edopoids and eryopoids-basal archegosauriforms, on the one hand, from dvinosaurs and dissorophoids, on the other. To evaluate the significance of this separation, however, a visual inspection of the rarefaction profiles may not suffice. Therefore, I performed a Kruskal–Wallis one-way analysis of variance test on the values of each disparity index obtained through the subsampling routine in RARE. Values in question are those that have been used to construct the rarefaction profiles. The Kruskal–Wallis test assesses the significance of differences among the medians of all groups. First, the values related to each group are ranked; second, the test operates on ranked data rather than on raw values. For each disparity index (Manhattan- and Euclidean-derived), the test seeks to determine if a null hypothesis of no differences among the rarefied profiles of these indexes can be rejected. If the null hypothesis is rejected, then nonparametric *post hoc* tests (pairwise Mann–Whitney tests employing Bonferroni correction) can be applied to identify those pair(s) of groups for which differences between profiles of the same index are significant. All statistical tests were carried out in PAST v. 1.79 (Hammer *et al.* 2001; http://www.folk.uio.no/ohammer/past) and using online statistical software from the following web site: http://www.chiryo.phar.nagoya-cu.ac.jp/javastat/JavaStat-e.htm.

The results of the Kruskal–Wallis test shows that the rarefied profiles of each disparity index differ significantly among major temnospondyl groups, regardless of which distance matrix is used to derive data. A *post hoc* test finds significant differences for each of the following pairs of clades: edopoids and dissorophoids (all indexes, except for Manhattan-based root-product of variances); eryopoids-basal archegosauriforms and dvinosaurs (for both Manhattan- and Euclidean-based sum and root-product of variances); edopoids and dvinosaurs (for both Manhattan- and Euclidean-based sum and root-product of variances); dissorophoids and eryopoids-basal archegosauriforms (only for Manhattan-based sum of ranges). All levels of significance for the Bonferroni-corrected pairwise Mann–Whitney tests are at p < 0.05.

TEXT-FIG. 5. Complete rarefaction profiles for the sum of ranges (A and E), root-product of ranges (B and F), sum of variances (C and G) and root-product of variances (D and H) based upon Manhattan (A–D) and Euclidean (E–H) intertaxon distances. In each graph, the circles, squares, triangles and rhombs represent edopoids, eryopoids-basal archegosauriforms, dvinosaurs and dissorophoids respectively (index abbreviations: s.r; p.r; s.v; p.v).

Principal coordinate analysis volume. The PCo volume normalized by taxon number requires very few comments. For both Manhattan and Euclidean distances, the value of this index for dvinosaurs greatly exceeds those for all other temnospondyl groups, showing that this clade occupies a greater portion of character-state space (see Table 1).

Pairwise distances. Overall disparity was further explored by comparing pairwise distances among taxa in each of the four main groups. To this end, the Manhattan and Euclidean distances were first distributed in four bins of values (one bin for each group) and then subjected to a Kruskal–Wallis test to assess similarities or differences among ranked intertaxon distances.

For Euclidean distances, the test returns significant results ($H = 60.152332$ corrected for ties; $\chi^2 = 7.815$). Bonferroni-corrected Mann–Whitney tests on all six pair-

wise comparisons among the four groups show five comparisons to be significant at $p < 0.05$. The only nonsignificant comparison is between eryopoids-basal archegosauriforms and dissorophoids. For Manhattan distances, the test is also significant ($H = 68.540474$ corrected for ties; $\chi^2 = 7.815$). The *post hoc* tests reveal that two of the comparisons are not significant (i.e. edopoids vs eryopoids-basal archegosauriforms; eryopoids-basal archegosauriforms vs dissorophoids), whereas the other four are ($p < 0.05$).

Character-state space

Using GINKGO, I visualized scree plots from the PCoA of Manhattan (Text-fig. 6A) and Euclidean (Text-fig. 6B) distances to derive the number of PCo axes that, together, explain at least 90 per cent of the total variance. The

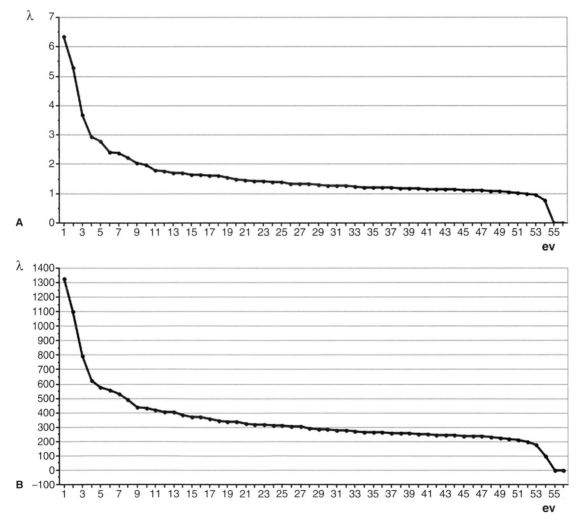

TEXT-FIG. 6. Scree plots of eigenvalues (λ) associated with PCo axes and based upon PCoA of Manhattan (A) and Euclidean (B) intertaxon distances.

simple visual inspection of such plots and the choice of the 90 per cent threshold value conform to some (e.g. Jolliffe 2002), but by no means all (e.g. Jackson 2003), of the current practice (see Cangelosi and Goriely 2007 for a novel mathematical treatment of axis selection). The plot of taxa using the first three PCo axes (Text-figs 7–14) provides a geometrically useful (albeit slightly simplified) depiction of morphospace occupation (e.g. Wills *et al.* 1994; Wills 2001*a, b*; Moyne and Neige 2007).

Both the disparity indexes (see above) and the graphic visualization of morphospace give simple indications of patterns of dispersal or aggregation. To test for differences among the groups' distributions (specifically, to test the null hypothesis of similar distributions), a one-way non-parametric multivariate ANOVA (npMANOVA) was used (see Anderson 2001; Hammer and Harper 2006). This test assesses statistically significant (or otherwise) differences among variances for the four temnospondyl groups, using the PCoA scores (real coordinates) on the first 45 and 47 axes for the Manhattan- and Euclidean-based calculations, respectively (see above). Levels of significance for overall differences among groups were evaluated through 10,000 permutations (default value) of group's species in PAST v. 1.79. The *post hoc* tests for significant differences between all pairs of groups (also provided by PAST, with and without Bonferroni correction for p values; Hammer and

Harper 2006) are reported in Table 2. The uncorrected p values (above the all-zero diagonal for each set of tabulated values) tend to be smaller than the Bonferroni-corrected p values (below the diagonal). All six pairwise comparisons among the four major groups show significant differences between the two temnospondyl groups that appear in each comparison, regardless of whether the PCoA scores used as input for npMANOVA were derived from Manhattan or Euclidean distances and regardless of the use of a Bonferroni correction. In Table 2, I only report comparisons among major temnospondyl groups. I point out that all outgroups (not shown in Table 2) are significantly separated from each of these groups, in all cases. Furthermore, *Balanerpeton* and the two species of *Dendrerpeton* have been lumped together and treated as an additional group (not shown in Table 2). This group is, however, far too small to provide any valid statistical inferences about patterns of dispersion, aggregation, overlap with, or separation from, other groups, and was considered in npMANOVA calculations exclusively for heuristic purposes. Without Bonferroni correction and for both Manhattan- and Euclidean-derived PCoA scores, the *Balanerpeton–Dendrerpeton* group is not significantly separated from any temnospondyl group, but it is significantly separated from the cluster of outgroups. However, with Bonferroni correction, the

TEXT-FIG. 7. Two-dimensional plot of taxa in a character-state morphospace defined by PCo axes 1 and 2, and based upon the PCoA of intertaxon Manhattan distances (for taxon abbreviations, see Appendix). Outgroup taxa are also labelled.

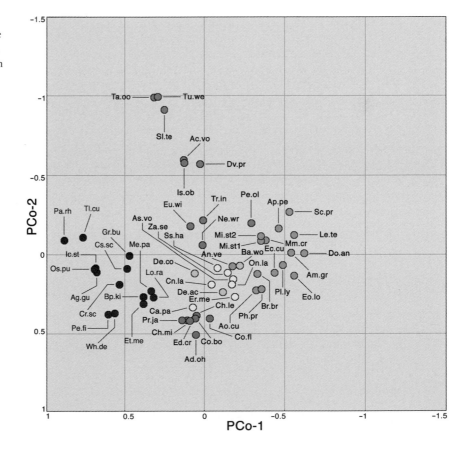

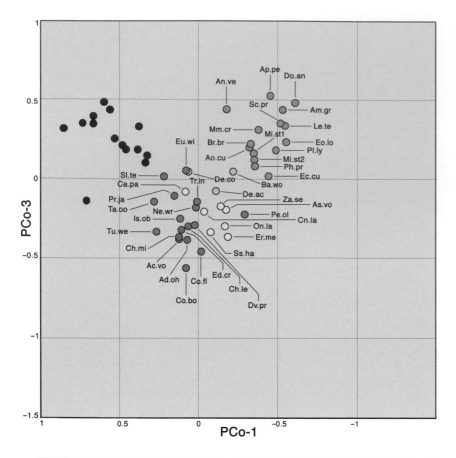

TEXT-FIG. 8. Two-dimensional plot of taxa in a character-state morphospace defined by PCo axes 1 and 3, and based upon PCoA of intertaxon Manhattan distances (for taxon abbreviations, see Appendix). Outgroup taxa are not labelled, for simplicity.

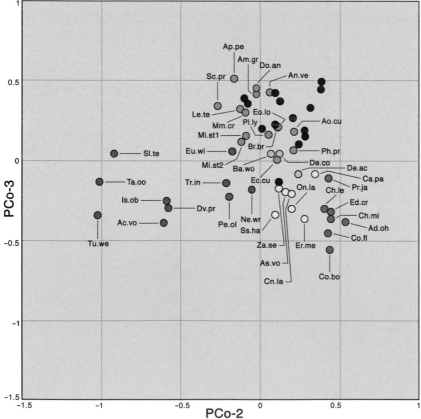

TEXT-FIG. 9. Two-dimensional plot of taxa in a character-state morphospace defined by PCo axes 2 and 3, and based upon PCoA of intertaxon Manhattan distances (for taxon abbreviations, see Appendix). Outgroup taxa are not labelled, for simplicity.

Balanerpeton–Dendrerpeton group is significantly separated from the outgroups and from all temnospondyls other than the eryopoids-basal archegosauriforms, in all cases. The levels of significance in the npMANOVA and in the *post hoc* pairwise comparison tests are also retrieved when only the first three PCo axes (i.e. a Cartesian representation of morphospace) are used.

Text-figures 7–14 permit additional observations on the distribution of taxa in morphospace. For simplicity, neither rectangular boxes nor convex hulls (e.g. Wills *et al.* 1994; Wills 2001*b*; Stayton and Ruta 2006; Pierce *et al.* 2008) have been drawn around different groups in the two-dimensional scatterplots (Text-figs 7–9, 11–13). However, the large dispersion of dvinosaurs relative to all other temnospondyls appears obvious. In particular, in the two-dimensional scatterplots constructed using the first two PCo axes (Text-figs 7, 11), dvinosaurs form an almost linear arrangement (*Perryella* is an outlier). A convex hull placed around dvinosaurs would thus appear narrow and almost wedge-like, and its axis of greater elongation would form a smaller angle with the positive direction of the PCo-2 axis than with the positive direction of the PCo-1 axis. In contrast, other temnospondyls plot out in a subelliptical area of morphospace. The axis of greater elongation of this subelliptical area forms an angle of *c.* 45 degrees with the negative directions of the PCo-1 and PCo-2 axes.

DISCUSSION

The Palaeozoic radiations of temnospondyls

Usually portrayed as a step-like arrangement of discrete radiations in order of increasing proximity to some or all of the crown lissamphibians (Milner 1988, 1990, 1993; Holmes 2000; Ruta *et al.* 2003; Anderson 2007; Anderson *et al.* 2008*a, b*; Ruta and Coates 2007; see Vallin and Laurin 2004 for a dissenting opinion), the higher-level interrelationships of temnospondyls may appear considerably less linear than previously thought (e.g. Laurin and Steyer 2000; Steyer and Laurin 2000; Yates and Warren 2000; Ruta and Bolt 2006 and this paper).

The phylogenetic hypothesis presented here can now be discussed in terms of the chronology of the main diversification events in the evolution of Palaeozoic temnospondyls. Available evidence (see Milner and Sequeira 1994) places the origin of the entire group minimally during the Viséan (Early Carboniferous). The

TEXT-FIG. 10. Three-dimensional plot of taxa in a character-state morphospace defined by PCo axes 1, 2 and 3, and based upon PCoA of intertaxon Manhattan distances (for taxon abbreviations, see Appendix). Outgroup taxa are not labelled, for simplicity.

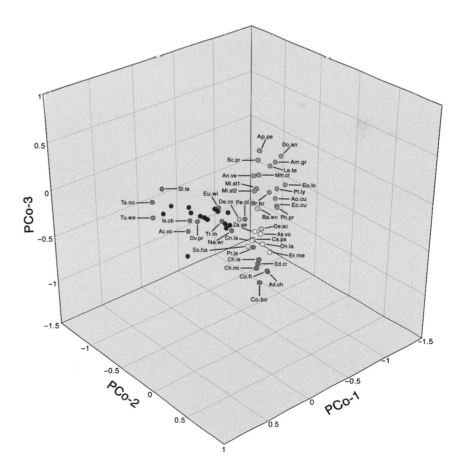

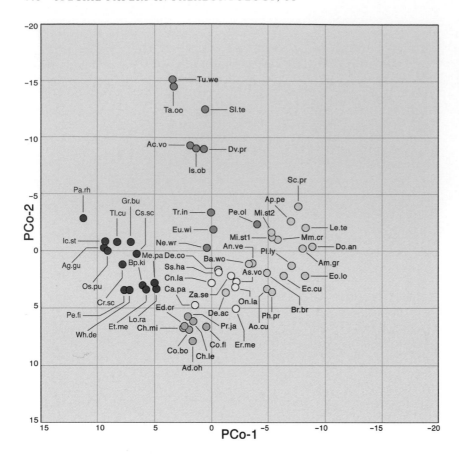

TEXT-FIG. 11. Two-dimensional plot of taxa in a character-state morphospace defined by PCo axes 1 and 2, and based upon PCoA of intertaxon Euclidean distances (for taxon abbreviations, see Appendix). Outgroup taxa are also labelled.

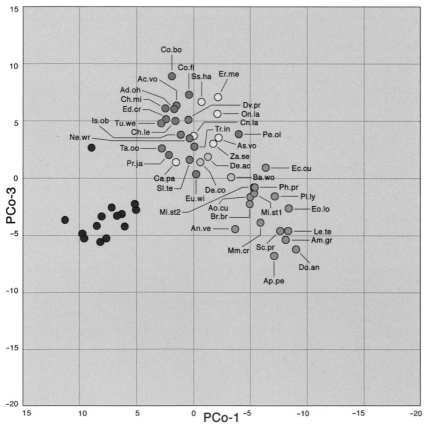

TEXT-FIG. 12. Two-dimensional plot of taxa in a character-state morphospace defined by PCo axes 1 and 3, and based upon PCoA of intertaxon Euclidean distances (for taxon abbreviations, see Appendix). Outgroup taxa are not labelled, for simplicity.

TEXT-FIG. 13. Two-dimensional plot of taxa in a character-state morphospace defined by PCo axes 2 and 3, and based upon PCoA of intertaxon Euclidean distances (for taxon abbreviations, see Appendix). Outgroup taxa are not labelled, for simplicity.

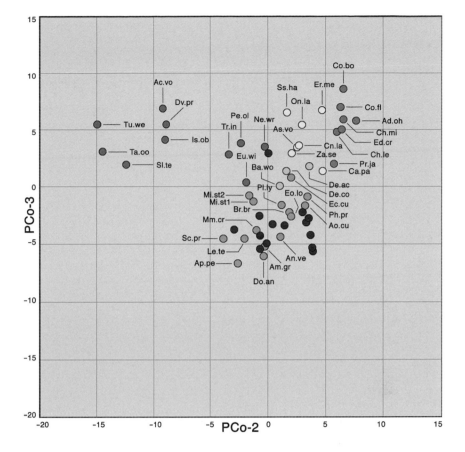

TEXT-FIG. 14. Three-dimensional plot of taxa in a character-state morphospace defined by PCo axes 1, 2 and 3, and based upon PCoA of intertaxon Euclidean distances (for taxon abbreviations, see Appendix). Outgroup taxa are not labelled, for simplicity.

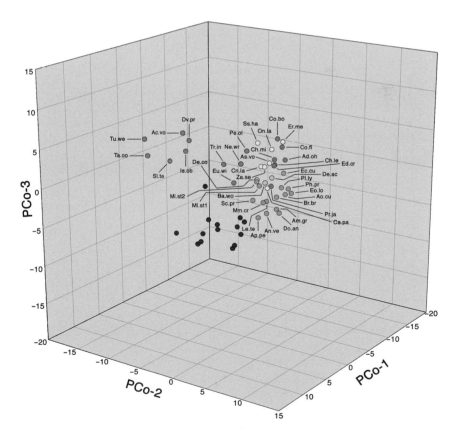

TABLE 2. Uncorrected (below diagonal) and Bonferroni corrected (above diagonal) p values for statistical comparisons among the four major temnospondyl groups in npMANOVA.

	EDO	ERY	DVI	DIS
EDO	0	0.0007	0	0
ERY	0.0105	0	0.0002	0
DVI	0	0.003	0	0
DIS	0	0	0	0

Abbreviations, see Text-figure 1. All p values are identical for both Manhattan- and Euclidean-based PCo scores.

earliest known temnospondyl, *Balanerpeton woodi*, does not occupy a basal position, but rather branches from within temnospondyls (see Milner and Sequeira 1994; Sequeira 2004; Laurin and Soler-Gijón 2006). This implies the presence of several ghost lineages and range extensions (*sensu* Norell 1992; Smith 1994) for various temnospondyl groups, as Milner and Sequeira (1998) noted.

In the following two stages (Serpukhovian; Bashkirian), various taxa pinpoint the earliest known records of edopoids and dvinosaurs (Romer 1930; Milner 1980, 1996; Hook and Baird 1986, 1988; Sequeira 1996; Ruta and Bolt 2006). By post-Viséan times, the evolutionary separation between the edopoids-basal archegosauriforms, on the one hand, and the dvinosaurs–dissorophoids, on the other, had already occurred. Given the position of *Balanerpeton* in the present paper, this separation may have dated minimally to the Viséan. Milner and Sequeira (1994) reported incomplete remains of a putative *Edops*-like taxon from the Viséan. However, the affinities or even the temnospondyl nature of these remains are uncertain (Ruta and Bolt 2006).

The highly variable placement of *Micropholis* in several recent studies is particularly interesting. The early Triassic (Induan) age of *Micropholis* entails a long, as yet largely unrecorded history for certain portions of the evolutionary history of dissorophoids. *Micropholis* is likely to represent a relict taxon, and is one of the last surviving members of a radiation that presumably originated well outside the South African Karoo, where this genus is recorded. A close relationship of *Micropholis* with older amphibamid taxa in some of the most recent studies (Huttenlocker *et al.* 2007; Anderson *et al.* 2008*a*) entails a long range extension for this Induan genus. However, the affinities of *Micropholis* are elusive. Thus, there are several similarities in the patterns of snout elongation and in the morphology of circumorbital bones between one of the two skull morphs of *Micropholis* described by Schoch and Rubidge (2005); long-snouted morph) and recently reported, mature individuals of *Micromelerpeton credneri*. In addition, the latter has been hypothesized to occupy a basal position among dissorophoids (Lillich and Schoch

2007). Remarkable similarities occur also between the long-snouted morph of *Micropholis* and the micromelerpetontid *Eimerisaurus* (Boy 1980, 2002; see also comments by Schoch and Rubidge (2005) and Huttenlocker *et al.* (2007)). Perhaps more significantly, *Eimerisaurus* is also similar to *Perryella*, though clearly distinct. Pending a revised phylogenetic analysis including all of the above-mentioned taxa, similarities among *Micropholis*, *Micromelerpeton* and *Eimerisaurus* may represent either plesiomorphic features of dissorophoids (e.g. Schoch and Rubidge 2005; Huttenlocker *et al.* 2007; Lillich and Schoch 2007; Anderson *et al.* 2008*a*), synapomorphies for micromelerpetontids, or characters of uncertain or mixed polarity (see also Milner and Sequeira 2003; Schoch and Milner 2008).

The dvinosaur–dissorophoid sister group relationship, perhaps the most unexpected feature of the cladistic analysis (see Steyer and Laurin 2000; Yates and Warren 2000; Ruta and Bolt 2006), requires further corroboration in the light of a more extensive taxon/character data sets, but presents novel implications for temnospondyl evolution in general (see Appendix for a list of characters supporting the dvinosaur–dissorophoid clade in this paper). A salient aspect of the palaebiology of dvinosaurs and dissorophoids is that both of these clades show widespread heterochronic features associated in various degrees with either neoteny (mostly among dvinosaurs; Boy and Sues 2000; Warren 2000) or paedomorphosis (in particular for dissorophoids; Milner 1990).

Although heterochrony is likely to have operated across a broad variety of temnospondyl groups, it is well studied and better understood in dvinosaurs and dissorophoids, several representatives of which display a mosaic of convergent juvenile features (e.g. Bolt 1969, 1977, 1979, 1991; Ricqlès 1975; Milner 1988, 1990, 1993; Laurin 1998; Schoch and Milner 2000, 2004; Steyer 2000; Schoch 1995, 2002*a*, *b*, 2006; Witzmann and Pfretzschner 2003; Schoch and Fröbisch 2006; Fröbisch *et al.* 2007). It would be interesting to assess whether heterochrony-driven speciation patterns in different groups are linked to key diversification episodes, such as are exemplified by unusually elevated net speciation rates (e.g. Ruta *et al.* 2006) and whether they occur preferentially on certain parts of the temnospondyl tree. As regards the latter point, if future cladistic analyses lend support to the close relationship between dvinosaurs and dissorophoids, then it would be possible to test the occurrence of concentrated heterochronic changes in these clades, i.e. testing for a nonrandom (or otherwise) distribution of such changes on the topology of the entire temnospondyl tree. Furthermore, it would be interesting to assess this distribution in relation to the origin of lissamphibians, which have long been hypothesized to be paedomorphic descendants of one or several families of dissorophoids (e.g. Bolt 1969, 1977,

1979, 1991; Milner 1988, 1990, 1993; Ruta *et al.* 2003; Schoch and Milner 2004; Anderson 2007; Anderson *et al.* 2008*b*; Ruta and Coates 2007).

In this respect, it has been noted that heterochronic processes underpin the origin of numerous crown radiations, including lungfish, tetrapods, birds and mammals (e.g. Bemis 1984; Long 1990; Bolt 1991; Padian *et al.* 2001; Clack 2002; Luo 2007). Perhaps more interesting is the observation that in many of the cases documented thus far, such crown radiations are 'grafted' onto groups that have themselves experienced heterochrony-driven models of diversification. In turn, these groups are phylogenetically close to clades in which heterochronic processes are either less widespread or less conspicuous. Some examples include but are by no means limited to: edopoids–eryopoids-basal archegosauriforms vs dvinosaurs–dissorophoids–crown lissamphibians; ornithischian dinosaurs vs saurischian dinosaurs–crown birds (but see also Weishampel *et al.* 2003; Rauhut and Fechner 2005).

Morphological diversity

An interesting feature of the tree topology is the fact that some of the oldest and morphologically most generalized temnospondyls (*Balanerpeton, Dendrerpeton* and *Capetus*) are not plesiomorphic, but occupy internal branches of the temnospondyl phylogeny. Thus, the tree suggests the possibility that the ancestry of all major temnospondyl groups might be linked to a terrestrial or semiterrestrial array of Carboniferous taxa exhibiting low overall disparity. At least some of these taxa occur close to the centroid in morphospace (Text-figs 7, 11).

Following their phylogenetic separation, the major Palaeozoic clades 'diffuse' quite rapidly into morphospace. In general, different groups occupy clearly distinct regions of morphospace and show hardly any overlap. Edopoids form a compact cluster, as do eryopoids-basal archegosauriforms and (to a lesser degree) dissorophoids, although for these two clades the dispersion in morphospace is marginally higher than in edopoids. Most peculiar is the pattern of dvinosaurs, which are not only dispersed as a group, but display a heterogeneous spread. Eobrachyopids (*Acroplous*; *Isodectes*), tupilakosaurids (*Slaugenhopia*; *Thabanchuia*; *Tupilakosaurus*) and dvinosaurids (*Dvinosaurus*) are distinctly separated from more basal dvinosaurs (*Eugyrinus*; *Perryella*) and from trimerorhachids (*Neldasaurus*; *Trimerorhachis*).

There is no clear separation in morphospace among the various dissorophoid families. Instead, species in these families tend to overlap to some degree. *Perryella* occupies a peripheral position relative to both dvinosaurs and dissorophoids. Using intertaxon PCoA distances from GINKGO (not the original Manhattan or Euclidean distances), it is possible to show that the mean distance of *Perryella* from all other dvinosaurs (sum of PCoA distances averaged out over nine comparisons) is slightly larger (26.428) than its mean distance from the 15 dissorophoids (25.944), although not in a significant way (p > 0.05; Mann–Whitney test). These figures refer to calculations deriving from the multivariate analysis of original Euclidean distances. As regards the PCoA of Manhattan distances, the mean PCoA distance of *Perryella* from remaining dvinosaurs (1.794) is, once again, marginally larger than its mean PCoA distance from dissorophoids (1.754), albeit not significantly so.

Concordance between morphology and phylogeny

The issue of concordance between phylogenetic and morphospace grouping of taxa (e.g. Stayton 2005, 2006; Stayton and Ruta 2006; Pierce *et al.* 2008) can now be explored briefly. Tests of the degree of concordance (or lack thereof) between phenetic and cladogenetic clustering are widely employed in geometric morphometric analyses. Though applicable to phylogeny-based studies of character-state space occupation, such tests may appear to be redundant because a certain amount of concordance is expected (M. A. Wills, pers. comm. 2008). After all, the characters that are used to reconstruct a phylogeny are also used to derive a character space (see Pie and Weitz 2005). Nevertheless, empirical studies conducted on various matrices (MR, unpublished data) show that, in fact, not only is the strength of the concordance highly variable, but concordance itself may be absent altogether. For this study, three separate intertaxon distance matrices were compared. The first is a matrix of patristic distances obtained in PAUP* under ACCTRAN (for brevity, I will only consider the patristic distances associated with the first of the most parsimonious trees, but results deriving from the use of other trees are identical). The patristic distance between any two taxa is the sum of character-state changes that occur along the branches that connect those two taxa along the shortest tree path. The patristic matrix summarizes data on character-state changes on the tree and yields information on the optimal distribution of these changes. Importantly, it also provides data on rates of character-state accumulation and transformation (e.g. Ruta *et al.* 2006; Wagner *et al.* 2006).

The second matrix consists of 'nodal' distances (once again, based upon the first tree). The nodal distance between any two taxa is the sum of nodes encountered along the shortest path that connects those two taxa on a tree. The nodal matrix does not include data on characters (i.e. branch lengths), but only the geometric relationships of species in a rooted network (although the latter is ultimately the product of parsimony analysis).

The third matrix considers PCo Euclidean distances. These are not the Euclidean distances from the original cladistic data matrix, but the Euclidean distances associated with ordination of species in the PCo character-state space. The PCo Euclidean distances can be conveniently regarded as the numerical representation of intertaxon distances in morphospace.

Tests of the congruence among the three distance matrices were performed with the program CADM (Legendre and Lapointe 2004; http://www.bio.umontreal.ca/Casgrain/en/labo/index.html). All three matrices are symmetric and were treated as having equal weights. Tests in CADM were carried out with 999 permutations and returned identical levels of significance. A global test (with null hypothesis of incongruence of all matrices and with independent permutations performed on each matrix; Friedman's χ^2 = 3796.7364; Kendall's W = 0.82234 with correction for tied ranks) indicates that the three matrices are congruent (p = 0.001). The same p value is retrieved in *post hoc* tests in which each matrix is evaluated for its congruence with the other two matrices in turn. The degree of resemblance among all three matrices was further assessed through a Mantel correlation test based upon ranks (again, with 999 permutations; p = 0.001). These results strongly support the correlation between proximity of taxa in morphospace and their phylogenetic relatedness.

Acknowledgements. This work honours Andrew Milner in recognition of his excellence in the study of early tetrapods. Andrew's unflagging enthusiasm for research and sheer passion for the natural world have exerted a truly contagious effect on generations of colleagues and students. Andrew cosupervised my PhD in a field far from his own expertise, and has offered me help and advice over many years. Those anecdotal, biographical, historical and palaeontological data that have enlivened our formal (and less formal) conversations continue to provide a substantial impetus for my own undertakings. I express my deepest gratitude to an enlightened intellect for having taught me the subtle, and often difficult, art of freeing one's mind of preconceived notions. As Andrew would quickly point out, more often than not, original ideas may not be good and good ideas may not be original. Dr Matthew Wills kindly made a copy of his MATRIX and RARE programs available to me. I also thank him for numerous informative exchanges and for the scientific inspiration that exudes from his work. Professor Charles Ciampaglio and Dr Oyvind Hammer saved me from some major howlers and assisted me with my many queries. Constructive comments from Drs Sebastien Steyer and Adam Yates improved considerably the quality of this paper. I am grateful to Dr Paul Barrett for his technical and editorial assistance and stylistic remarks. I thank Messrs. Massimo Bernardi, Marco Brandalise, Graeme Lloyd and Mark Young for allowing me to carry out some analyses on their computers. I remain solely responsible for omissions and inaccuracies. Research is funded by NERC (ARF NE/F014872/1).

REFERENCES

ADAMOWICZ, S. J., PURVIS, A. and WILLS, M. A. 2008. Increasing morphological complexity in multiple parallel lineages of the Crustacea. *Proceedings of the National Academy of Sciences, USA*, **105**, 4786–4791.

ADRAIN, J. M., EDGECOMBE, G. D. and LIEBERMAN, B. S. 2001. *Fossils, phylogeny, and form: an analytical approach.* Plenum Publishers, New York, 402 pp.

AHLBERG, P. E., CLACK, J. A., LUKSEVICS, E., BLOM, H. and ZUPINS, I. 2008. *Ventastega curonica* and the origin of tetrapod morphology. *Nature*, **453**, 1199–1204.

ANDERSON, M. J. 2001. A new method for non-parametric multivariate analysis of variance. *Austral Ecology*, **26**, 32–46.

ANDERSON, J. S. 2007. Incorporating ontogeny into the matrix: a phylogenetic evaluation of developmental evidence for the origin of modern amphibians. 182–227. In ANDERSON, J. S. and SUES, H.-D. (eds). *Major transitions in vertebrate evolution*. Indiana University Press, Bloomington, IN, 417 pp.

——HENRICI, A. C., SUMIDA, S. S., MARTENS, T. and BERMAN, D. S. 2008*a*. *Georgenthalia clavinasica*, a new genus and species of dissorophoid temnospondyl from the Early Permian of Germany, and the relationships of the family Amphibamidae. *Journal of Vertebrate Paleontology*, **28**, 61–75.

—— REISZ, R. R., SCOTT, D., FRÖBISCH, N. B. and SUMIDA, S. S. 2008*b*. A stem batrachian from the Early Permian of Texas and the origin of frogs and salamanders. *Nature*, **453**, 515–518.

ANGIELCZYK, K. D. and SHEETS, H. D. 2007. Investigation of simulated tectonic deformation in fossils using geometric morphometrics. *Paleobiology*, **33**, 125–148.

ARTHUR, W. 1997. *The origin of animal body plans.* Cambridge University Press, Cambridge, 357 pp.

BEMIS, W. E. 1984. Paedomorphosis and the evolution of the Dipnoi. *Paleobiology*, **10**, 293–307.

BOLT, J. R. 1969. Lissamphibian origins: possible protolissamphibian from the Lower Permian of Oklahoma. *Science*, **166**, 888–891.

—— 1977. Dissorophoid relationships and ontogeny, and the origin of the Lissamphibia. *Journal of Paleontology*, **51**, 235–249.

——1979. *Amphibamus grandiceps* as a juvenile dissorophid: evidence and implications. 529–563. *In* NITECKI, M. H. (ed.). *Mazon Creek fossils.* Academic Press, New York, 581 pp.

—— 1991. Lissamphibian origins. 194–222. *In* SCHULTZE, H.-P. and TRUEB, L. (eds). *Origins of the higher groups of tetrapods: controversy and consensus.* Cornell University Press, Ithaca, 724 pp.

BOTHA, J. and ANGIELCZYK, K. D. 2007. An integrative approach to distinguishing the Late Permian dicynodont species *Oudenodon bainii* and *Tropidostoma microtrema* (Therapsida: Anomodontia). *Palaeontology*, **50**, 1175–1209.

BOY, J. A. 1980. Die Tetrapodenfauna (Amphibia, Reptilia) des saarpfälzischen Rotliegenden (Unter-Perm; SW-Deutschland) 2. *Tersomius graumanni* n. sp. *Mainzer Geowissenschaftliche Mitteilungen*, **8**, 17–30.

—— 1981. Zur Anwendung der Hennigschen Methode in der Wilbertierpaläontologie. *Paläontologische Zeitschrift*, **55**, 87–107.

—— 2002. Über die Micromelerpetontidae (Amphibia: Temnospondyli) 3. *Eimerisaurus* n. g. *Neues Jahrbuch für Geologie und Paläontologie, Abhandlungen*, **225**, 425–452.

—— and SUES, H.-D. 2000. Branchiosaurs: larvae, metamorphosis and heterochrony in temnospondyls and seymouriamorphs. 1150–1197. *In* HEATWOLE, H. and CARROLL, R. L. (eds). *Amphibian biology. Volume 4. Palaeontology. The evolutionary history of amphibians.* Surrey Beatty and Sons, Chipping Norton, NSW, 524 pp.

BRIGGS, D. E. G., FORTEY, R. A. and WILLS, M. A. 1992. Morphological disparity in the Cambrian. *Science*, **256**, 1670–1673.

CÁCERES, M. DE, FONT, X., OLIVA, F. and VIVES, S. 2007. GINKGO, a program for non-standard multivariate fuzzy analysis. *Advances in Fuzzy Sets and Systems*, **2**, 41–56.

CAILLIEZ, F. 1983. The analytical solution of the additive constant problem. *Psychometrika*, **48**, 305–308.

CANGELOSI, R. and GORIELY, A. 2007. Component retention in principal component analysis with application to cDNA microarray data. *Biology Direct*, **2**, 1–21.

CIAMPAGLIO, C. N., KEMP, M. and MCSHEA, D. V. 2001. Detecting changes in morphospace occupation patterns in the fossil record: a characterization and analysis of measures of disparity. *Paleobiology*, **27**, 695–715.

CLACK, J. A. 2002. *Gaining ground: the origin and evolution of tetrapods.* Indiana University Press, Bloomington, IN, 369 pp.

CLARK, C. and CURRAN, D. J. 1986. Outgroup analysis, homoplasy, and global parsimony: a response to Maddison, Donoghue, and Maddison. *Systematic Zoology*, **35**, 422–426.

DAMIANI, R., SIDOR, C. A., STEYER, J. S., SMITH, R. M. H., LARSSON, H. C. E., MAGA, A. and IDE, O. 2006. The vertebrate fauna of the Upper Permian of Niger. V. The primitive temnospondyl *Saharastega moradiensis*. *Journal of Vertebrate Paleontology*, **26**, 559–572.

EBLE, G. J. 2000*a*. Contrasting evolutionary flexibility in sister groups: disparity and diversity in Mesozoic atelostomate echinoids. *Paleobiology*, **26**, 56–79.

—— 2000*b*. Theoretical morphology: state of the art. *Paleobiology*, **26**, 520–528.

ELEWA, A. M. T. 2004. *Morphometrics: applications in biology and paleontology.* Springer-Verlag, Berlin, 263 pp.

ENGLEHORN, J., SMALL, B. J. and HUTTENLOCKER, A. 2008. A redescription of *Acroplous vorax* (Temnospondyli: Dvinosauria) based on new specimens from the Early Permian of Nebraska and Kansas, U.S.A. *Journal of Vertebrate Paleontology*, **28**, 291–305.

ERWIN, D. H. 2007. Disparity: morphological pattern and developmental context. *Palaeontology*, **50**, 57–73.

FOOTE, M. 1991. Morphologic patterns of diversification: examples from trilobites. *Palaeontology*, **34**, 461–485.

—— 1992. Rarefaction analysis of morphological and taxonomic diversity. *Paleobiology*, **18**, 1–16.

—— 1993. Contributions of individual taxa to overall morphological disparity. *Paleobiology*, **19**, 403–419.

—— 1996. Models of morphological diversification. 62–86. *In* JABLONSKI, D., ERWIN, D. H. and LIPPS, J. H. (eds). *Evolutionary paleobiology.* University of Chicago Press, Chicago, IL, 492 pp.

—— 1997. The evolution of morphological diversity. *Annual Review of Ecology and Systematics*, **28**, 129–152.

—— and MILLER, A. J. 2007. *Principles of paleontology.* W. H. Freeman and Co., New York, 354 pp.

FORTEY, R. A., BRIGGS, D. E. G. and WILLS, M. A. 1996. The Cambrian evolutionary 'explosion': decoupling cladogenesis from morphological disparity. *Biological Journal of the Linnean Society*, **57**, 13–33.

FRÖBISCH, N. B., CARROLL, R. L. and SCHOCH, R. R. 2007. Limb ossification in the Paleozoic branchiosaurid *Apateon* (Temnospondyli) and the early evolution of preaxial dominance in tetrapod limb development. *Evolution and Development*, **9**, 69–75.

—— and REISZ, R. R. 2008. A new Lower Permian amphibamid (Dissorophoidea, Temnospondyli) from the fissure fill deposits near Richards Spur, Oklahoma. *Journal of Vertebrate Palentology*, **28**, 1015–1030.

GOTELLI, N. and COLWELL, R. K. 2001. Quantifying biodiversity: procedures and pitfalls in the measurement and comparison of species richness. *Ecology Letters*, **4**, 379–391.

GOULD, S. J. 1989. *Wonderful life: the Burgess Shale and the nature of history.* W. W. Norton and Co., New York, 347 pp.

HAMMER, Ø. and HARPER, D. 2006. *Paleontological data analysis.* Blackwell Publishing, Oxford, 351 pp.

—— and RYAN, P. D. 2001. PAST: paleontological statistics software package for education and data analysis. *Palaeontologia Electronica*, **4**, 9 pp.

HARPER, D. A. T. and GALLAGHER, E. 2001. Diversity, disparity and distributional patterns amongst the orthide brachiopod groups. *Journal of the Czech Geological Society*, **46**, 87–94.

HOLMES, R. B. 2000. Palaeozoic temnospondyls. 1081–1120. *In* HEATWOLE, H. and CARROLL, R. L. (eds). *Amphibian biology. Volume 4. Palaeontology. The evolutionary history of amphibians.* Surrey Beatty and Sons, Chipping Norton, NSW, 524 pp.

HOOK, R. W. and BAIRD, D. 1986. The diamond coal mine of Linton, Ohio, and its Pennsylvanian-age vertebrates. *Journal of Vertebrate Paleontology*, **6**, 174–190.

—— —— 1988. An overview of the Upper Carboniferous fossil deposit at Linton, Ohio. *Ohio Journal of Science*, **88**, 55–60.

HUTTENLOCKER, A. K., PARDO, J. D. and SMALL, B. J. 2007. *Plemmyradytes shintoni*, gen. et sp. nov., an Early Permian amphibamid (Temnospondyli: Dissorophoidea) from the Eskridge Formation, Nebraska. *Journal of Vertebrate Paleontology*, **27**, 316–328.

JACKSON, J. E. 2003. *A user's guide to principal components.* Wiley-Interscience, New York, 592 pp.

JOLLIFFE, I. T. 2002. *Principal component analysis.* Springer, New York, 502 pp.

KRILOFF, A., GERMAIN, D., CANOVILLE, A., VINCENT, P., SACHE, M. and LAURIN, M. 2008. Evolution of bone microanatomy of the tetrapod tibia and its use in palaeobiological inference. *Journal of Evolutionary Biology*, **21**, 807–826.

LAURIN, M. 1998. The importance of global parsimony and historical bias in understanding tetrapod evolution. Part II – vertebral centrum, costal ventilation, and paedomorphosis. *Annales des Sciences Naturelles, Zoologie, Paris, 13e Série*, **19**, 99–114.

—— 2004. The evolution of body size, Cope's Rule and the origin of amniotes. *Systematic Biology*, **53**, 594–622.

—— and SOLER-GIJÓN, R. 2006. The oldest known stegocephalian (Sarcopterygii: Temnospondyli) from Spain. *Journal of Vertebrate Paleontology*, **26**, 284–299.

—— and STEYER, J.-S. 2000. *Phylogeny and apomorphies of temnospondyls.* http://www.tolweb.org/accessory/Phylogeny_and_Apomorphies_of_Temnospondyls?acc_id=582.

LEFEBVRE, B., EBLE, G. J., NAVARRO, N. and DAVID, B. 2006. Diversification of atypical Paleozoic echinoderms: a quantitative survey of patterns of stylophoran disparity, diversity, and geography. *Paleobiology*, **32**, 483–510.

LEGENDRE, P. and ANDERSON, M. J. 1998. *Program DISTPCOA.* Département de Sciences Biologiques, Université de Montréal, 10 pp.

—— and LAPOINTE, F.-J. 2004. Assessing the congruence among distance matrices: single malt Scotch whiskies revisited. *Australian and New Zealand Journal of Statistics*, **46**, 615–629.

—— and LEGENDRE, L. 1998. *Numerical ecology.* Elsevier Science, Amsterdam, 853 pp.

LILLICH, R. and SCHOCH, R. 2007. Finally grown up – the significance of adult *Micromelerpeton. Journal of Vertebrate Paleontology*, **27** (3 Suppl.), 106A.

LINGOES, J. C. 1971. Some boundary conditions for a monotone analysis of symmetric matrices. *Psychometrika*, **36**, 195–203.

LONG, J. A. 1990. Heterochrony and the origin of tetrapods. *Lethaia*, **23**, 157–166.

LUO, Z.-X. 2007. Transformation and diversification in early mammal evolution. *Nature*, **450**, 1011–1019.

MACLEOD, N. and FOREY, P. L. 2002. *Morphology, shape and phylogeny.* Taylor and Francis, London, 308 pp.

MADDISON, W. P., DONOGHUE, M. J. and MADDISON, D. R. 1984. Outgroup analysis and parsimony. *Systematic Zoology*, **33**, 83–103.

—— and MADDISON, D. R. 2000. *MacClade 4: analysis of phylogeny and character evolution (CD-ROM).* Sinauer Associates, Sunderland, Massachusetts.

MARJANOVIC, D. and LAURIN, M. 2007. Fossils, molecules, divergence times, and the origin of lissamphibians. *Systematic Biology*, **56**, 369–388.

MCGHEE, G. R. Jr 1999. *Theoretical morphology. The concept and its applications.* Columbia University Press, New York, 316 pp.

MCGOWAN, A. J. 2004. Ammonoid taxonomic and morphologic recovery patterns after the Permian-Triassic. *Geology*, **32**, 665–668.

—— 2007. Does shape matter? Morphological diversity and differential survivorship among Triassic ammonoid genera. *Historical Biology*, **19**, 157–171.

—— and SMITH, A. B. 2007. Ammonoids across the Permian/Triassic boundary: a cladistic perspective. *Palaeontology*, **50**, 573–590.

MILLER, A. I. and FOOTE, M. 1996. Calibrating the Ordovician Radiation of marine life: implications for Phanerozoic diversity trends. *Paleobiology*, **22**, 304–309.

MILNER, A. R. 1980. The temnospondyl amphibian *Dendrerpeton* from the Upper Carboniferous of Ireland. *Palaeontology*, **23**, 125–141.

—— 1988. The relationships and origin of living amphibians. 59–102. *In* BENTON, M. J. (ed.). *The phylogeny and classification of the tetrapods. Volume 1. Amphibians, reptiles, birds.* Clarendon Press, Oxford, 392 pp.

—— 1990. The radiations of temnospondyl amphibians. 321–349. *In* TAYLOR, P. D. and LARWOOD, G. P. (eds). *Major evolutionary radiations.* Clarendon Press, Oxford, 437 pp.

—— 1993. The Paleozoic relatives of lissamphibians. *Herpetological Monographs*, **7**, 8–27.

—— 1996. A revision of the temnospondyl amphibians from the Upper Carboniferous of Joggins, Nova Scotia. 81–103. *In* MILNER, A. R. (ed.) *Studies on Carboniferous and Permian vertebrates. Special Papers in Palaeontology*, **52**.

—— and SEQUEIRA, S. E. K. 1994. The temnospondyl amphibians from the Viséan of East Kirkton, West Lothian, Scotland. *Transactions of the Royal Society of Edinburgh: Earth Sciences*, **84**, 331–361.

—— —— 1998. A cochleosaurid temnospondyl amphibian from the Middle Pennsylvanian of Linton, Ohio, U.S.A. *Zoological Journal of the Linnean Society*, **122**, 261–290.

—— —— 2003. On a small *Cochleosaurus* described as a large *Limnogyrinus* (Amphibia, Temnospondyli) from the Upper Carboniferous of the Czech Republic. *Acta Palaeontologica Polonica*, **48**, 143–147.

—— —— 2004. *Slaugenhopia texensis* (Amphibia: Temnospondyli) from the Permian of Texas is a primitive tupilakosaurid. *Journal of Vertebrate Paleontology*, **24**, 320–325.

MOYNE, S. and NEIGE, P. 2007. The space-time relationship of taxonomic diversity and morphological disparity in the Middle Jurassic ammonite radiation. *Palaeogeography, Palaeoclimatology, Palaeoecology*, **248**, 82–95.

NAVARRO, N. 2003. MDA: a MATLAB-based program for morphospace-disparity analysis. *Computers and Geosciences*, **29**, 655–664.

NIXON, K. C. and CARPENTER, J. M. 1993. On outgroups. *Cladistics*, **9**, 413–426.

NORELL, M. A. 1992. Taxic origin and temporal diversity: the effect of phylogeny. 88–118. *In* NOVACEK, M. J. and WHEELER, Q. D. (eds). *Extinction and phylogeny.* Columbia University Press, New York, 253 pp.

PADIAN, K., RICQLÈS, A. J. DE and HORNER, J. R. 2001. Dinosaurian growth rates and bird origins. *Nature*, **412**, 405–408.

PATTERSON, C. 1994. Null or minimal models. 173–192. *In* SCOTLAND, R. W., SIEBERT, D. J. and WILLIAMS, D. M. (eds). *Models in phylogeny reconstruction.* Clarendon Press, Oxford, 376 pp.

PIE, M. R. and WEITZ, J. S. 2005. A null model of morphospace occupation. *The American Naturalist*, **166**, E1–E13 [Electronic article, DOI: 10.1086/430727].

PIERCE, S. E., ANGIELCZYK, K. D. and RAYFIELD, E. J. 2008. Patterns of morphospace occupation and mechanical performance in extant crocodilian skulls: a combined geometric morphometric and finite element modeling approach. *Journal of Morphology*, **269**, 840–864.

QUICKE, D. L. J., TAYLOR, J. and PURVIS, A. 2001. Changing the landscape: a new strategy for estimating large phylogenies. *Systematic Biology*, **50**, 60–66.

RAUHUT, O. W. M. and FECHNER, R. 2005. Early development of the facial region in a non-avian theropod dinosaur. *Proceedings of the Royal Society of London: Series B*, **272**, 1179–1183.

RAUP, D. M. 1975. Taxonomic diversity estimation using rarefaction. *Paleobiology*, **1**, 333–342.

RICHTSMEIER, J. T., DELEON, V. B. and LELE, S. R. 2002. The promise of geometric morphometrics. *Yearbook of Physical Anthropology*, **45**, 63–91.

RICQLÈS, A. DE. 1975. Quelques remarques paléo-histologiques sur le problème de la néoténie chez les stégocéphales. *Problèmes Actuels de Paléontologie: Evolution des Vertébrés. Colloque International du Centre National de Recherche Scientifique*, **218**, 351–363.

ROMER, A. S. 1930. The Pennsylvanian tetrapods of Linton, Ohio. *Bulletin of the American Museum of Natural History*, **59**, 77–147.

RUNNEGAR, B. 1987. Rates and modes of evolution in the Mollusca. 39–60. *In* CAMPBELL, K. S. W. and DAY, M. F. (eds). *Rates of evolution*. Allen and Unwin, London, 314 pp.

RUTA, M. and BOLT, J. R. 2006. A reassessment of the temnospondyl amphibian *Perryella olsoni* from the Lower Permian of Oklahoma. *Transactions of the Royal Society of Edinburgh: Earth Sciences*, **97**, 113–165.

—— and COATES, M. I. 2007. Dates, nodes and character conflict: addressing the lissamphibian origin problem. *Journal of Systematic Palaeontology*, **5**, 69–122.

—————— and QUICKE, D. L. J. 2003. Early tetrapod relationships revisited. *Biological Reviews of the Cambridge Philosophical Society*, **78**, 251–345.

——PISANI, D., LLOYD, G. T. and BENTON, M. J. 2007. A supertree of Temnospondyli: cladogenetic patterns in the most species-rich group of early tetrapods. *Proceedings of the Royal Society of London, Series B*, **274**, 3087–3095.

——WAGNER, P. J. and COATES, M. I. 2006. Evolutionary patterns in early tetrapods. I. Rapid initial diversification followed by decrease in rates of character change. *Proceedings of the Royal Society of London, Series B*, **273**, 2107–2111.

SCHOCH, R. R. 1995. Heterochrony in the development of the amphibian head. 107–124. *In* MCNAMARA, K. J. (ed.). *Evolutionary change and heterochrony*. Wiley, Chichester, 286 pp.

——2002a. The early formation of the skull in extant and Paleozoic amphibians. *Paleobiology*, **28**, 278–296.

——2002b. The evolution of metamorphosis in temnospondyls. *Lethaia*, **35**, 309–327.

—— 2006. Skull ontogeny: developmental patterns of fishes conserved across major tetrapod clades. *Evolution and Development*, **8**, 524–536.

—— and FRÖBISCH, N. B. 2006. Metamorphosis and neoteny: alternative pathways in an extinct amphibian clade. *Journal of Vertebrate Paleontology*, **25**, 502–522.

—— and MILNER, A. R. 2000. *Handbuch der Paläoherpetologie. Teil 3B. Stereospondyli*. Verlag Friedrich Pfeil, Munich, 220 pp.

—————— 2004. Structure and implications of theories on the origin of lissamphibians. 345–377. *In* ARRATIA, G., WILSON, M. V. H. and CLOUTIER, R. (eds). *Recent advances in the origin and early radiation of vertebrates*. Verlag Dr. Friedrich Pfeil, Munich, 703 pp.

—————— 2008. The intrarelationships and evolutionary history of the temnospondyl family Branchiosauridae. *Journal of Systematic Palaeontology*, **6**, 409–431.

—— and RUBIDGE, B. S. 2005. The amphibamid *Micropholis* from the *Lystrosaurus* Assemblage Zone of South Africa. *Evolution*, **60**, 1467–1475.

SEQUEIRA, S. E. K. 1996. A cochleosaurid amphibian from the Upper Carboniferous of Ireland. 65–80. *In* MILNER, A. R. (ed.) *Studies on Carboniferous and Permian vertebrates*. *Special Papers in Palaeontology*, **52**.

—— 2004. The skull of *Cochleosaurus bohemicus* Fric, a temnospondyl from the Czech Republic (Upper Carboniferous) and cochleosaurid interrelationships. *Transactions of the Royal Society of Edinburgh: Earth Sciences*, **94**, 21–43.

—— 2009. The postcranum of *Cochleosaurus bohericus* Frič, a primitive Upper Carboniferous temnospondyl from the Czech Republic. 137–153. *In* RUTA, M., CLACK, J. A. and MILNER, A. C. (eds). *Patterns and Processes in Early Vertebrate Evolution. Special Papers in Palaeontology*, **81**.

—— 2002a. The early formation of the skull in extant and Paleozoic amphibians. *Paleobiology*, **28**, 278–296.

SHEN, B., DONG, L., XIAO, S. and KOWALEWSKI, M. 2008. The Avalon explosion: evolution of Ediacara morphospace. *Science*, **319**, 81–84.

SIDOR, C. A., O'KEEFE, F. R., DAMIANI, R., STEYER, J. S., SMITH, R. M. H., LARSSON, H. C. E., SERENO, P. C., IDE, O. and MAGA, A. 2005. Permian tetrapods from the Sahara show climate-controlled endemism in Pangaea. *Nature*, **434**, 886–889.

SIMPSON, G. G. 1944. *Tempo and mode in evolution*. Columbia University Press, New York, 237 pp.

SMITH, A. B. 1994. *Systematics and the fossil record: documenting evolutionary patterns*. Blackwell Scientific Publications, London, 223 pp.

STAYTON, C. T. 2005. Morphological evolution of the lizard skull: a geometric morphometrics survey. *Journal of Morphology*, **263**, 47–59.

—— 2006. Testing hypotheses of convergence with multivariate data: morphological and functional convergence among herbivorous lizards. *Evolution*, **60**, 824–841.

—— 2008. Is convergence surprising? An examination of the frequency of convergence in simulated datasets. *Journal of Theoretical Biology*, **252**, 1–14.

—— and RUTA, M. 2006. Geometric morphometrics of the skull roof of stereospondyls (Amphibia: Temnospondyli). *Palaeontology*, **49**, 307–337.

STEYER, J.-S. 2000. Ontogeny and phylogeny in temnospondyls: a new method of analysis. *Zoological Journal of the Linnean Society*, **130**, 449–467.

——DAMIANI, R., SIDOR, C. A., O'KEEFE, F. R., LARSSON, H. C. E., MAGA, A. and IDE, O. 2006. The vertebrate fauna of the Upper Permian of Niger. IV. Nigerpeton ricqlesi (Temnospondyli: Cochleosauridae), and the edopoid colonization of Gondwana. *Journal of Vertebrate Paleontology*, **26**, 18–28.

—— and LAURIN, M. 2000. *Temnospondyli*. http://www.tolweb.org/Temnospondyli/15009/2000.12.11.

STOCKMEYER LOFGREN, A., PLOTNICK, R. E. and WAGNER, P. J. 2003. Morphological diversity of Carboniferous arthropods and insights on disparity patterns of the Phanerozoic. *Paleobiology*, **29**, 350–369.

SWOFFORD, D. L. 2002. PAUP*: phylogenetic analysis using parsimony (and other methods) 4.0 beta. Sinauer Associates, Sunderland, Massachusetts.

THOMPSON, D. A. W. 1942. *On growth and form*. Cambridge University Press, Cambridge, 368 pp.

VALENTINE, J. W. 2004. *On the origin of phyla*. University of Chicago Press, Chicago, IL, 614 pp.

VALLIN, G. and LAURIN, M. 2004. Cranial morphology and affinities of *Microbrachis*, and a reappraisal of the phylogeny and lifestyle of the first amphibians. *Journal of Vertebrate Paleontology*, **24**, 56–72.

VAN VALEN, L. 1974. Multivariate structural statistics in natural history. *Journal of Theoretical Biology*, **54**, 235–247.

VILLIER, L. and EBLE, G. 2004. Assessing the robustness of disparity estimates: the impact of morphometric scheme, temporal scale, and taxonomic level in spatangoid echinoids. *Paleobiology*, **30**, 652–665.

—— and KORN, D. 2004. Morphological disparity of ammonoids and the mark of Permian mass extinctions. *Science*, **306**, 264–266.

WAGNER, P. J. 1996. Patterns of morphological diversification during the initial radiation of the 'Archaeogastropoda'. 161–169. *In* TAYLOR, J. D. (ed.). *Origin and evolutionary radiation of the Mollusca*. Oxford University Press, Oxford, 392 pp.

—— 1997. Patterns of morphologic diversification among the Rostroconchia. *Paleobiology*, **23**, 115–150.

—— RUTA, M. and COATES, M. I. 2006. Evolutionary patterns in early tetrapods. II. Differing constraints on available character space among clades. *Proceedings of the Royal Society of London: Series B*, **273**, 2113–2118.

WARREN, A. 2000. Secondarily aquatic temnospondyls of the Upper Permian and Mesozoic. 1121–1149. *In* HEATWOLE, H. and CARROLL, R. L. (eds). *Amphibian biology. Volume 4. Palaeontology. The evolutionary history of amphibians*. Surrey Beatty and Sons, Chipping Norton, NSW, 524 pp.

WEBSTER, M. 2007. A Cambrian peak in morphological variation within trilobite species. *Nature*, **317**, 499–502.

WEISHAMPEL, D. B., JIANU, C.-M., CSIKI, Z. and NORMAN, D. B. 2003. Osteology and phylogeny of *Zalmoxes* (n. g.), an unusual euornithopod dinosaur from the latest Cretaceous of Romania. *Journal of Systematic Palaeontology*, **1**, 65–123.

WILLS, M. A. 1998a. Cambrian and recent disparity: the picture from priapulids. *Paleobiology*, **24**, 177–199.

—— 1998b. Crustacean disparity through the Phanerozoic: comparing morphological and stratigraphic data. *Biological Journal of the Linnean Society*, **65**, 455–500.

—— 2001a. Morphological disparity: a primer. 55–144. *In* ADRAIN, J. M., EDGECOMBE, G. D. and LIEBERMAN, B. S. (eds). *Fossils, phylogeny, and form – an analytical approach*. Kluwer Academic, New York, 402 pp.

—— 2001b. Disparity vs. diversity. 495–500. *In* BRIGGS, D. E. G. and CROWTHER, P. R. (eds). *Palaeobiology II*. Blackwell Science, London, 608 pp.

—— BRIGGS, D. E. G. and FORTEY, R. A. 1994. Disparity as an evolutionary index: a comparison of Cambrian and Recent arthropods. *Paleobiology*, **20**, 93–130.

WITZMANN, F. 2006a. Cranial morphology and ontogeny of the Permo-Carboniferous temnospondyl *Archegosaurus decheni* Goldfuss, 1847 from the Saar-Nahe Basin, Germany. *Transactions of the Royal Society of Edinburgh: Earth Sciences*, **96**, 131–162.

—— 2006b. Developmental patterns and ossification sequence in the Permo-Carboniferous temnospondyl *Archegosaurus decheni* (Saar-Nahe Basin, Germany). *Journal of Vertebrate Paleontology*, **26**, 7–17.

—— and PFRETZSCHNER, H.-U. 2003. Larval ontogeny of *Micromelerpeton credneri* (Temnospondyli: Dissorophoidea). *Journal of Vertebrate Paleontology*, **23**, 750–768.

—— and SCHOCH, R. R. 2006. The postcranium of *Archegosaurus decheni*, and a phylogenetic analysis of temnospondyl postcrania. *Palaeontology*, **49**, 1211–1235.

—— —— MILNER, A. R. 2007. The origin of the Dissorophoidea – an alternative perspective. *Journal of Vertebrate Paleontology*, **27** (3 Suppl.), 167A.

—— and SCHOLZ, H. 2007. Morphometric study of allometric skull growth in the temnospondyl *Archegosaurus decheni* from the Permian/Carboniferous of Germany. *Geobios*, **40**, 541–554.

YATES, A. M. and WARREN, A. A. 2000. The phylogeny of the 'higher' temnospondyls (Vertebrata: Choanata) and its implications for the monophyly and origins of the Stereospondyli. *Zoological Journal of the Linnean Society of London*, **128**, 77–121.

ZELDITCH, M. L., SWIDERSKI, D. L., SHEETS, H. D. and FINK, W. L. 2004. *Geometric morphometrics for biologists: a primer*. Elsevier Academic Press, San Diego, 443 pp.

APPENDIX

Modifications of Ruta and Bolt's (2006) Data Matrix

New character

NAS 7. Absence (0) or presence (1) of condition: lateral margins of nasals diverging abruptly in their anterior parts. Remarks: The derived condition is seen in many dissorophoids. In the derived state, the anterior tracts of the lateral margins of the nasals diverge anterolaterally, so that these bones increase rapidly in width anteriorly. This condition is not linked to the size of the external nostrils. Below, I report in square brackets the state of this character for all taxa in Ruta and Bolt's (2006) data matrix.

Acanthostomatops vorax [0], *Acheloma cumminsi* [0], *Acroplous vorax* [0], *Adamanterpeton ohioensis* [0], *Amphibamus grandiceps* [1], *Anconastes vesperus* [0], *Apateon pedestris* [1], *Balanerpeton woodi* [1], *Broiliellus brevis* [0], *Capetus palustris* [0], *Cheliderpeton latirostre* [0], *Chenoprosopus lewisi* [0], *Chenoprosopus milleri* [0], *Cochleosaurus bohemicus* [0], *Cochlesaurus florensis* [0], *Dendrerpeton acadianum* [0], *Dendrerpeton confusum* [0], *Doleserpeton annectens* [1], *Dvinosaurus primus* [0], *Ecolsonia cutlerensis* [0], *Edops craigi* [0], *Eoscopus lockardi* [0], *Eryops megacephalus* [0], *Eugyrinus wildi* [?], *Isodectes obtusum* [0], *Leptorophus tener* [1], *Micromelerpeton credneri* [0], *Micropholis stowi* (long-skulled morph) [0], *Micropholis stowi* (short-skulled morph) [0], *Neldasaurus wrightae* [0], *Onchiodon labyrinthicus* [0], *Perryella olsoni* [?], *Phonerpeton pricei* [0], *Platyrhinops lyelli* [1], *Procochleosaurus jarrowensis* [?], *Schoenfelderpeton prescheri* [1], *Sclerocephalus haeuseri* [0], *Slaugenhopia texensis* [1], *Thabanchuia oomiae* [?], *Tupilakosaurus wetlugensis* [0], *Trimerorhachis insignis* [0], *Zatrachys serratus* [0], *Acanthostega gunnari* [0], *Baphetes kirkbyi* [0], *Colosteus scutellatus* [0], *Crassigyrinus scoticus* [0], *Eucritta melanolimnetes* [0], *Greererpeton burkemorani* [0], *Ichthyostega stensioei* [0], *Loxomma rankini* [0], *Megalocephalus pachycephalus* [0], *Ossinodus pueri* [?], *Panderichthys rhombolepis* [?], *Pederpes finneyae* [0], *Tulerpeton curtum* [?], *Whatcheeria deltae* [0].

Character coding for the two Micropholis morphs using Ruta and Bolt's (2006) expanded data matrix. The 'a' coding for character 56 in the long-skulled morph indicates poymorphism (a = 0&1)

Micropholis stowi 1 (long-skulled morph)

1001100100 0?00000000 0000000002 1002112111 00000001??
?0001a0100 0111000100 0110001021 0021100002 014?001101
1000001001 1110000100 0100100000 1010111100 011????210
0000300000 0?1???1001 10???????? ?????????1 1111000010
?0001?0111 11?10111?1 0002100001 1101001??1 001?40

Micropholis stowi 2 (short-skulled morph)

1001?00100 0?00000000 0000000002 1002112111 00000001??
?000000100 0111000100 0110001011 0000100002 014?001111
1000?01001 1110000100 0100100000 101011100? ?11????210
0000300000 0?1???1001 10???????? ?????????1 1111000110
?0001?0111 11?10111?1 0002100001 1101001??1 001?40

List of character-states supporting the dvinosaur–dissorophoid clade in all minimal trees. For each tree, the list includes the character number followed, in brackets, by its consistency index and state change. All state changes are described after the trees, and refer to the modified version of the Ruta and Bolt (2006) data matrix. Note that the newly introduced character (see above) occupies the tenth column in Ruta and Bolt's (2006) modified data matrix. Therefore, all characters preceding it have the same number as in Ruta and Bolt's (2006) original data matrix (characters 1–9), whereas all characters following it are shifted by one column (characters 11–246). For instance, character 210 in the modified matrix corresponds to character 209 in the original matrix.

Tree 1: 50 (0.25; 0 $\Rightarrow$ 1); 51 (0.333; 0 $\Rightarrow$ 1); 54 (0.2; 1 $\rightarrow$ 0); 84 (0.222; 1 $\Rightarrow$ 0); 85 (0.222; 2 $\Rightarrow$ 1); 107 (0.143; 0 $\Rightarrow$ 1); 135 (0.2; 0 $\Rightarrow$ 1); 210 (0.333; 0 $\rightarrow$ 1).

Tree 2: 50 (0.25; 0 $\Rightarrow$ 1); 51 (0.333; 0 $\Rightarrow$ 1); 54 (0.2; 1 $\rightarrow$ 0); 84 (0.222; 1 $\Rightarrow$ 0); 85 (0.25; 2 $\Rightarrow$ 1); 107 (0.143; 0 $\Rightarrow$ 1); 135 (0.2; 0 $\Rightarrow$ 1); 210 (0.333 0 $\rightarrow$ 1).

Tree 3: 12 (0.167; 1 $\rightarrow$ 0); 53 (0.077; 1 $\rightarrow$ 0); 103 (0.167; 1 $\rightarrow$ 0); 111 (0.2; 0 $\Rightarrow$ 1); 116 (0.167; 1 $\Rightarrow$ 0); 131 (0.2; 0 $\rightarrow$ 1); 175 (0.5; 0 $\Rightarrow$ 1); 182 (0.333; 0 $\rightarrow$ 1); 210 (0.5; 0 $\Rightarrow$ 1); 211 (0.25; 0 $\rightarrow$ 1); 218 (0.2; 0 $\rightarrow$ 1); 219 (0.5; 0 $\rightarrow$ 1).

Tree 4: 12 (0.167; 1 $\rightarrow$ 0); 31 (0.25; 0 $\rightarrow$ 1); 48 (0.143; 0 $\rightarrow$ 1); 53 (0.077; 1 $\rightarrow$ 0); 103 (0.167; 1 $\rightarrow$ 0); 111 (0.2; 0 $\Rightarrow$ 1); 116 (0.167; 1 $\Rightarrow$ 0); 131 (0.2; 0 $\Rightarrow$ 1); 175 (0.5; 0 $\Rightarrow$ 1); 182 (0.333; 0 $\rightarrow$ 1); 210 (0.5; 0 $\Rightarrow$ 1); 211 (0.25; 0 $\rightarrow$ 1); 218 (0.2; 0 $\Rightarrow$ 1).

Tree 5: 50 (0.25; 0 $\Rightarrow$ 1); 51 (0.333; 0 $\Rightarrow$ 1); 54 (0.2; 1 $\rightarrow$ 0); 84 (0.25; 1 $\Rightarrow$ 0); 85 (0.25; 2 $\Rightarrow$ 1); 107 (0.143; 0 $\Rightarrow$ 1); 135 (0.2; 0 $\Rightarrow$ 1); 210 (0.333; 0 $\rightarrow$ 1).

Tree 6: 12 (0.167; 1 $\rightarrow$ 0); 31 (0.25; 0 $\rightarrow$ 1); 48 (0.125; 0 $\rightarrow$ 1); 53 (0.077; 1 $\rightarrow$ 0); 103 (0.167; 1 $\rightarrow$ 0); 111 (0.2; 0 $\Rightarrow$ 1); 116 (0.167; 1 $\Rightarrow$ 0); 131 (0.2; 0 $\Rightarrow$ 1); 175 (0.5; 0 $\Rightarrow$ 1); 182 (0.333; 0 $\rightarrow$ 1); 210 (0.5; 0 $\Rightarrow$ 1); 211 (0.25; 0 $\rightarrow$ 1); 218 (0.2; 0 $\Rightarrow$ 1).

Tree 7: 50 (0.25; 0 $\Rightarrow$ 1); 51 (0.333; 0 $\Rightarrow$ 1); 54 (0.2; 1 $\rightarrow$ 0); 84 (0.222; 1 $\Rightarrow$ 0); 85 (0.222; 2 $\Rightarrow$ 1); 107 (0.143; 0 $\Rightarrow$ 1); 135 (0.2; 0 $\Rightarrow$ 1); 210 (0.333; 0 $\rightarrow$ 1).

Tree 8: 12 (0.167; 1 $\rightarrow$ 0); 31 (0.25; 0 $\rightarrow$ 1); 48 (0.125; 0 $\rightarrow$ 1); 53 (0.077; 1 $\rightarrow$ 0); 103 (0.167; 1 $\rightarrow$ 0); 111 (0.2; 0 $\Rightarrow$ 1); 116 (0.167; 1 $\Rightarrow$ 0); 131 (0.2; 0 $\Rightarrow$ 1); 175 (0.5; 0 $\Rightarrow$ 1); 182 (0.333; 0 $\rightarrow$ 1); 210 (0.5; 0 $\Rightarrow$ 1); 211 (0.25; 0 $\rightarrow$ 1); 218 (0.2; 0 $\Rightarrow$ 1).

12: in dorsal aspect, lateralmost point of prefrontal–postfrontal suture situated at level of anterior half of orbit length (measured parallel to skull midline); **31**: frontal contributing to orbit margin; **48**: intertemporal absent as a separate ossification; **50**: intertemporal less than half as broad as supratemporal; **51**: intertemporal subquadrangular and approximately as long as wide; **53**: supratemporal–squamosal contact smooth; **54**: absence of conical posterior process of tabular projecting backward from the bone and occurring ventral to the level of the ornamented surface of the skull table; **84**: orbit centre closer to anterior extremity of premaxillae than to posterior margin of skull roof; **85**: orbit centre occupying mid-length between anterior extremity of premaxillae and posterodorsal margin of squamosal; **103**: vomerine fangs aligned subparallel to marginal tooth row; **107**: vomer extending posteriorly along lateral margins of anterior extremity of cultriform process; **111**: palatine contributing to interpterygoid vacuity; **116**: ectopterygoid without denticles; **131**: absence of suture between vomer and palatal ramus of pterygoid; **135**: maximum combined width of interpterygoid vacuities greater than their length; **175**: presence of prearticular–surangular contact; **182**: absence of prearticular–anterior coronoid contact; **210**: absence of entepicondylar foramen; **211**: absence of ectepicondylar foramen; **218**: width of entepicondyle smaller than half of humerus length; **219**: length of part of humerus shaft lying proximal to entepicondyle greater than maximum width of humerus head.

Taxon abbreviations for the character-state morphospaces is given in Text-figures 7–14

Edopoids

Ad.oh, *Adamanterpeton ohioensis*; Ch.le, *Chenoprosopus lewisi*; Ch.mi, *Chenoprosopus milleri*; Co.bo, *Cochleosaurus bohemicus*; Co.fl, *Cochleosaurus florensis*; Ed.cr, *Edops craigi*; Pr.ja, *Procochleosaurus jarrowensis*.

Eryopoids-basal archegosauriforms

As.vo, *Acanthostomatops vorax*; Ca.pa, *Capetus palustris*; Cn.la, *Cheliderpeton latirostre*; Er.me, *Eryops megacephalus*; On.la, *Onchiodon labyrinthicus*; Ss.ha, *Sclerocephalus haeuseri*; Za.se, *Zatrachys serratus*.

Dvinosaurs

Ac.vo, *Acroplous vorax*; Dv.pr, *Dvinosaurus primus*; Eu.wi, *Eugyrinus wildi*; Is.ob, *Isodectes obtusus*; Ne.wr, *Neldasaurus wrightae*; Pe.ol, *Perryella olsoni*; Sl.te, *Slaugenhopia texensis*; Ta.oo, *Thabanchuia oomie*; Tr.in, *Trimerorhachis insignis*; Tu.we, *Tupilakosaurus wetlugensis*.

Dissorophoids

Am.gr, *Amphibamus grandiceps*; An.ve, *Anconastes vesperus*; Ao.cu, *Acheloma cumminsi*; Ap.pe, *Apateon pedestris*; Br.br, *Broiliellus brevis*; Do.an, *Doleserpeton annectens*; Ec.cu, *Ecolsonia cutlerensis*; Eo.lo, *Eoscopus lockardi*; Le.te, *Leptorophus tener*; Mi.st1, *Micropholis stowi* long-skulled morphotype; Mi.st2, *Micropholis stowi* short-skulled morphotype; Mm.cr, *Micromelerpeton credneri*; Ph.pr, *Phonerpeton pricei*; Pl.ly, *Platyrhinops lyelli*; Sc.pr, *Schoenfelderpeton prescheri*.

Other temnospondyls

Ba.wo, *Balanerpeton woodi*; De.ac, *Dendrerpeton acadianum*; De.co, *Dendrerpeton confusum*.

Outgroups

Ag.gu, *Acanthostega gunnari*; Bp.ki, *Baphetes kirkbyi*; Cr.sc, *Crassigyrinus scoticus*; Cs.sc, *Colosteus scutellatus*; Et.me, *Eucritta melanolimnetes*; Gr.bu, *Greererpeton burkemorani*; Ic.st, *Ichthyostega stensioei*; Lo.ra, *Loxomma rankini*; Me.pa, *Megalocephalus pachycephalus*; Os.pu, *Ossinodus pueri*; Pa.rh, *Panderichthys rhombolepis*; Pe.fi, *Pederpes finneyae*; Tl.cu, *Tulerpeton curtum*; Wh.de, *Whatcheeria deltae*.

[Special Papers in Palaeontology 81, 2009, pp. 121–136]

THE TEMNOSPONDYL *GLANOCHTHON* FROM THE LOWER PERMIAN MEISENHEIM FORMATION OF GERMANY

by RAINER R. SCHOCH* *and* FLORIAN WITZMANN†

*Staatliches Museum für Naturkunde, Rosenstein 1, D-70191 Stuttgart, Germany; e-mail: schoch.smns@naturkundemuseum-bw.de
†Humboldt Universität zu Berlin, Museum für Naturkunde, Invalidenstraße 43, D-10015 Berlin, Germany; e-mail: florian.witzmann@museum.hu-berlin.de

Typescript recieved 25 June 2008; accepted in revised form 2 September 2008

Abstract: New material of the up to 1 m long temnospondyl 'Cheliderpeton' latirostre from the Lower Permian Saar-Nahe Basin in SW Germany is described and attributed to the new genus *Glanochthon*. The material falls into two distinct species, which were stratigraphically and ecologically separated: the classical *G. latirostre* from the ironstone concretions of the Lebach region, and *G. angusta* from the Humberg Black Shale. Autapomorphies of the new genus are the postnarial constriction of the preorbital region, with a conspicuously convex lateral margin of the premaxilla in adults, and the long preorbital region that has 2.0–2.3 times the length of posterior skull table in adults. *G. latirostris* is characterized by a predominantly radial dermal sculpture on skull roof and interclavicle, and the quadrate condyles that are located only slightly posterior to the occipital condyles. In contrast, *G. angusta* has a clear polygonal dermal sculpture, and the quadrate condyles are situated well posterior to the occipital region in juveniles and adults. A cladistic analysis of the two *Glanochthon* species and 17 further temnospondyl taxa with 54 characters finds *Glanochthon* to be monophyletic. The genus is nested within the basal stereospondylomorphs, above *Sclerocephalus* and below *Intasuchus*, *Memonomenos*, *Archegosaurus* and the stereospondyl crown. In general, *Memonomenos* and *Intasuchus* are much more similar to *Glanochthon* than *Cheliderpeton* is. However, the position of *Cheliderpeton* is unclear because of poor preservation and the unknown morphology of its palate. The results of the analysis further suggest that eryopids and stereospondylomorphs are sister-groups, what contradicts the Euskelia hypothesis that implies that eryopids are more closely related to dissorophoids and zatracheids.

Key words: Autunian, Permian, phylogeny, Rotliegend, Stereospondylomorpha.

THE Permo-Carboniferous lake deposits of Central Europe rank among the richest aquatic tetrapod *lagerstätten* in the late Palaeozoic. Among these, the largest and most locality-rich region forms the Franco-German Saar-Nahe-Lorraine Basin (SNLB), which yielded numerous well-preserved specimens of temnospondyls in the last 165 years (von Meyer 1844, 1857, 1858; Boy 1972, 1976, 2007; Boy and Sues 2000). Apart from the abundant branchiosaurids, which comprise the bulk of the finds in most localities, the occurrence of large tetrapods such as the 1.5 m long *Sclerocephalus haeuseri* (Goldfuss, 1847; Boy 1988) and *Archegosaurus decheni* (von Meyer, 1857; Witzmann 2006a) has been recorded in numerous localities (Boy 1987). Here we focus on the least-known of the larger temnospondyls, 'Cheliderpeton' latirostre, after the other taxa have been revised recently (Boy 2007; Witzmann 2006a, b; Witzmann and Schoch 2006a; Schoch and Witzmann in press).

'Cheliderpeton' latirostre was discovered in the ironstone concretions of Lebach, in the southwestern corner of the SNLB. It was first considered a smaller, broad-snouted species of *Archegosaurus* (i.e. Jordan 1849; Burmeister 1850; von Meyer 1857). After it had been referred to the genus *Actinodon* by Romer (1947) and known from the French Autun basin, it was later attributed to *Sclerocephalus* by Boy (1987) or *Cheliderpeton* (Boy, 1993a; Schoch and Milner, 2000). In a recent revision of the type species of *Cheliderpeton* from the Czech Republic (*C. vranyi*), Werneburg and Steyer, (2002) showed that 'C.' latirostre was quite distinct from the Czech taxon, warranting the creation of a new generic name for the German taxon. Here, we revise the complete set of available material of 'Cheliderpeton' latirostre from the SNLB, focusing on the following objectives: (1) an analysis of how many taxa the heterogenous sample of 'C.' latirostre actually include; (2) the morphology of the adult skull and the postcranium, which have been only partially described so far and (3) the phylogenetic position(s) of the represented taxa within the temnospondyls.

Krätschmer (2006) described a new species from the Kappeln Horizon of the Meisenheim Formation, which at Odernheim underlies the Humberg Black Shale. He referred the new species to *Cheliderpeton*. This taxon or *C. lellbachae* is truly different from both of the two species described here and does not share the autapomorphies of *C. latirostre*. In most respects, it resembles *Sclerocephalus haeuseri*. Since the paper of Krätschmer, (2006), several additional specimens have appeared, including large adults. We have not considered this taxon in our cladistic analysis, because the complete material needs a detailed description, before phylogenetic conclusions may be drawn from it.

MATERIAL

All material investigated here is collected from the Lower Autunian (Lower Permian) Meisenheim Formation (L-O 10). The Meisenhem Formation comprises the lower Jeckenbach and upper Odernheim members, the latter of which includes the Humberg Black Shale horizon (Text-Fig. 1). Two different lithofacies are developed in the top member of the Meisenheim Formation, which are black shale horizons or the Humberg Black Shale and the overlying Lebach sideritic concretions (local name = Lebacher Knollen). The black shale is developed as poorly laminated black mudstone (=Schwarzschiefer) and thinly laminated paper shale mudstone (=Papierschiefer).

The Lebach concretions have yielded the classical specimens of *C. latirostris*, which were collected in the midnineteenth century. Most of the specimens from the Lebach concretions were prepared during the 19th century, which was a time where preparation technique

was less advanced. Therefore, the classical Lebach collection has suffered from crude preparation despite the often well-preserved material. The poorly laminated black shale preserves very good and only gently crushed specimens, whereas the paper shale usually yield poorly preserved and crushed bones.

A total of 59 specimens identified as '*Cheliderpeton*' *latirostre* have been examined (see lists in the next section). Since the last revision of the material (Boy 1993*a*), the number of available specimens has nearly doubled. The material falls into two sets: (1) the classical taxon from Lebach, comprising 15 specimens examined in public collections; and (2) a separate taxon, here described as a new species, of which 44 specimens are now accessible to research.

From the basal part of the Humberg Black Shale, *Sclerocephalus haeuseri* is rarely found together with '*C.*' *latirostre*. The following specimens of this species were examined: GMBS 177, GPIM-N 400, 1166, MB.Am. 1345 and POL-54Gre1.

In the present paper, we refer to skulls up to c. 80 mm as larvae because they exhibit the larval branchial dentition (see below), and those beyond 90 mm as adults (see Boy 1993*a*). The adult skulls (90–149 mm) are characterized by proportionally longer preorbital regions and a differentiated lateral margin (see diagnosis).

Institutional abbreviations. GMBS, Geologische Sammlung der Saarbergwerke, Saarbrücken, Germany; GPIM-N, Institut für Geowissenschaften, Universität Mainz, Germany; GPIT, Geologisch-Paläontologisches Institut und Museum, Universität Tübingen, Germany; MB, Museum für Naturkunde, Humboldt Universität zu Berlin, Germany; MCZ, Museum of Comparative Zoology, Harvard University, Cambridge/MA, USA; MSN, Museum Stapf, Nierstein, Germany; PIB, Paläontologisches Institut der Universität Bonn, Germany; POL, Pollichia Collection, Geoskop, Thallichtenberg, Germany; ROM, Royal Ontario Museum, Toronto, Canada; SMNS, Staatliches Museum für Naturkunde, Stuttgart, Germany.

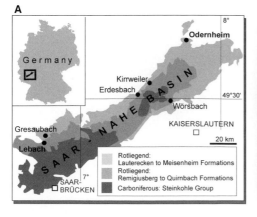

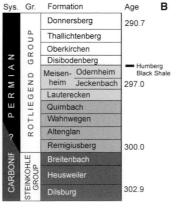

TEXT-FIG. 1. A, map of Saar-Nahe-Lorraine Basin with outcropping Lower Rotliegend (Autunian) areas shaded and localities yielding *Glanochthon* indicated. B, stratigraphy of the Rotliegend section in the Saar-Nahe Basin with the position of the Humberg Black Shale indicated. The age is given in million years. Sys., system; Gr., group.

SYSTEMATIC PALAEONTOLOGY

TEMNOSPONDYLI Zittel, 1888
STEREOSPONDYLOMORPHA Yates and Warren, 2000

Genus GLANOCHTHON gen. nov

Derivation of name. *Glanum*, the aboriginal Celtic name for the Glan river at Odernheim, and *he chthôn* (fem.), Greek for 'native soil, region, people' (Menge 1985).

Diagnosis

1. Autapomorphies, as contrasted with eryopids, sclerocephalids and archegosaurids:
   (i) Preorbital region constricted posterior to naris, with lateral margin of premaxilla markedly convex (adults).
   (ii) Preorbital region measuring 2.0–2.3 times the length of posterior skull table (adults).
   (iii) Posterior end of orbits rounded-triangular (only large adults).
2. Characters shared with some other stereospondylomorph temnospondyls:
   (iv) Palatine and ectopterygoid bearing a continuous row of teeth (like in *Intasuchus* and *Archegosaurus*) which are markedly larger than the maxillary teeth (unlike *A. decheni*) and the ectopterygoid arcade having no tusk pair (unlike *Sclerocephalus*).
   (v) Locally differentiated teeth in the snout, with larger teeth producing lateral excursions of margin (*Intasuchus*, *Sclerocephalus*).
   (vi) Orbits elongated oval (*Intasuchus*, *Memonomenos*, *Archegosaurus*).

Glanochthon latirostris (Jordan, 1849)

1849   *Archegosaurus latirostris* Jordan, p. 78, pl. 4, fig. 2.
1850   *Archegosaurus latirostris*; Burmeister, p. 69, pl. 2, figs 3–4.
1857   *Archegosaurus latirostris*; Meyer, p. 211, pl. 1–2.
1858   *Archegosaurus latirostris*; Meyer, p. 219, fig. 1.
1939   *Actinodon latirostris*; Romer, 1939, p. 758, fig. 4.
1947   *Actinodon latirostris*; Romer, p. 136.
1976   '*Actinodon*' *latirostris*; Boy, p. 54, fig. 27.
1977   '*Actinodon*' *latirostris*; Boy, 1977, p. 134, fig. 8.
1987   *Sclerocephalus latirostris*; Boy, p. 758.
1993a   *Cheliderpeton latirostre*; Boy, pp. 123–141, figs 1–5.
1993b   *Cheliderpeton latirostre*; Boy, pp. 155–169, fig. 1.
2000   *Cheliderpeton latirostre*; Schoch and Milner, p. 53, fig. 44.

2008   *Cheliderpeton latirostre*; Kriwet *et al.* pp. 182–183, fig. 1.

Holotype. PIB-4, a skull fragment, figured by Jordan (1849, pl. 4, fig. 2) and Boy (1993a, fig. 1e).

Type locality. 'Lebach', probably the region between Rümmelbach and Gresaubach, one of the various ironstone localities, Saarland, Germany (Text-fig. 1).

Type horizon. Top of Humberg Black Shale, Odernheim member, Meisenheim Formation, Autunian (Lower Rotliegend), lowermost Permian (DSK (Deutsche Stratigraphische Kommission) 2002). Correlates with L-O 10 of Boy and Fichter (1982).

Referred material. MCZ 1272; GMBS 228, 244/24, 308; GPIT-Am-34, 673; MB.Am.113, 163, 211, 215, 211, 224; PIB-1, PIB-2; ROM 5735.

Diagnosis

1. Autapomorphies, as contrasted with *G. angusta*:
   (i) Ornament consisting of numerous fine pits in central regions of elements and longitudinal radial ridges in the periphery (small specimens and adults).
   (ii) Jaw joint only slightly posterior to the posterior margin of the postparietals in adults. (Text-fig. 2C, F–G, J–K).
2. Characters shared with some other stereospondylomorph temnospondyls, but not with *G. angusta*:
   (iii) Lateral line sulci are clearly impressed on the dermal skull roof.
   (iv) Interpterygoid vacuities slightly wider than in *C. angusta*.

Glanochthon angusta sp. nov.

1987   *Sclerocephalus latirostris* Boy, p. 758 (*partim*).
1993a   *Cheliderpeton latirostre*; Boy, pp. 123–141 (*partim*), figs 1–5.
2000   *Cheliderpeton latirostre*; Schoch and Milner, p. 53, fig. 44.
2007   '*Cheliderpeton*' *latirostre*; Boy, p. 266, fig. 3b.

Holotype. GPIM-N 1217, a complete skull in dorsal exposure (136 mm).

Type locality. Odernheim, Rheinland-Pfalz, Germany (Text-fig. 1).

Type horizon. Black shale proper and Paper Shale (*Papierschiefer*) horizons of Humberg Black Shale, Odernheim Member,

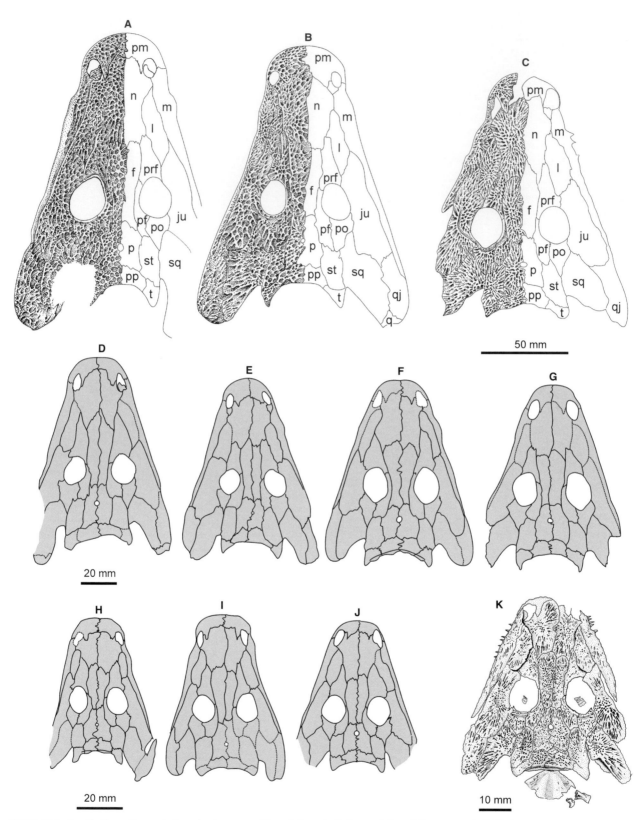

TEXT-FIG. 2. Skull roof morphology in the genus *Glanochthon*. A–B, large adults of *G. angusta*, A, POL-F 1994-1. B, GPIM-N 1217 (type specimen). C, *G. latirostris*, large adult, MB.Am 224. D–E, *G. angusta*, small adults. D, POL-F 002 and E, SMNS 91003, GPIT.Am.34. F–G, *G. latirostris*, small adults. F, GPIT.Am.673 and G, GPIT.Am.34. H–I, *G. angusta*, juveniles. H, MB.Am. 1310 and I, MB.Am.1272. J–K, *G. latirostris*, juveniles. J, MB.Am.113 and K, MB.Am.211a.

Meisenheim Formation, Autunian (Lower Rotliegend), lowermost Permian. Correlates with L-O 10 of Boy and Fichter (1982).

Geographical range. Gresaubach (black shale), Erdesbach ('*Papierschiefer*', black shale), Rathsweiler (black shale), Wörsbach (black shale), Odernheim ('*Papierschiefer*', black shale) (Text-fig. 1).

Referred material. GMBS-Gre 24, 85, 93, 174, 179, 282, 250, 307, 348, 393-394; GPIM-N 793, 1177, 1180, 1220, 1843, 1844; MB.Am.1272, 1274–1276, 1281–1283, 1286, 1287, 1289, 1293, 1296, 1306, 1310, 1312; POL-F-002, -1994-1, -1989-3, -1991-5, -1991-6, 1992-3, Gre-17, Gre-148; SMNS 58784, 90697, 91000, 91003.

Adult specimens. Large adults, on which the focus of this paper lies, are now available from the Humberg Black Shale. The best specimens are as follows, with skull length (SL) given in parentheses. POL-F 1994-1 (148 mm) -F-002 (115 mm, with a good portion of the postcranium, slightly disarticulated), SMNS 90697 (108 mm, with parts of postcranium). GPIM-N 1844 (large postcranium, including interclavicle, scapulocoracoid and humerus). The largest specimen of *Glanochthon angusta* (MSN 1) preserves almost the complete trunk and appendicular skeleton regions. There is a range of still larger specimens (150–175 mm) housed in private collections, which unfortunately could not be considered; these have slender snouts but share the autapomorphies listed in the diagnosis.

Remarks. Boy (1993*a*) mentioned and figured two morphs but was cautious in separating them into two species, because at that time the number of specimens from the Humberg Black Shale was limited. Since then, the material has almost doubled, and the differences between the original Lebach taxon and the Humberg taxon are now apparent. According to our observations, the two species never occurred together and are separated both stratigraphically and locally. *Glanochthon latirostris* is confined to the ironstone facies at the top of the Humberg Black Shale whereas *G. angusta* is only found at the base of this unit i.e. in black shales and finely laminated paper shales (Boy 1987, 1993*a*, and pers. comm. 2008).

Diagnosis

1. Autapomorphies, as contrasted with *G. latirostris*:
   (i) Lateral margin of skull straight or weakly concave (Text-fig. 2A–B, D–E, H).
   (ii) Ornament consisting of mainly large polygons.
   (iii) Generally more heavily ossified than *G. latirostris*, with sutures obscured by intense ornament in places.
   (iv) Braincase ossifies earlier than in other taxa, but this is probably also the case in *G. latirostris*.

2. Characters shared with some other stereospondylomorph temnospondyls, but not with *G. latirostris*:
   (v) Lateral line sulci largely masked by the taller ornamenting ridges and the larger polygons as compared to *C. latirostris*.
   (vi) Jaw joint well posterior to occiput not only in adults, but also in juveniles.

Description. In the following description, we focus on hitherto unknown characters as well as features that differ between the two newly-distinguished species. The study focus on the largest adult specimens, which were discovered after Boy (1993*a*) description.

Identification. Boy (1993*a*) noted that is not easy to distinguish small specimens (10–50 mm skull length) of *Sclerocephalus* from those of *Glanochthon*. This is not a problem in the Lebach concretions, because *G. latirostris* is accompanied only by the much more frequent *Archegosaurus decheni* (Boy, 1987). The preorbital region narrows conspicuously anterior to the orbits to form a rostrum of the latter and it has a jugal width lateral to the orbit, which does not reach half the orbital width (Witzmann 2006*a*). In the black shale facies, *Sclerocephalus haeuseri* is rare and confined to the basal beds of the section (Boy 1987), but present with both juveniles and adults. It is known from both the Humberg and Gresaubach localities, and from the latter a range of good specimens is available (GMBS 177; MB.Am.1345; POL-54Gre1). According to these specimens and to Schoch (2003), small skulls of *Sclerocephalus* have: (1) a longer basipterygoid branch of the pterygoid giving a wider interpterygoid vacuity; (2) a wider basal plate of the parasphenoid, further adding to the width of the skull; and (3) the markedly later ossification of braincase elements and quadrate relative to *Glanochthon*. Adult specimens of *Sclerocephalus* differ from *Glanochthon* by the longer posterior skull table in relation to the snout (*Sclerocephalus*: SNL:POL = 1.7–1.8; *Glanochthon*: 2.0–2.3), the poor differentiation of the humerus, the stout shaft and broad dorsal end of the ilium, and the less pronounced uncinate processes in the anterior thoracal ribs.

Skull roof. The adult skull of *Glanochthon* is very characteristic by its elongated, laterally convex and anteriorly blunt snout. This is shared by both species and constitutes their most robust synapomorphy. *Intasuchus* (Upper Permian of European Russia) and *Cheliderpeton vranyi* (Lower Permian of Czech Republic) have the most similar cranial outlines compared with *Glanochthon*, but both lack the preorbital bulge and blunt tip and also differ by having more slender oval orbits (Konzhukova 1956; Gubin 1991, 1997). However, these taxa all share the slender jugals with preorbital regions at least twice as long as the postorbital skull table. In turn, *Sclerocephalus* and the eryopids differ from *Glanochthon* in having a wider skull, a parabolic preorbital region, and an extended tip of the snout anterior to the naris.

The two *Glanochthon* species differ from one another in the width of the cheek, with the wide-cheeked *G. latirostre* having a triangular skull, whereas that of *G. angusta* is much narrower

(Text-fig. 2A–C). Further, the ornament differs clearly between the two species: while *G. latirostris* has a fine ornament of numerous thin, radial ridges and few central polygons, the ornament of *G. angusta* consists of large rounded polygons formed by tall and robust ridges; these are hardly elongated except for the lateral region of the squamosal and small areas of the nasal. In these features, the ornament of *G. angusta* is much more similar to that of *Sclerocephalus haeuseri* than to *G. latirostris*.

The following description applies to both *Glanochthon* species except otherwise noted (Text-fig. 2). The naris is small, located at the lateral margin and bordered by a substantial septomaxilla posteriorly. The nasal is only slightly wider than the frontal, and the lacrimal is rather short and narrow, quite like in *Sclerocephalus*. Lacrimal and septomaxilla are always separated by a long maxilla-nasal suture. The prefrontal is large and has a variable shape within both species, sometimes the anterior end being rounded, but mostly ending in a point. In adults, the maxilla has one or more lateral projections that bear larger teeth.

The contact between pre and postfrontal is more slender than in *Sclerocephalus*, but always present. The orbit is elongated oval, often with pointed anterior or posterior ends; a lateral emargination like in large *Sclerocephalus haeuseri* (Schoch and Witzmann in press) is found in some large *G. angusta*. A large specimen (POL-F 1994-1) has two orbits of different size, a phenomenon also known from *Mastodonsaurus* (Schoch 1999). Among the two *Glanochthon* species, the jugal is slightly wider in adult *G. latirostris*, but generally more slender than in *Sclerocephalus*.

Similar to *Sclerocephalus*, the posterior skull table is slender with long parietals and supratemporals. The tabular forms a well-defined posterior horn, clearly offset from the posterior skull margin (Text-fig. 2). This feature is most pronounced in adult *G. angusta*. Both *Glanochthon* species have a posteroventrally directed tabular extension, as described for *Archegosaurus* (Witzmann 2006a). In adult *G. latirostris*, the occipital margin is not as strongly concave as in *G. angusta*, a variational range that is also known from *Sclerocepahlus* (Schoch and Witzmann in press). The most conspicuous difference between the two *Glanochthon* species is the proportion and length of the cheek, with adult *G. angusta* having quadrate condyles much further posterior than *G. latirostris*. Interestingly, even some small juveniles have reached that condition, which in *Sclerocephalus* and *Archegosaurus* was only established in large adult specimens (Boy 1988; Witzmann 2006a).

The 3d-structure of the skull roof is well-preserved in MB.Am.113 (*G. latirostris*), showing parasagittal ridges running along: (1) the lateral corners of the posterior skull table (postorbital-supratemporal-tabular); and (2) a line starting from the orbit and ending in the alary process of the premaxilla (prefrontal-nasal-premaxilla). Similar ridges are known from well-preserved skulls of other temnospondyls (*Sclerocephalus*: Boy, 1988; *Eryops, Onchiodon*: Werneburg, 2008; *Archegosaurus*: Witzmann, 2006a). In *G. angusta*, stronger compaction has probably obscured these ridges. The lateral line sulci are more clearly impressed in *G. latirostris*, especially along a line running through the prefrontal, postfrontal, postorbital, and jugal. In *G. angusta*, they are largely masked by the taller ornamenting ridges and the larger polygons. In general, the sulci are deeper and more conspicuous in adult skulls than in small specimens.

Palate. The best-preserved palates are from *G. angusta*: a small juvenile (POL-F 1989-3, Text-fig. 3A) and small adult (GPIM-N 1180, Text-figs 3D, 4). The two species differ little in the structure of the palate, except for the slightly wider interpterygoid vacuities in *G. latirostris*, which correlate with the wider posterior half of the skull in that species. The dentition, outline of the choana, and the slenderness of all palate bones form major differences between *Glanochthon* and *Sclerocephalus*. As in many Permian temnospondyls, the posteriormost premaxillar teeth are enlarged, with their sockets exceeding the size of the vomerine tusks. The tusks of the vomer are arranged transversely, and the element is covered by numerous tubercles that could represent vestigial denticles. The maxillar teeth become continuously smaller posteriorly, but they all have in common a strong base and a marked medial curvature. By far the largest tusks are found on the palatine, which may form a pair or three large teeth in succession, followed by 3–4 smaller teeth. The ectopterygoid, reaching almost twice the length of the palatine, houses 9–11 teeth that become smaller posteriorly, and generally rank in size between those of the palatine (larger) and maxilla (smaller). The shape of the sockets ranges from round to transversely oval and is overall variable.

The vomer is longer than in *Sclerocephalus*, with the anterior portion extended. There, the elongated anterior palatal depressions are always present, even in juveniles (Text-figs 3–4). The choana forms an elongated slit-like opening, about as long but much narrower than the palatine. Unlike in *Sclerocephalus*, it has parallel medial and lateral margins; in some specimens, they are not straight but slightly curved. The palatine and the ectopterygoid are underplated by a broad shelf of the maxilla, in turn being overlapped by the pterygoid medially (POL-F 1989-3; Text-fig. 3A). The pterygoid has a very slender palatine ramus which, like in *Sclerocephalus, Archegosaurus*, and eryopids, curves medially to suture with the vomer; it always excludes the palatine and ectopterygoid from the margin of the interpterygoid vacuity. The basipterygoid ramus is shorter than in *Sclerocephalus*, and the basal plate of the parasphenoid is anteriorly more slender, correlating with the narrower skull. The basal plate has a triangular denticle field in larvae and generally resembles the condition in *Micromelerpeton* (Boy 1995), while in adults the plate is much narrower and longer without a clearly defined denticle field. The dorsal surface of the parasphenoid bears two curved crests on each side, as in *Archegosaurus* (Witzmann 2006a): the anterior, transversely orientated parapterygoid crest, and the posterolateral paroccipital crest, separated from each other by the sulcus intercristatus. Between the left and right paraoccipital crest, the fossa basioccipitalis is present.

Braincase. A very clear difference to *Sclerocephalus* and the eryopids is the early ossification of endocranial elements in *Glanochthon angusta*. The quadrates, exoccipitals, and the basisphenoid are fully formed in POL-F 1989-3 (44 mm SL, Text-fig. 3A). In POL-F 1989-3, the basisphenoid is fully ossified and firmly attached to the dorsal side of the parasphenoid. It forms a symmetrical unpaired bone with dorsally raised lateral wings and a central depression that houses a pair of opening on each side. The exoccipitals are short and poorly differentiated. In *G. latirostris*, the case is not so clear, although the exoccipitals are

TEXT-FIG. 3. Palatal morphology in the genus *Glanochthon*. A–B, small *G. angusta*, POL-F 1989-3. C, small *G. latirostris*, MB.Am.211b. D, adult *G. angusta*, GPIM-N 1180. E, adult *G. latirostris*, GPIT.Am.34.

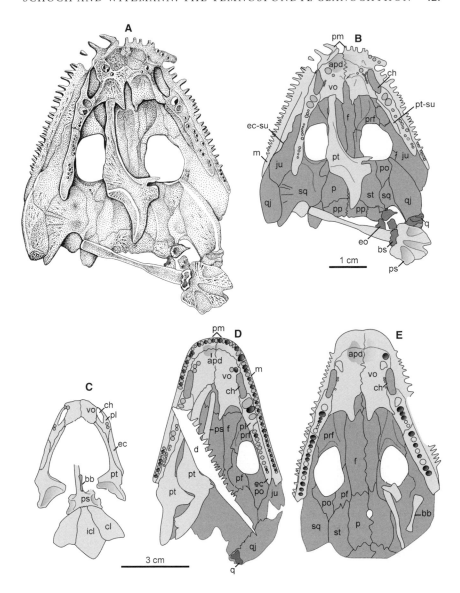

probably present in MB.Am.211 (SL = 55 mm) and in ROM 5735 (SL = 58 mm). There is no trace of either sphenethmoid or epipterygoid in any specimen.

Altogether, the much earlier ossification of exoccipitals, quadrates, and the basisphenoid is a feature shared at least between *Glanochthon angusta* and *Archegosaurus decheni* (Witzmann 2006a, b). The early ossification of exoccipitals and quadrates is further a common feature with crown stereospondyls like *Benthosuchus* (Bystrow and Efremov 1940) and *Watsonisuchus* (Steyer 2003). The late ossification or complete absence, respectively, of the sphenethmoid and epipterygoids in *Glanochthon* is shared with *Sclerocephalus*, *Archegosaurus*, and *Benthosuchus*.

Visceral skeleton. Only the stapes and basibranchial are ossified in the visceral skeleton. Unlike in *Sclerocephalus* and *Archegosaurus* (Schoch 2002a; Witzmann 2006a), the stapes is more similar to that of stereospondyls in having a slender shaft that lacks a quadrate process. Consistent with most other temnospondyl stapes, the footplate and ventral process are present albeit not clearly separated. A round stapedial foramen is located near the point where the footplate merges into the shaft. The basibranchial was described and figured by Boy (1993a) and is often not adequately preserved. We know only two good specimens, both of *G. latirostris*, in which the basibranchial is fully exposed. In the adult GPIT.Am.34, the basibranchial measures about the length of the orbit and is only slightly expanded at both ends, in contrast to the basibranchial of *Sclerocephalus* (Boy 1988; Werneburg 1992). Its anterior third is broader than the rest of the element. In adults of *Glanochthon angusta*, the element has not been observed. In *G. angusta*, the basibranchial is first visible in a specimen of 48 mm SL (GPIM-N 1286), and in *G. latirostris* in a specimen of 55 mm (MB.Am.211b). If this is not a an artefact of preservation, this would mean that the basibranchial started to ossify late compared to *Sclerocephalus* (17 mm, Schoch 2003) and *Onchiodon* (20 mm, Witzmann 2005), but comparable to *Archegosaurus* (45 mm, Witzmann 2006b).

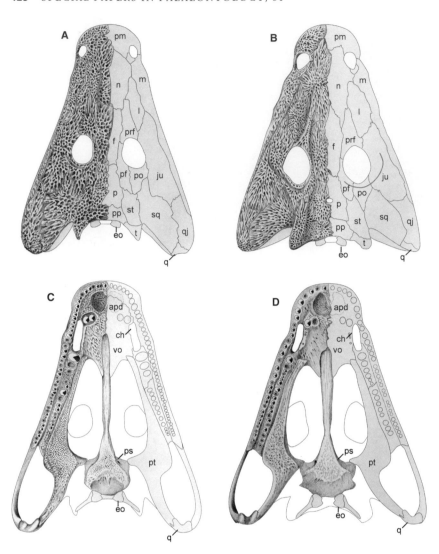

TEXT-FIG. 4. Reconstruction of the skull in the genus *Glanochthon*. A–B, dorsal view. C–D, ventral view. A, C, *G. angusta*. B, D, *G. latirostris*.

A well-developed branchial dentition is preserved in both *Glanochthon* species. In *G. angusta*, branchial denticles can be demonstrated at least until 65 mm SL (MB.Am.1312), and in *G. latirostris* at least until 76 mm SL (MB.Am.224). Similar to *Archegosaurus* (Witzmann 2004) and the stereospondyl *Uranocentrodon* (Schoch 2002b), the branchial platelets are elongate and bear six to more than 20 small teeth, in contrast to *Sclerocephalus* in which the plates are smaller, have a more roundish or rounded squarish shape and bear less denticles. This seems to be the plesiomorphic condition in temnospondyls since it is found also in eryopids (Witzmann 2005), micromelerpetontids (Boy 1972) and in saurerpetontids (pers. obs.). The elongated branchial platelets of *Glanochthon* covered the ceratobranchials proper, which themselves were not ossified but remained cartilaginous. In MB.Am.224 (*G. latirostris*), the branchial platelets are arranged in four rows. Between them, three carbonized, rod-like elements are preserved that can be interpreted as imprints of the cartilaginous ceratobranchials, similar to a large larva of *Archegosaurus* (Witzmann 2006a). The long retention of branchial denticles in *Glanochthon* is shared with *Archegosaurus* (Witzmann 2004) and *Uranocentrodon* (van Hoepen 1915; Schoch 2002b).

Mandible. The lower jaw of *Glanochthon latirostris* was described and reconstructed in detail by Boy (1993a). In *G. angusta*, the mandible is almost identical, the only difference is that the articular was ossified in small specimens (SL 44 mm) and the posterior coronoid dentigerous.

Postcranium. Most of the postcranial material present belongs to *Glanochthon angusta*, whereas the postcranial remains are very scarce in *G. latirostris*. Therefore, the following description of the postcranium refers to *G. angusta*, unless otherwise indicated.

Vertebrae. Based on MB.Am.1271 (SL = 58 mm), the presacral count is estimated as 24–25, similar to *Archegosaurus* (24 presacrals, Witzmann and Schoch 2006a) and *Sclerocephalus* (24–25, Schoch and Witzmann in press). In the other specimens, the presacral vertebral column is incompletely preserved. The atlas-axis complex is never exposed, with the possible exeption of MB.Am.1271 in which a slender element located immediately posterior to the skull table might represent the atlas neural arch. As typical for temnospondyl larvae (Boy 1974), the ossification and differentiation of the vertebrae proceeded very slowly in *G. angusta*

TEXT-FIG. 5. Postcranial morphology in *Glanochthon angusta*. A, larval skeleton, MB.Am.1275. B, adult interclavicle, ventral view, POL-F 002. C, large adult, interclavicle, dorsal view, GPIM-N 1844. D, scapulocoracoid, POL-F 002. E, humerus, GPIM-N 1844. F, humerus, POL-F 002. G, vertebrae and ribs of the anterior trunk, POL-F 002. H, small ilium, MB.Am.1275. I, large ilium, GPIM-N 1844.

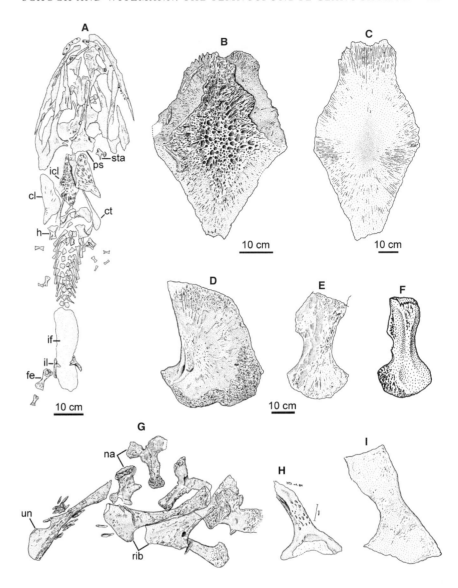

(Text-fig. 5A). At a SL of c. 60 mm, the vertebral centra are still cartilaginous, and the ossified neural arches are still paired. The neural spines are low dorsal outgrowths, and the zygapophyses are rudimentarily developed. On their medial side, the neural arches have inwardly curved anterior and posterior margins, indicating that the arches represent perichondral ossifications that enclosed a cartilaginous core. Transverse processes are not visible. At this size stage, ribs are short and straight. In the thoracic region they are longer and broader, and have distally slightly expanded ends. In MB.Am.1271, the sacral ribs are exposed. They have a short shaft and a widened distal end for articulation with the ilium. Their morphology resembles closely that of similar size stages of *Archegosaurus* (Witzmann and Schoch 2006a), and they are less stout than in larval/juvenile *Onchiodon* (Witzmann 2005) and *Acanthostomatops* (Witzmann and Schoch 2006b).

In adult specimens, the neural arches possess tall neural spines whose dorsal portions are thickened and have a rugose surface (Text-fig. 5G). The lateral sides of the spines may bear tuberosities. The neural canal is ventrally not closed by a bony bridge and might indicate that the largest investigated animals were not

fully grown. In their general proportions, the neural arches resemble those of *Archegosaurus* (Witzmann and Schoch 2006a). The horseshoe shaped intercentra are rather thin and leave large space for the notochord and cartilage, respectively. Pleurocentra are never well exposed, but seem to have been rather small and paired. The thoracal ribs are proportionally longer than in the small specimens and slightly curved. The ribs terminate proximally in a flattened rib head consisting of capitulum and tuberculum without a clear incisure between them. As common in middle- and large sized temnospondyls, the thoracal ribs bear uncinate processes in their distal parts. The processes are blade-like and bear foramina and notches at their posterior margin, that gives the blades a sometimes bizarre shape, similar to the processes in *Sclerocephalus* (Schoch and Witzmann in press) and the early tetrapod *Pederpes* (Clack and Finney 2005). Compared to *G. angusta*, the blades of *Sclerocephalus* are proportionally distinctly larger and more pronounced, whereas those of *Archegosaurus* are less well developed (Witzmann and Schoch 2006a). The outline of the sacral rib in adults corresponds to that described above for larval specimens.

Pectoral girdle. The interclavicle is elongated rhomboidal with a pointed posterior end (Text-fig. 5B, C). It is longer than wide in all growth stages (see also Boy 1993a). The width constitutes less than the width of the posterior skull table in all known growth stages. There is some variation in interclavicle width, but not as large as in *Sclerocephalus* or *Onchiodon* (Boy 1988, 1990). In small larvae, its length constitutes slightly less than half the skull length, in adults it is slightly more than that value. On the ventral surface, the interclavicle bears a pronounced sculpture or ornament in all known size stages (Text-fig. 5A–B). Most of the dermal sculpture consists of coarse pits, whereas the anterior portion of the interclavicle bears a rugose and irregular, but basically radial sculpture of furrows and ridges. In the postero-lateral parts of the ventral surface, radiating ridges and furrows are also present, but are much shallower. The anterolateral facets for the clavicular blades are conspicuous and bear faint striae. They do not meet in the ventral midline, and in the small specimen MB.Am.1275 (SL = 37 mm), they are framed medially by a distinct bulge (Text-fig. 5A). The dorsal surface of the interclavicle is concave and bears a faint, radial striation (Text-fig. 5C). The shape of the interclavicle in *G. latirostris* corresponds to that of *G. angusta*. In contrast, the dermal sculpture on the ventral surface in *G. latirostris* shows the predominantly radial pattern, similar to the sculpture of the skull roof.

The clavicles have long and slender ventral blades, corresponding to the long facets on the interclavicle (Text-fig. 5A). Their dermal sculpture consists mainly of radial ridges and furrows, whereas polygonal pits are present in the region of curvature into the shaft. The cleithrum possesses a slender dorsal blade, comparable to the situation in *Archegosaurus* (Witzmann and Schoch 2006a), and is more slender than in *Onchiodon* (Witzmann 2005) and *Sclerocephalus* (Meckert 1993; Schoch 2003). The scapula starts to ossify at a SL of 28 mm (MB.Am.1281), earlier than in *Archegosaurus* (almost 40 mm SL) and later than in *Sclerocephalus* (23 mm SL, Witzmann 2006b). In this size stage, only the semilunar, posteroventral portion of the element is ossified. At a SL of almost 40 mm, the ossified portion of the scapula is reniform with a convex anterior and concave posterior margin (Text-fig. 5A). Posteriorly, the supraglenoidal buttress is already visible and tapers dorsally. In the adult specimen GPIM 1844, the scapular blade has a straight, posterodorally directed posterior edge (Text-fig. 5D). The large supraglenoid foramen is visible posterolateral to the supraglenoidal buttress. The ventral margin is irregular and indicates the continuation in the still cartilaginous coracoid portion with the glenoid. Compared to *Archegosaurus* and especially *Sclerocephalus*, the scapulocaoracoid of *G. angusta* is poorly ossified.

Pelvic girdle. In small specimens, the ilium has a slender, posterodorsally inclined shaft with a ventral, triangular base. The lateral surface of the shaft is pitted, and a dorsoventrally directed groove is visible at its posterior edge (Text-fig. 5H). The acetabulum differentiates earlier than in *Archegosaurus* (Witzmann and Schoch 2006a). In adult specimens, the shaft is proportionally stouter, and its dorsal end is expanded (Text-fig. 5I). The pubis and ischium have not been identified in any specimen. As in *Archegosaurus*, there is no co-ossification with the ischium, and the pubis remained cartilaginous.

Limbs. In small specimens around 20–40 mm SL, the humerus is poorly ossified, short and undifferentiated (Text-fig. 5A). Its length corresponds approximately to the dorsoventral extension of the scapular ossification. Radius and ulna measure c. 70% of the length of the humerus, and the femur is 1.6 times longer than the humerus. At about 60 mm SL, the humerus is proportionally longer and has a slender shaft. The femur has 1.4 times the length of the humerus. In adults, the humerus has a well developed deltopectoral crest, but a supinator process is absent (Text-fig. 5E–F). Although more robust in *G. angusta*, the humerus resembles the poorly differentiated element of *Archegosaurus*, whereas the humerus of *Sclerocephalus* is much more differentiated (Meckert 1993; Schoch and Witzmann in press). Radius and ulna are equally long and poorly differentiated, the ulna lacking a distinct olecranon. The manus consists of elongated and slender metacarpals and phalanges, the count is not preserved. The leg is poorly preserved and generally similar to that of *Sclerocephalus haeuseri*.

Dermal scales. In both species of *Glanochthon*, the elongate, gastral scales arranged en chevron, and the dorsal, round-oval scales are present, as in most Palaeozoic temnospondyls (Witzmann 2007). In both species, the gastral scales of all known growth stages are slender and spindular in outline, as in *Archegosaurus*, and represent the 'juvenile' type rather than the 'adult' thomboid type *sensu* Witzmann (2007). Gastral scales of small larvae are thin and bear concentric growth rings, and during ontogeny, they get increasingly more slender. In larger specimens, the gastral scales of the anterior trunk region are thick and well ossified. In contrast, those of the posterior half are thinner and bear growth rings like the gastral scales of the small larvae. In *G. latirostris*, the round-oval scales of the flanks and the back of the trunk do not mutually overlap and have even no point contact. This resembles closely the situation in *Archegosaurus*, whereas larvae and juveniles of *Sclerocephalus* have imbricating, closely set round-oval scales, a pattern that represents the plesiomorphic condition in temnospondyls (Witzmann 2007). A further similarity with *Archegosaurus* are the round-oval scales in the region of the pectoral girdle, that are enlarged compared to those of the remaining trunk.

Palaeobiology

The presence of lateral line sulci throughout ontogeny and the poor degree of postcranial ossification including the weakly differentiated humerus strongly suggests that both *Glanochthon* species were primarily water-dwelling animals. This is further supported by the presence of branchial denticles (and thus open gill slits) in specimens of up to at least 76 mm SL. This suggests that the hyobranchial skeleton of adults was quite similar to that of small specimens and no remodeling took place during ontogeny. Like *Archegosaurus*, *G. latirostris* inhabited the deep water region of the 80 km long Humberg lake, whereas *G. angusta* was confined to an earlier, more shallow phase of the same lake.

Glanochthon probably mainly fed on acanthodian prey (Boy 1993*b*), which is indicated by remnants of prey. Spines of acanthodians are preserved immediately posterior to the skull table in MB.Am.1276 (*G. angusta*, 44 mm SL) and MB.Am.120 (*G. latirostris*, 91 mm SL). Also Kriwet *et al.* (2008) reported acanthodian remains in the digestive tract of a *G. latirostris* with c. 55 mm SL. Apart from these remnants of acanthodians, intestine fillings consisting of unidentifiable prey are present in MB.Am. 1275 (SL 37 mm; Text-fig. 5A) and MB.Am.1286, SL = 48 mm (both *G. angusta*).

PHYLOGENETIC ANALYSIS

The distinction of a second species within the sample formerly referred to '*Cheliderpeton*' *latirostre* requires a phylogenetic analysis to test whether the new genus *Glanochthon* actually forms a monophylum. Further major questions to be addressed by this analysis are: (1) what is the relationship of *Glanochthon* to *Sclerocephalus* as the most basal taxon of the Stereospondylomorpha; (2) are the close similarities between *Glanochthon* and *Intasuchus* highlighted by earlier papers supported; and (3) is the intermediate position of *Glanochthon* between *Sclerocephalus* and *Archegosaurus* corroborated, and what are the positions of the less well known taxa *Memonomenos (Archegosaurus) dyscriton* and *Cheliderpeton vranyi*?

Taxon sample. We restricted the analysis to Carboniferous and Permian temnospondyls, as the phylogeny of the Triassic stereospondyls is beyond the scope of this study. To address the four above-listed questions, we have included the following taxa: as outgroups, *Balanerpeton woodi* (Milner and Sequeira 1993), *Dendrerpeton acadianum* (Holmes *et al.* 1998), and *Cochleosaurus bohemicus* (Sequeira 2004). As ingroups, we chose *Iberospondylus schultzei* (Laurin and Soler-Gijón 2001, 2006), *Eryops megacephalus* (Sawin 1941), *Onchiodon labyrinthicus* (Boy 1990; Witzmann 2005), *Onchiodon (Actinodon) frossardi* (Werneburg 1997; Werneburg and Steyer 1999), *Micropholis stowi* (Schoch and Rubidge 2005); *Acanthostomatops vorax* (Witzmann and Schoch 2006*b*), *Sclerocephalus haeuseri* (Boy 1988; Schoch and Witzmann in press), *Glanochthon latirostris* and *G. angusta* (present study), *Intasuchus silvicola* (Konzhukova 1956; Gubin 1997), *Cheliderpeton vranyi* (Werneburg and Steyer 2002), *Memonomenos dyscriton* (Milner 1978), *Archegosaurus decheni* (Witzmann 2006*a, b*; Witzmann and Schoch 2006*a*), *Platyoposaurus stuckenbergi* (Gubin 1991), *Australerpeton cosgriffi* (Barberena 1998; Dias and Schultz 2003). The characters used here derive to a large part from a recent analysis centered at *Sclerocephalus* (Schoch and Witzmann

in press). They are listed and defined in the Appendix. A phylogenetic analysis of 54 characters and 18 taxa was performed in the branch-and-bound mode of PAUP 3.1 (Swofford 1991), with all characters unordered and three outgroups. This gave 12 most parsimonious trees (95 steps, CI: 0.632, RI: 0.821, RC: 0.519). If all characters except for the two multistates 53 and 54 are coded as ordered, four trees are found requiring 96 steps. In this alternative, the stereospondylomorphs are completely resolved, with *Intasuchus* and *Memonomenos* forming a clade.

Results. The obtained consensus tree is not completely resolved, with a basal polytomy formed by the three outgroups and the ingroup and two further trichotomies within the ingroup. The following nodes were found (Text-fig. 6).

Postedopoid temnospondyls: Dissorophoidea + Zatracheidae, *Iberospondylus*, Eryopidae, Stereospondylomorpha. Synapomorphies: 22, 48. Support: 84% Bootstrap, 2 steps Bremer.

Dissorophoidea + Zatracheidae: Synapomorphies: 7, 47, 50. Homoplasies: 23, 46. Support: 83% Bootstrap, 2 steps Bremer.

Eryopidae + Stereospondylomorpha: Synapomorphies: 16, 24, 26, 28, 39, 51. Homoplasy: 4. Support: 100% Bootstrap, 5 steps Bremer.

Eryopidae: Synapomorphies: 41. Homoplasies: 19, 38, 46. Support: 87% Bootstrap, 2 steps Bremer.

Onchiodon + *Eryops*: Synapomorphies: 14, 31. Homoplasy: 23. Support: 98% Bootstrap, 4 steps Bremer.

Stereospondylomorpha: Synapomorphies: 20, 25, 43. Support: 83% Bootstrap, 2 steps Bremer.

Post*Sclerocephalus* stereospondylomorphs: Synapomorphies: none. Support: 1 step Bremer.

Long-snouted stereospondylomorphs: Synapomorphies: 33, 35, 37, 44, 52. Homoplasy: 2. Support: 59% Bootstrap, 3 steps Bremer.

Genus *Glanochthon*: Synapomorphies: 5, 9. Support: 90% Bootstrap, 2 steps Bremer.

Post*Glanochthon* stereospondylomorphs: Synapomorphy: 32. Support: 1 step Bremer.

Archegosaurid plesion: Synapomorphies: 6, 29, 36. Support: 64% Bootstrap, 1 step Bremer.

Platyoposaurus + *Australerpeton*: Synapomorphies: 30. Support: 91% Bootstrap, 2 steps Bremer.

An equivocal node, reflected by some trees, is a clade formed by *Iberospondylus*, eryopids, and stereospondylomorphs, supported by four homoplasies (39, 45, 51, 54). Within the outgroup, *Cochleosaurus* forms the most primitive temnospondyl, followed by *Balanerpeton* and finally *Dendrerepeton*. The latter two share with the ingroup two synapomorphies (1, 40).

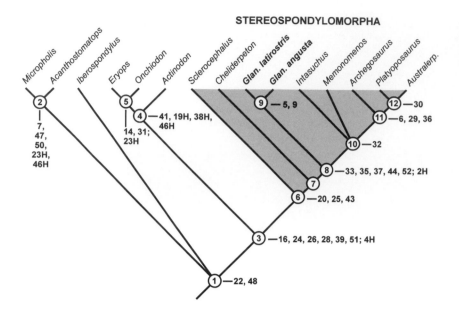

TEXT-FIG. 6. Phylogeny based on the present analysis, nodes consecutively numbered (white circles) and supporting synapomorphies listed (see Appendix).

Discussion. The results suggest that eryopids and stereospondylomorphs probably form a clade, which contradicts the Euskelia hypothesis of Yates and Warren (2000). Further, *Sclerocephalus* is found at the base of the Stereospondylomorpha, and *Glanochthon* holding a position between *Sclerocephalus* and the archegosaurid plesion. The support for the two *Glanochthon* species to form a clade is strong, but the support for the post*Sclerocephalus* clade rather weak; the major problem is the position of *Cheliderpeton*, whose palate is entirely unknown. *Glanochthon*, *Intasuchus*, *Memonomenos*, and the archegosaurid plesion are well-supported by four synapomorphies and three steps Bremer, but their Bootstrap support is low (59%).

Major steps in resolving the open questions indicated by the cladistic analysis would be to more fully prepare the Czech material of *Cheliderpeton vranyi* and *Memonomenos dyscriton* in order to get access to the palate. Another logical step will be to describe and analyse the material from stratigraphically lower horizons than *Glanochthon*, particularly the Klauswald sample, which has yielded some specimens that appear as perfect intermediates between *Glanochthon* and *Sclerocephalus*. This will have to involve cladistics as well as multivariate analysis.

CONCLUSIONS

After the revision of the type species *Cheliderpeton vranyi* from the Czech Republic (Werneburg and Steyer 2002) it became clear that '*Cheliderpeton*' *latirostre* from the Saar-Nahe-Lorraine Basin does not share the autapomorphies of that genus and is generally rather different. Consequently, we propose the new generic name *Glanochthon* for '*Cheliderpeton*' *latirostre*.

The material previously assigned to *Glanochthon* ('*Cheliderpeton*') *latirostre* falls into two distinct species, which exclude one another stratigraphically: (1) the classic taxon from the Lebach concretions (*G. latirostris* proper); and (2) a slightly older, more slender-skulled and coarsely ornamented taxon here referred to the new species *G. angusta*.

The two species of *Glanochthon* share unambiguous synapomorphies not found in any other clade; cladistic analysis of 19 temnospondyl taxa and 54 characters finds these two species to be monophyletic.

Glanochthon nests within the stereospondylomorphs, above *Sclerocephalus* and below *Intasuchus*, *Memonomenos*, *Archegosaurus*, and the stereospondyl crown. In general, *Memonomenos* and *Intasuchus* are much more similar to *Glanochthon* than *Cheliderpeton* is, but many parts of their skeletons are unknown. The position of *Cheliderpeton* is unclear because of poor preservation and the unknown morphology of its palate; while cladistic analysis finds it to nest between *Sclerocephalus* and *Glanochthon*, it may as well hold a more basal position.

Acknowledgements. We are both pleased to contribute this paper to a Festschrift for Andrew Milner, as we have benefitted much from his knowledge and generosity in our careers. The senior author first met him at a conference in Bad Dürkheim in 1990, where Andrew was one of the few scientists who talked to the would-be student. We have always been impressed by Andrew's encyclopaedic knowledge that made him the leading authority on temnospondyls and beyond. In addition to his meticulous collection of data, which is always up-to-date, he is full of questions and ideas. Certainly influenced by his own teaching of all aspects of biology, Andrew has always an eye on functional morphology, development, and ecology in addition to his main fields of morphology and phylogeny. Like many other projects, the present paper developed from discussions we had with Andrew

Milner, who is an expert not only on lower tetrapod material from the vast range of Commonwealth countries, but also very much so from Central Europe.

We thank Jürgen Boy, Klaus Krätschmer, Andrew Milner, Marcello Ruta, Jean-Sébastien Steyer, and Ralf Werneburg for fruitful discussions, and Jennifer Clack and Tomasz Sulej for their helpful reviews. Thomas Bach, Jürgen Boy, Rudolf Becker, Eberhard Frey, Oliver Hampe, Michael Maisch, Michael Maus, Dieter Schweiss, Kevin Seymour, Harald Stapf, and Ralf Werneburg are acknowledged for access to collections. Marit Kamenz carried out a meticulous preparation of a delicate palate, and Stefanie Klug prepared silikon casts, which is gratefully acknowledged.

REFERENCES

BARBERENA, M. C. 1998. *Australerpeton cosgriffi* n.g. n.sp., a Late Permian rhinesuchoid amphibian from Brazil. *Anais de Academia Brasileira de Ciências*, **70**, 125–137.

BOY, J. A. 1972. Palökologischer Vergleich zweier berühmter Fossillagerstätten des deutschen Rotliegenden (Unter-Perm; Saar-Nahe-Gebiet). *Notizblatt des hessischen Landesamtes für Bodenforschung*, **100**, 46–59.

—— 1974. Die Larven der rhachitomen Amphibien (Amphibia: Temnospondyli; Karbon – Trias). *Paläontologische Zeitschrift*, **48**, 236–282.

—— 1976. Überblick über die Fauna des saarpfälzischen Rotliegenden. *Mainzer geowissenschaftliche Mitteilungen*, **5**, 13–85.

—— 1977. Typen und Genese jungpaläozoischer Tetrapoden-Lagerstätten. *Palaeontographica A*, **156**, 111–167.

—— 1987. Die TetrapodenLokalitäten des saarpfälzischen Rotliegenden (?OberKarbon–UnterPerm; SWDeutschland) und die Biostratigraphie der RotliegendTetrapoden. *Mainzer geowissenschaftliche Mitteilungen*, **16**, 31–65.

—— 1988. Über einige Vertreter der Eryopoidea (Amphibia: Temnospondyli) aus dem europäischen Rotliegend (?höchstes Karbon–Perm) 1. *Sclerocephalus*. *Paläontologische Zeitschrift*, **62**, 107–132.

—— 1990. Über einige Vertreter der Eryopoidea (Amphibia: Temnospondyli) aus dem europäischen Rotliegend (?höchstes Karbon–Perm). 3. *Onchiodon*. *Paläontologische Zeitschrift*, **64**, 287–312.

—— 1993*a*. Über einige Vertreter der Eryopoidea (Amphibia: Temnospondyli) aus dem europäischen Rotliegend (?höchstes Karbon–Perm) 4. *Cheliderpeton latirostre*. *Paläontologische Zeitschrift*, **67**, 123–143.

—— 1993*b*. Synopsis of the tetrapods from the Rotliegend (Lower Permian) in the Saar-Nahe Basin (SW-Germany). 155–169. *In* HEIDTKE, U. H. J. (ed.). *New research on permo-carboniferous faunas*. Pollichia-Buch vol 29. Pollichia, Bad Dürkheim, 199 pp.

—— 1995. Über die Micromelerpetontidae (Amphibia: Temnospondyli). 1. Morphologie und Paläoökologie des *Micromelerpeton credneri* (Unter-Perm; SW-Deutschland). *Paläontologische Zeitschrift*, **69**, 429–457.

—— 2007. Als die Saurier noch klein waren: Tetrapoden im Permokarbon. 258–286. *In* SCHINDLER, T. and HEID-

TKE, U. H. J. (eds). *Kohlesümpfe, Seen und Halbwüsten*. Pollichia, Bad Dürkheim, 318 pp.

—— and FICHTER, J. 1982. Zur Stratigraphie des saarpfälzischen Rotliegenden (?Ober-Karbon–Unter-Perm, SW-Deutschland). *Zeitschrift der deutschen geologischen Gesellschaft*, **133**, 607–642.

—— and SUES, H.-D. 2000. Branchiosaurs: larvae, metamorphosis and heterochrony in temnospondyls and seymouriamorphs. 1150–1197. *In* HEATWOLE, H. and CARROLL, R. L. (eds). *Amphibian biology, volume 4: palaeontology*. Surrey Beatty, Chipping Norton, 523 pp.

BURMEISTER, H. 1850. *Die Labyrinthodonten aus dem Saarbrücker Steinkohlengebirge*. Reimer, Berlin, 74 pp.

BYSTROW, A. P. and EFREMOV, I. A. 1940. *Benthosuchus sushkini* Efremov. A labyrinthodont from the Eotriassic of Sharzenga River. *Travaux de l'institut paléontologique académie de sciences de l'USSR*, **10**, 1–152. [in Russian].

CLACK, J. A. and FINNEY, S. 2005. *Pederpes finneyae*, an articulated tetrapod from the Tournaisian of western Scotland. *Journal of Systematic Palaeontology*, **2**, 311–346.

DIAS, E. V. and SCHULTZ, C. L. 2003. The first Paleozoic temnospondyl postcranial skeleton from South Africa. *Revista Brasileira de Paleontologia*, **6**, 29–42.

DSK (DEUTSCHE STRATIGRAPHISCHE KOMMISSION) 2002. *Stratigraphische Tabelle von Deutschland*. Stein, Potsdam.

GOLDFUSS, G. A. 1847. *Beiträge zur vorweltlichen Fauna des Steinkohlengebirges*. Naturhistorischer Verein für die preussischen Rheinlande, Bonn, 30 pp.

GUBIN, Y. M. 1991. Permian archegosauroid amphibians from the USSR. *Trudy Paleontologiceskogo Instituta Akademiya Nauk SSSR*, **249**, 1–138. [In Russian].

—— 1997. Skull morphology of *Archegosaurus decheni* Goldfuss (Amphibia, Temnospondyli) from the Early Permian of Germany. *Alcheringa*, **21**, 103–121.

van HOEPEN, E. C. N. 1915. Stegocephalia of Senekal, O. F. S. *Annals of the Transvaal Museum*, **5**, 124–149.

HOLMES, R. B., CARROLL, R. L. and REISZ, R. R. 1998. The first articulated skeleton of *Dendrerpeton acadianum* (Temnospondyli, Dendrerpetontidae) from the Lower Pennsylvanian locality of Joggins, Nova Scotia, and a review of its relationships. *Journal of Vertebrate Paleontology*, **18**, 64–79.

JORDAN, H. 1849. Ergänzende Beobachtungen zu der Abhandlung von Goldfuss über die Gattung *Archegosaurus*. *Verhandlungen des naturhistorischen Vereins für die Rheinlande und Westphalen*, **6**, 76–81.

KONZHUKOVA, E. D. 1956. The Intan Lower Permian fauna of the northern Ural region. *Trudy Paleontologiceskogo Instituta Akademiya Nauk SSSR*, **62**, 5–50. [In Russian].

KRÄTSCHMER, K. 2006. Neue temnospondyle Amphibien aus dem Rotliegend des südwestdeutschen Saar-Nahe-Beckens Teil 1. *Geowissenschaftliche Beiträge zum Saarpfälzischen Rotliegenden*, **4**, 3–46.

KRIWET, J., WITZMANN, F., KLUG, S. and HEIDTKE, U. H. J. 2008. First evidence of a vertebrate three-level trophic chain in the fossil record. *Proceedings of the Royal Scociety, Series B*, **275**, 181–186.

LAURIN, M. and SOLER-GIJÓN, R. 2001. The oldest stegocephalian from the Iberian Peninsula: evidence that temnospondyls were euryhaline. *Comptes Rendus Academie des Sciences de la vie*, **324**, 495–501.

—— —— 2006. The oldest known stegocephalian (Sarcopterygii: Temnospondyli) from Spain. *Journal of Vertebrate Paleontology*, **26**, 284–299.

MECKERT, D. 1993. Der Schultergürtel des *Sclerocephalus haeuseri* GOLDFUSS, 1847 im Vergleich mit *Eryops* Cope, 1877 (Eryopoidea, Amphibia, Perm). *Palaeontographica A*, **229**, 113–140.

MENGE, H. 1985. *Altgriechisch. Langenscheidts Taschenwörterbuch*. New edition. Langenscheidt, Berchtesgaden, 547 pp.

von MEYER, H. 1844. Briefliche Mitteilung an Professor Bronn. *Neues Jahrbuch für Mineralogie*, **1844**, 336–337.

—— 1857. Reptilien aus der Steinkohlen-Formation in Deutschland. *Palaeontographica*, **6**, 59–218.

—— 1858. Nachtrag zu den Reptilien der Steinkohlen-Formation in Deutschland, insbesondere zu *Archegosaurus latirostris*. *Palaeontographica*, **6**, 219–220.

MILNER, A. R. 1978. A reappraisal of the early Permian amphibians *Memonomenos dyscriton* and *Cricotillus brachydens*. *Palaeontology*, **21**, 667–686.

—— and SEQUEIRA, S. E. K. 1993. The temnospondyl amphibians from the Viséan of East Kirkton, West Lothian, Scotland. *Transactions of the Royal Society of Edinburgh Earth Sciences*, **84**, 331–361.

ROMER, A. S. 1939. Notes on branchiosaurs. *American Journal of Science*, **237**, 748–761.

—— 1947. Review of the Labyrinthodontia. *Bulletin of the Museum of comparative Zoology Harvard*, **99**, 1–368.

SAWIN, H. J. 1941. The cranial antomy of *Eryops megacephalus*. *Bulletin of the Museum of Comparative Zoology Harvard*, **88**, 407–463.

SCHOCH, R. R. 1999. Comparative osteology of *Mastodonsaurus giganteus* (Jaeger, 1828) from the Lettenkeuper (Ladinian: Longobardian) from Germany (Baden-Württemberg, Bayern, Thüringen). *Stuttgarter Beiträge zur Naturkunde, B*, **278**, 1–178.

—— 2002a. The stapes and middle ear of the Permo-Carboniferous temnospondyl *Sclerocephalus*. *Neues Jahrbuch für Geologie und Paläontologie Monatshefte*, **2002**, 671–680.

—— 2002b. The evolution of metamorphosis in temnospondyls. *Lethaia*, **35**, 309–327.

—— 2003. Early larval ontogeny of the Permo-Carboniferous temnospondyl *Sclerocephalus*. *Palaeontology*, **46**, 1055–1072.

—— and MILNER, A. R. 2000. *Stereospondyli. Handbuch der Paläoherpetologie. Part 3B*. Verlag Dr. Friedrich Pfeil, München, 203 pp.

—— and RUBIDGE, B. S. 2005. The amphibamid *Micropholis stowi* from the Lystrosaurus Assemblage Zone of South Africa. *Journal of Vertebrate Paleontology*, **25**, 502–522.

—— and WITZMANN, F. in press. Osteology and relationships of the temnospondyl genus *Sclerocephalus*. *Zoological Journal of the Linnean Society*.

SEQUEIRA, S. E. K. 2004. The skull of *Cochleosaurus bohemicus* Frič, a temnospondyl from the Czech Republic (Upper Carboniferous) and cochleosaurid interrelationships. *Transactions of the Royal Society of Edinburgh Earth Sciences*, **94**, 21–43.

STEYER, J. S. 2003. A revision of the early Triassic 'capitosaurs' (Stegocephali, Stereospondyli) from Madagascar, with remarks on their comparative ontogeny. *Journal of Vertebrate Paleontology*, **23**, 544–555.

SWOFFORD, D. 1991. PAUP: phylogenetic analysis using parsimony, version 3.1. Illinois Natural History Survey, Champaign, Illinois.

WERNEBURG, R. 1992. *Sclerocephalus jogischneideri* n. sp. (Eryopoidea, Amphibia) aus dem Unterrotliegenden (Unterperm) des Thüringer Waldes. *Freiberger Forschungshefte, C*, **445**, 29–48.

—— 1997. Der Eryopide *Onchiodon* (Amphibia) aus dem Rotliegend des Beckens von Autun (Frankreich). *Freiberger Forschungshefte, C*, **466**, 167–181.

—— 2008. Der 'Manebacher Saurier' – ein neuer großer Eryopide (*Onchiodon*) aus dem Rotliegend (Unter-Perm) des Thüringer Waldes. *Veröffentlichungen des Naturhistorischen Museums Schleusingen*, **22**, 3–40.

—— and STEYER, J.-S. 1999. Redescription of the holotype of *Actinodon frossardi* (Amphibia, Temnospondyli) from the Lower Permian of the Autun Basin (France). *Geobios*, **32**, 599–607.

—— —— 2002. Revision of *Cheliderpeton vranyi* Fritsch, 1877 (Amphibia, Temnospondyli) from the Lower Permian of Bohemia (Czech Republic). *Paläontologische Zeitschrift*, **76**, 149–162.

WITZMANN, F. 2004. The external gills of Palaeozoic amphibians. *Neues Jahrbuch für Geologie und Paläontologie, Abhandlungen*, **232**, 375–401.

—— 2005. Hyobranchial and postcranial ontogeny of the temnospondyl *Onchiodon labyrinthicus* (Geinitz, 1861) from Niederhäslich (Döhlen Basin, Autunian, Saxony). *Paläontologische Zeitschrift*, **79**, 479–492.

—— 2006a. Cranial anatomy and ontogeny of the Permo-Carboniferous temnospondyl *Archegosaurus decheni* from the Saar-Nahe Basin, Germany. *Transactions of the Royal Society of Edinburgh Earth Sciences*, **96**, 131–162.

—— 2006b. Developmental patterns and ossification sequence in the Permo-Carboniferous temnospondyl *Archegosaurus decheni* (Saar-Nahe Basin, Germany). *Journal of Vertebrate Paleontology*, **26**, 7–17.

—— 2007. The evolution of the scalation pattern in temnospondyl amphibians. *Zoological Journal of the Linnean Society*, **150**, 815–834.

—— and SCHOCH, R. R. 2006a. The postcranium of *Archegosaurus decheni* and the phylogeny of temnospondyl postcrania. *Palaeontology*, **49**, 1211–1235.

—— —— 2006b. Skeletal development of the temnospondyl *Acanthostomatops vorax* from the Lower Permian Döhlen Basin of Saxony. *Transactions of the Royal Society of Scotland Earth Sciences*, **96**, 365–385.

YATES, A. and WARREN, A. A. 2000. The phylogeny of the 'higher' temnospondyls (Vertebrata: Choanata) and its implications for the monophyly and origins of the Stereospondyli. *Zoological Journal of the Linnean Society*, **128**, 77–121.

ZITTEL, K. 1888. *Handbuch der Palaeontologie. 1. Abtheilung. Palaeozoologie. Vol. III. Vertebrata (Pisces, Amphibia, Reptilia, Aves)*. Rudolf Oldenbourg, Berlin, 900 pp.

APPENDIX

Anatomical abbreviations

apd, anterior palatal depression; bb, basibranchial; bs, basisphenoid; ch, choana; cl, clavicle; ct, cleithrum; d, dentary; ec, ectopterygoid; eo, exoccipital; f, frontal; fe, femur; h, humerus; icl, interclavicle; if, intestine filling; il, ilium; ju, jugal; l, lacrimal; m, maxilla; n, nasal; na, neural arch; p, parietal; pf, postfrontal; pl, palatine; pm, premaxilla; po, postorbital; pp, postparietal; prf, prefrontal; prsp, presplenial; ps, parasphenoid; pt, pterygoid; q, quadrate; qj, quadratojugal; sc, scapulocoracoid; sh-m, shelf of maxilla; sh-pe, shelf of palatine and ectopterygoid; SL, skull length; sq, squamosal; st, supratemporal; sta, stapes; su-f, supraglenoid foramen; t, tabular; ul, ulna; un, uncinate process; vo, vomer.

Character-taxon matrix

De, Dendrerpeton; Ba, Balanerpeton; Co, Cochleosauru; Mi, Micropholis; Ac, Acanthostomatops; Ib, Iberospondylus; Er, Eryops; On, Onchiodon; Ac, Actinodon; Sh, Sclerocephalus haeuseri; Gl, Glanochthon latirostris; Ga, G. angusta; In, Intasuchus; Me, Memonomenos; Ch, Cheliderpeton vranyi; Ar, Archegosaurus; Pl, Platyoposaurus; Au, Australerpeton.

	1	2	3	4	5	6	7	8	9	10	11	12	13	14	15	16	17	18	19	20	21	22	23	24	25	26	27	28
De	1	0	0	0	0	0	0	0	0	0	0	0	0	0	0	0	0	0	0	0	0	0	0	0	0	0	0	0
Ba	1	0	0	0	0	0	0	0	0	0	0	0	0	0	0	0	0	0	0	0	0	0	0	0	0	0	0	0
Co	0	1	0	1	0	0	0	0	0	0	0	0	0	0	0	0	0	0	1	0	0	0	0	0	0	0	0	0
Mi	1	0	0	0	0	0	1	0	0	0	0	0	0	0	0	0	0	0	0	0	0	1	1	0	0	0	0	0
Ac	1	1	0	0	0	0	1	0	0	0	0	0	0	0	0	1	0	0	0	1	0	0	1	1	0	0	0	0
Ib	1	0	0	0	0	0	0	0	0	0	0	0	0	0	0	1	0	0	0	0	0	0	1	0	0	0	0	0
Er	1	0	1	1	0	0	0	0	0	1	0	0	0	1	1	1	1	1	1	0	0	1	1	1	0	1	0	1
On	1	0	1	1	0	0	0	0	0	1	0	0	0	1	1	1	1	0	1	0	0	1	1	1	0	1	0	1
Ac	1	0	1	1	0	0	0	0	0	0	0	0	0	0	1	1	0	0	1	0	0	1	0	1	0	1	0	1
Sh	1	0	1	1	0	0	0	0	0	1	1	1	1	0	0	1	0	0	0	0,1	2	1	0	1	1	1	0	1
Gl	1	1	1	1	1	0	0	0	1	1	1	1	1	0	1	0	1	0	0	0	1	1	1	0	1	1	1	1
Ga	1	1	1	1	1	0	0	0	1	1	1	1	0,1	0	1	1	0	0	0	1	2	1	0	1	1	1	0	1
In	1	1	0	1	0	0	0	1	?	0	0	0	0	0	1	1	0	1	0	?	0	1	0	1	?	1	1	1
Me	1	1	1	1	0	0	0	1	0	0	1	0	0	0	1	1	0	1	0	?	1	1	0	1	?	1	0	?
Ch	1	0	0	1	0	0	0	0	0	0	?	0	0	0	0	1	0	1	0	1	2	1	0	1	?	1	0	?
Ar	1	1	1	1	0	1	0	1	0	0	1	0	0	0	1	1	0	1	0	1	2	1	0	1	1	1	1	1
Pl	1	1	1	1	0	1	0	0	0	0	1	0	0	0	1	1	0	1	0	1	2	1	0	1	1	1	0	1
Au	1	1	1	1	0	1	0	0	0	0	1	0	0	0	1	1	0	1	0	1	2	1	0	1	1	1	0	1

	29	30	31	32	33	34	35	36	37	38	39	40	41	42	43	44	45	46	47	48	49	50	51	52	53	54
De	0	0	0	0	0	0	0	0	0	0	0	1	0	0	0	0	0	0	0	0	0	?	0	0	0	0
Ba	0	0	0	0	0	0	0	0	0	0	0	1	0	0	0	0	0	0	0	0	0	0	0	0	0	0
Co	0	0	0	0	0	0	0	0	0	0	0	0	0	0	0	0	0	0	0	0	1	0	0	0	0	0
Mi	0	0	0	0	0	0	0	0	0	0	0	1	0	0	0	0	0	1	1	1	1	1	0	0	1	0
Ac	0	0	0	0	0	0	0	0	0	0	0	1	0	0	0	0	0	1	1	1	0	1	0	1	2	0
Ib	0	0	0	0	0	0	0	0	0	0	?	?	0	0	?	?	?	?	?	?	?	?	?	?	?	1
Er	0	0	1	0	0	0	0	0	0	1	1	1	1	0	0	0	0	1	0	1	0	0	1	0	2	2
On	0	0	1	0	0	0	0	0	0	1	1	1	1	0	0	0	0,1	1	0	1	0	0	1	0	2	2
Ac	0	0	0	0	0	0	0	0	0	1	1	1	1	?	0	0	1	1	0	1	0	?	1	0	?	2
Sh	0	0	0	0	0	0	0	0	0	0	1	1	0	1	1	0	1	0	0	1	0	0	1	0	1	2
Gl	0	0	0	0	1	0	1	0	1	0	1	1	0	0	1	1	1	0	0	1	1	0	1	1	1	1
Ga	0	0	0	0	1	0	1	0	1	0	1	1	0	0	1	1	1	0	0	1	1	0	1	1	1	1
In	0	0	0	1	1	1	1	0	1	0	1	1	0	?	?	?	?	?	?	?	?	?	?	?	?	?
Me	?	?	?	?	?	?	?	?	?	?	?	?	?	?	?	1	1	?	0	?	?	1	?	?	?	?
Ch	?	?	?	?	?	?	?	?	?	?	?	?	?	?	1	0	1	0	0	1	1	0	1	1	2	1

	29	30	31	32	33	34	35	36	37	38	39	40	41	42	43	44	45	46	47	48	49	50	51	52	53	54
Ar	1	0	0	1	1	1	1	1	1	0	1	1	0	1	1	1	1	0	0	1	1	0	1	1	1	1
Pl	1	1	0	1	1	1	1	1	1	0	1	1	0	0	1	1	1	0	0	1	1	?	1	1	1	1
Au	1	1	0	1	1	1	1	1	1	1	1	1	0	0	1	1	1	0	0	1	0	?	1	1	1	1

List of characters

1. Premaxilla (alary process). Absent (0), or present (1).
2. Premaxilla (prenarial portion). Short (0), or expanded anteriorly by about the length of the naris (1).
3. Premaxilla (outline). Parabolically rounded (0), or box-like, anteriorly blunt (1).
4. Snout (internarial distance). Narrower than interorbital distance (0), or wider (1).
5. Snout (margin). Straight (0), or laterally constricted at level of naris (1).
6. Rostrum. Absent (0), or present (1).
7. Internarial fenestra. Absent (0), or present (1).
8. Orbits. Round to slightly oval (0), or elongated oval (1).
9. Orbits. Ends rounded (0), or pointed (1).
10. Maxilla (anterior margin). Straight (0), or laterally convex due to enlarged teeth (1).
11. Maxilla (contact to nasal). Absent, separated by lacrimal (0), or present (1).
12. Nasal (lateral margin). Straight (0), or stepped, with lateral excursion anterior to prefrontal, accommodating narrower lacrimal (1).
13. Lacrimal (length). As long as nasal (0), shorter than nasal (1).
14. Lacrimal (width). Lateral suture parallels medial one (0), or lateral suture posterolaterally expanded to give broader preorbital region (1).
15. Preorbital region (length). Less than twice the length of posterior skull table (0), or more (1).
16. Prefrontal-jugal (contact). Absent (0), or present (1).
17. Prefrontal (anterior end). Pointed (0), or wide and blunt (1).
18. Frontal-nasal (length). Frontal as long or longer than nasal (0), or shorter (1).
19. Interorbital distance. Narrower than orbital width (0), or wider (1).
20. Lateral line (sulci). Absent in adults (0), or present (1).
21. Posterior skull table (length). Less than 0.7 times the width (0), or 0.7-0.8 (1), or larger than 0.8 (2).
22. Intertemporal. Present (0), or absent (1).
23. Postorbital. Long triangular, wedged deeply between squamosal and supratemporal (0), or short (1).
24. Squamosal embayment (size). Wide, giving semilunar flange on squamosal (0), or slit-like, with thin flange on squamosal (1).
25. Tabular (ventral crest). Absent (0), or present (1).
26. Jugal (preorbital expansion). Absent in adults (0), or present (1).
27. Ornament. Polygons and short ridges (0), long ridges arranged radially (1).

28. Vomer. Smooth (0), or with paired depressions anteriorly (1).
29. Vomerine tusks. Anterolateral to choana, transverse row (0), or well anterior to choana, sagittal row (1).
30. Anterior palatal openings. Absent (0), or present (1).
31. Choana (width). Elongated oval or slit-like (0), or round (1).
32. Premaxilla. Borders choana (0), or not (1).
33. Palatine, ectopterygoid (continuous tooth row). Absent (0), or present (1).
34. Palatine. Fangs and no more than 3-4 extra teeth (0), or 5 and more extra teeth (1).
35. Ectopterygoid (tusks). Present (0), or absent (1).
36. Parasphenoid. Denticle field on plate triangular (0), or round (1).
37. Basipterygoid ramus (length). Transverse, rod-like (0), or short, without medial extension (1).
38. Basicranial articulation. Moveable overlap (0), or tightly sutured (1).
39. Carotid foramina (entrance). Anteromedial on basal plate, close to cultriform process (0), or at posterolateral corner of plate (1).
40. Vomer. Separated by pterygoid from interpterygoid vacuity (0), or bordering that opening (1).
41. Cultriform process (width). Throughout of similar width (0), or posteriorly expanding abruptly to about twice the width (1).
42. Stapes (quadrate process). Absent (0), or present (1).
43. Interclavicle (adult shape). As wide as long (0), or longer than wide (1).
44. Interclavicle (width). As wide or wider than posterior skull table (0), or narrower (1).
45. Interclavicle (size). Shorter than posterior skull table (0), or longer than half of skull length (1).
46. Interclavicle (posterior margin). Triangular, pointed (0), or rounded to blunt (1).
47. Interclavicle (outline). Rhomboid (0), or quadrangular to pentagonal (1).
48. Humerus (entepicondylar foramen). Present (0), or absent (1).
49. Humerus (supinator). Present (0), or absent (1).
50. Humerus. Short with slow growth rate in larvae (0), or long due to rapid growth (1).
51. Femur. Intercondylar fossa on dorsodistal surface forming deep trough (0), or shallow groove (1).
52. Pubis. Ossified (0), or unossified (1).
53. Ilium. Shaft kinked, posteriorly directed (0), or shaft straight and dorsal with broadened end (1), or shaft straight posterodorsally directed (2). Unordered.
54. Ribs. Short (0), or long rod-like with small uncinates (1), or long with blade-like uncinates (2). Unordered.

[Special Papers in Palaeontology 81, 2009, pp. 137–153]

THE POSTCRANIUM OF *COCHLEOSAURUS BOHEMICUS* FRIČ, A PRIMITIVE UPPER CARBONIFEROUS TEMNOSPONDYL FROM THE CZECH REPUBLIC

by SANDRA E. K. SEQUEIRA

Department of Palaeontology, The Natural History Museum, Cromwell Road, London SW7 5BD, UK; e-mail: s.sequeira@btinternet.com

Typescript received 29 April 2008; accepted in revised form 4 June 2008

Abstract: The postcranium of the Upper Carboniferous basal temnospondyl *Cochleosaurus bohemicus* Frič is comprehensively described and figured for the first time, using data from both small and larger specimens. The unspecialized postcranial skeleton of *C. bohemicus* lacks apomorphies and is only slightly modified during ontogeny. *C. bohemicus* has an estimated 24 rhachitomous presacral vertebrae which are undifferentiated posterior to the atlas complex; ribs lack uncinate processes. The scapulocoracoid is partly cartilaginous, the pubis is unossified and femora lack specific modifications. Small individuals have poorly-ossified and unfused neural arches, their humerus retains an entepicondylar foramen, carpals are unossified and the dorsal blade of the ilium is very broad. Ontogenetic changes include apparent loss of the entepicondylar foramen in larger humeri, ossification of carpals in a probable *C. bohemicus* manus, and elongation and narrowing of the iliac dorsal blade. Incomplete ossification of the pectoral and pelvic girdles implies that *C. bohemicus* was mainly amphibious. The postcranial data supports the sister-group relationship of the Cochleosauridae to *Balanerpeton*, *Dendrerpeton* and the trimerorhachoids. Cladistic integrity of the Edopoidea and the phylogenetic position of *Edops* remains problematic due to difficulties in identifying size-independent characters for that genus.

Key words: Temnospondyli, *Cochleosaurus*, postcranium, ontogeny, Edopoid phylogeny, amphibians.

T EMNOSPONDYL amphibians comprise the largest group of non-amniote tetrapods recorded from Lower Carboniferous to Mid-Cretaceous fossil deposits. Despite their relative abundance in the fossil record, most temnospondyls have been described only from isolated skulls, since postcranial bones are rarely retained postmortem. Even where taphonomic conditions permit complete preservation, skeletons may be dissociated by scavenging, geological and fluvial transport or collector bias. Collectors have commonly ignored dissociated postcranium, selecting only attractive and easily recognizable skulls.

Those factors have hindered our understanding of temnospondyl postcranial anatomy and many phylogenetic analyses now derive almost exclusively from datasets of cranial characters. Some recent studies of *Balanerpeton woodi* (Milner and Sequeira 1994), *Dendrerpeton acadianum* (Holmes *et al.* 1998), *Sclerocephalus haeuseri* (Lohmann and Sachs 2001), *Archegosaurus decheni* (Witzmann and Schoch 2006) and *Trimerorhachis insignis* (Pawley 2007) have begun to address the paucity of recorded postcranial anatomy. However, those taxa are all more derived than the basal temnospondyl *Cochleosaurus bohemicus* (Ruta *et al.* 2003), which is an unspecialized member of the plesiomorphic Edopoidea (= *Edops* + Cochleosauridae). The Cochleosauridae currently comprises Euro-American taxa (*Procochleosaurus*, *Adamanterpeton*, *Cochleosaurus bohemicus*, *Cochleosaurus florensis*, *Chenoprosopus*), and the West African species, *Nigerpeton ricqlesi*.

This is the first comprehensive description of the postcranial skeleton of *Cochleosaurus bohemicus*. It provides a reassessment of previously poorly documented, small specimens, and includes new data from larger, undescribed, individuals. Understanding the postcranial anatomy of such a primitive temnospondyl is of potential value for future cladistic analyses.

HISTORY

The exceptionally productive Upper Carboniferous (Asturian) assemblage at Nýřany, in the northwestern Czech

Republic, has yielded over 2000 vertebrates, including numerous fossils of the superficially crocodyliform temnospondyl, *Cochleosaurus bohemicus*. During the Palaeozoic, Nýřany was an intermontane, lacustrine and swampy environment (Pešek 1974; Milner 1980) and *C. bohemicus* was a common faunal component at the locality, reaching an estimated maximum length of about 2 m. Several tens of specimens of *C. bohemicus* were collected there during the late nineteenth century; most derive from a richly fossiliferous 30 cm band of laminated coal and mudstone matrix (= Plattelkohle), which is estimated to have accumulated in less than 100 years (Milner 1980; extrapolating from Skoček 1968). Despite their relative abundance at Nýřany, very few *C. bohemicus* specimens retain any articulated postcranium, and those which do so, are mainly small individuals.

The Czech naturalist Antonin Frič provided the first formal report of the Nýřany fossil fauna in a preliminary, unillustrated, paper (Frič 1876). Frič published under the German form of his name (= Anton Fritsch) after 1876, and that convention is followed herein. He initially (Frič 1876) reported the cochleosaur material as *Melosaurus bohemicus*, a Permian temnospondyl from Russia, but later (Fritsch 1885) described the species fully and made it the type of the new genus *Cochleosaurus*. This reattribution was made in a lavishly-illustrated, 15-part, monograph (Fritsch 1879–1901) describing the Nýřany vertebrates. However, Fritsch assigned the cochleosaur specimens therein to several taxa (*Cochleosaurus falax, Dendrerpeton pyriticum, Dendrerpeton ?deprivatum, Gaudrya latistoma, Nyrania trachystoma*) which are now synonymised as *Cochleosaurus bohemicus* (Steen 1938; Romer 1947; Sequeira and Milner 1993; Sequeira 2004). Although Fritsch also included some fragmentary postcranial material in his illustrations of *C. bohemicus* (Fritsch 1885, figs 7, 9, 10–13, table 61), none of it can be certainly associated with *C. bohemicus* skulls. The only determinate *C. bohemicus* postcranium in his report is in a photograph of a plaster cast from Leoben Museum (Fritsch 1899, fig. 391), but this was not illustrated further by Fritsch.

Broili later described four more Nýřany specimens, some with associated postcranium (Broili 1905, figs 1–3, tables 1–2,), as *C. bohemicus*. However, one is not a cochleosaur (Broili 1905, fig. 3a, table 1). Two of the other three *C. bohemicus* specimens were destroyed in World War 2, and only AMNH23363, counterpart BMNH R2823 (Broili 1905, fig. 2, table 2) remains. His illustrations of the postcranium associated with these two small *C. bohemicus* skulls (Broili, figs 1–2, table 2) contain little detail.

There were no subsequent descriptions of the Nýřany *C. bohemicus* specimens until Steen (1938) reviewed the fossil amphibians from the locality. Once again, the study focused entirely on small *C. bohemicus* specimens, with minimal postcranial detail for two small individuals (Steen 1938, pl. 6, text-figs 29, 31c). One of these (Steen 1938, text-fig. 29) was the counterpart to the specimen previously illustrated by Broili (Broili 1905, fig. 2, table 2), but little extra information was discernible in Steen's publication.

No further progress was made until Klembara (1985, figs 8–9) and Hook (1993, fig. 3A–C) described small amounts of cochleosaurid postcranium from two Euro-American sites. The material from Florence, Nova Scotia, Canada, which Klembara described, is probably *Cochleosaurus florensis*, since it was found at the same locality as the type specimen described by Rieppel (1980). Hook (1993) erected a new species of the cochleosaurid genus *Chenoprosopus*, namely *C. lewisi*, based on a small skull and associated postcranium from North Central Texas, USA. More recently, Reisz *et al.* (2005) described a new specimen of *Chenoprosopus milleri* from New Mexico, USA and reassessed data from Langston's (1953) earlier description of that species. The new data shows that Hook's characters previously distinguishing *C. lewisi* from *C. milleri* are actually due to differences in preservation, ontogeny and biological variation and that the type of *C. lewisi* is a juvenile specimen of *C. milleri*. That synonymy is followed in this text. Fragmentary postcranium has been recovered with another new Upper Permian cochleosaurid, *Nigerpeton riqclesi*, from Niger (Sidor *et al.* 2005; Steyer *et al.* 2006), but those three presacral vertebrae, associated ribs and an isolated femur remain unfigured.

Despite cochleosaurids having been documented for over a century, our knowledge of their postcranial anatomy remains incomplete. Although the monophyly of the Cochleosauridae is unequivocal, their association with the very much larger, single specimen of *Edops* in the Edopoidea is weakly supported. Romer and Witter (1942) considered *Edops* to be closely related to *C. bohemicus* and *Dendrerpeton* on the basis of similar cranial anatomy and the Edopoidea is still defined by skull characters which may be size-linked. Data from this study now provides a standard against which to assess the postcranium of *Edops* and other basal temnospondyls for the first time in detail.

SYSTEMATIC PALAEONTOLOGY

Institutional abbreviations. AMNH, American Museum of Natural History, New York, USA; BMNH, Department of Palaeontology, Natural History Museum, London, UK; MB, Museum für Naturkunde, Humboldt Universität, Berlin, Germany; 'MM', herein, an informally designated, unregistered specimen from the Geologisches Museum, Marburg, Germany; NMP, Národní Museum, Prague, Czech Republic; NMW, Naturhistorisches Museum, Vienna; ZP, Západočéské Museum, Plzeň, Czech Republic.

Text-figure abbreviations. act, acetabulum; add, adductor crest; at, atlas vertebra; ax, axis vertebra; c1–c4, centralia 1–4; cla, clavicle; cth, cleithrum; d1–d4, distal carpals 1–4; ent.for., entepicondylar foramen; fem, femur; fib, fibula; gas, gastralia; hum, humerus; ic, intercentrum; icl, interclavicle; il, ilium; in, intermedium; isc, ischium; na, neural arch; ns, neural spine; ole, olecranon process; pc, pleurocentrum; ph, phalanx; pro, proatlas; r, rib; ra, radius; rdl, radiale; sc, scapulocoracoid; sr, sacral rib; tib, tibia; tp, terminal phalanx; ul, ulna; uln, ulnare.

AMPHIBIA Linnaeus, 1758
TEMNOSPONDYLI von Zittel, 1888
EDOPOIDEA Romer, 1945 (as EDOPSOIDEA)
emended Langston, 1953
COCHLEOSAURIDAE von Zittel and Broili, 1923

COCHLEOSAURUS Frič, 1885

Type species. Cochleosaurus bohemicus Frič, 1885.

Holotype. NMP M540/NMP M1268, part and counterpart of a skull in dorsal view. NMP M540 retains a small amount of bone anterior to the orbits, but the posterior half of the specimen is a plaster cast. NMP M1268 is a natural mould in coal of the type skull, lacking the anterior snout region.

Locality and horizon. The Humboldt Mine, Nýřany, Czech Republic. Plattelkohle (= laminated canneloid shales and mudstones) from the base of the Nýřany Gaskohle series, which is part of the Nýřany Member of the Lower Grey Beds of the Plzeň limnic basin. Upper Carboniferous, uppermost Asturian (=Westphalian D).

Diagnosis. Cochleosaurus bohemicus has two autapomorphic cranial characters: postorbital covered in fine pits on the ventral surface; postparietal lappets large and spatulate with both dorsal and ventral dermal ornament (Sequeira 2004).

No autapomorphic postcranial characters have been identified in this study.

Cochleosaurus bohemicus Frič, 1885
Text-figures 1–12

Material. The following specimens form the basis of the description herein, with some reference to other, unfigured individuals. Most specimens are described from silicone or latex peels of natural impressions in a coal matrix, except where otherwise indicated.

The postcranial material is associated with determinate *Cochleosaurus bohemicus* specimens, mainly individuals of skull length less than 100 mm. Unfortunately, there are far fewer large *C. bohemicus* individuals from Nýřany and consequently adult postcranial remains are very rare.

Although the isolated temnospondyl manus (NMPM1270) described herein cannot be certainly attributed to *C. bohemicus*,

that is its most likely source, given the relative abundance of the genus in the Nýřany fauna.

Midline skull length (msl) is given below.

Specimens of skull length less than 100 mm
AMNH23363/BMNH R2823. (Text-figs 1, 2A–B; Broili 1905, fig. 2, table 2; Steen 1938, pl. 6, text-fig. 29). (msl 67 mm). Part and counterpart. Skull and postcranium in dorsal and ventral aspect. Partial pectoral and pelvic girdles, forelimbs and hindlimbs; anterior vertebrae and ribs, gastralia.

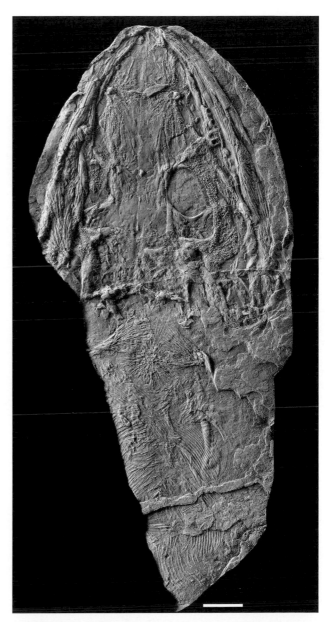

TEXT-FIG. 1. *Cochleosaurus bohemicus* Frič, BMNH R2823 (cpt AMNH23363). Photograph of a cast of skull and postcranium in ventral view. Scale bar represents 10 mm.

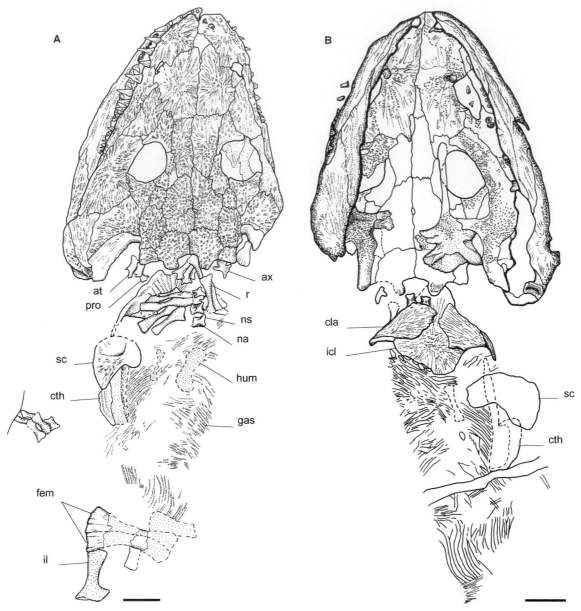

TEXT-FIG. 2. *Cochleosaurus bohemicus* Frič. A, AMNH23363 (cpt BMNH R2823). Skull and postcranium in dorsal view, including proatlas-atlas-axis complex. B, BMNH R2823. Ventral view. Scale bar represents 10 mm.

MBAm25. (Text-figs 6A–C, 8A–B). (msl 36 mm). Skull and postcranium in dorsal aspect. Partial pectoral and pelvic girdle, incomplete forelimbs and hindlimbs, gastralia.

MBAm50. (Text-figs 4A–B, 9A–B, 10A, 12). (msl 22 mm). Skull and anterior postcranium in dorsal aspect. Partial pectoral girdle, forelimbs, anterior vertebrae and gastralia.

'MM1'. (Text-fig. 3A). (msl 24 mm). An unregistered specimen in the Geologisches Museum, Marburg, informally designated as MM1 herein. Skull and anterior postcranium. Partial pectoral girdle, ribs, anterior vertebrae and gastralia. Redrawn (with permission) from an original unpublished drawing by A. R. Milner.

NMPM1270. (Text-figs 10B–C, 12). (msl 80 mm est.). Isolated temnospondyl left manus, probably *Cochleosaurus bohemicus* or

Capetus palustris. Redrawn (with permission) from an original unpublished drawing by A. R. Milner.

ZP1367. (Text-fig. 3B). (msl 80 mm est.). Skull (posterior skull table), partial pectoral girdle and humeri, anterior ribs and vertebrae.

Specimens of skull length more than 100 mm
MBAm81. (Text-figs 3C–D, 10A). (msl 110 mm). Partial skull in dorsal aspect, incomplete forelimb and hindlimb, anterior vertebrae, gastralia.

MBAm83.1/MBAm83.2 (Text-figs 7A–B, 8D, 11A–C). (msl 105 mm). Part and counterpart, comprising some original bone and natural moulds in coal matrix. Skull in dorsal aspect, interclavicle, rib, partial hindlimb and gastralia.

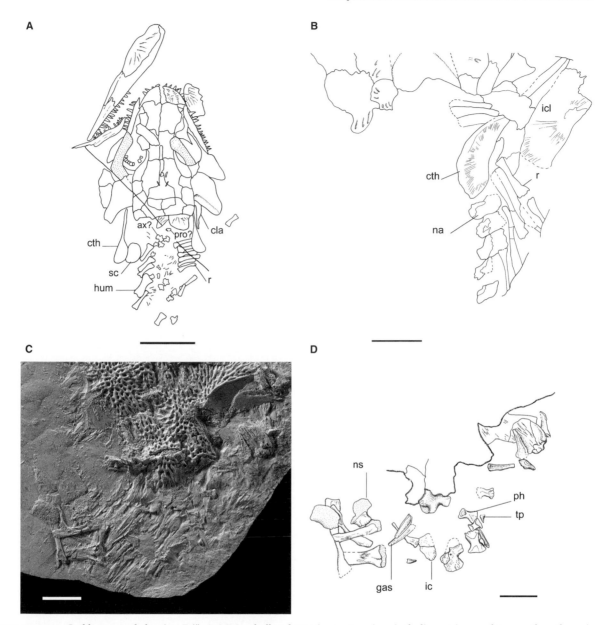

TEXT-FIG. 3. *Cochleosaurus bohemicus* Frič. A, MM1, skull and anterior postcranium including an incomplete proatlas-atlas-axis complex. Redrawn (with permission) from an unpublished illustration by A. R. Milner. B, ZP1367, anterior vertebrae and partial pectoral girdle. C, MBAm81, intercentrum and disarticulated phalanges. D, MBAm81, interpretive drawing. Scale bar represents 10 mm.

NMPM631. (Text-figs 5A–B, 7C, 8C, 12; Fritsch 1879–1901, fig. 391). (msl 172 mm est.). Plaster cast of a posterior skull table in dorsal aspect, incomplete pectoral girdle and forelimb, vertebrae associated with ribs, ilium, gastralia.

NMW 1899 III 11a/b. Unfigured herein. (120 mm est.). Part and counterpart; partial skull and dissociated forelimb.

Description

Vertebrae
The rhachitomous trunk vertebrae of *C. bohemicus* are structurally similar to those of other basal temnospondyls, includ-ing *Balanerpeton* and *Dendrerpeton*. Each *C. bohemicus* vertebra comprises a well-ossified neural arch associated with a single crescentic intercentrum and smaller, paired pleurocentra.

Proatlas-atlas-axis complex. Certain identification of the proat-las-atlas complex is difficult, since the anteriormost cervical vertebrae are only adequately preserved in one specimen (AMNH23363/ BMNH R2823, Text-figs 1, 2A–B). Broili (1905, fig. 1) illustrated a small skull with 8–9 cervical vertebrae, but unfortunately that specimen was destroyed in World War 2 and the drawing is not very informative.

In AMNH23363/ BMNH R2823, the most likely interpre-tation is that the complex is comprised of a series of three associated bones, lying adjacent to the left tabular. The smallest of these is a single, saddle-shaped proatlas, with a small neural spine. It is similar in outline to the proatlas of *Balanerpeton woodi* (Milner and Sequeira 1994, fig. 9) and *Archegosaurus decheni* (Witzmann and Schoch 2006, Text-fig. 1A–B). An isolated proatlas may also be present in MM1 (Text-fig. 3A).

The atlas neural arches are also incompletely preserved, with some bone missing along their ventral margins. The neural arches are paired and unfused; they extend posterodorsally as tall and rather broad neural spines. The prezygapophyses are well-defined, but the postzygapophyses cannot be seen, being are either poorly developed or damaged.

From its size and position, the robust and relatively large vertebra underlying the right tabular in AMNH23363 (Text-fig. 2A) appears to be the displaced axis. It is unpaired, suggesting that it comprised fused neural arches. Although partly obscured, its neural spine is broad and is at least as tall as wide. The anterior zygapophysis is clearly differentiated, but this is somewhat exaggerated by the backwards displacement or possible fragmentation of the ventral section of the neural arch. In a much smaller individual (MM1, Text-fig. 3A), the anterior cervical vertebrae, including the presumed axis, are paired and not co-ossified.

Trunk vertebrae. Five small specimens retain a few vertebrae near their pectoral girdles (Text-figs 1, 2A–B, 3A–D, 4A–B); a sixth, larger, individual (Text-fig. 5A–B) has an almost complete vertebral series reaching from the pectoral girdle to the ilium. Unfortunately, this specimen is only known from a plaster cast, which retains indistinct impressions of their detailed structure.

Neural arches are approximately rectangular in lateral view, with broad neural spines which occupy about half their total vertebral height. There is little, if any, variation in the dimensions of the neural arches in the available specimens and none of the four postaxial vertebrae (AMNH23363/BMNH R2823, Text-figs 1, 2A–B) seems to have a reduced neural spine. This contrasts with several other temnospondyls (*Balanerpeton*, *Trimerorhachis*, *Eryops* (Moulton 1974), *Archegosaurus*) in which the neural spine of the fourth trunk vertebra (or fifth vertebra, in *Balanerpeton*) is shorter than that of adjacent vertebrae. In the two smallest individuals, with skulls less than 25 mm long (MM1, MBAm50, Text-figs 3A, 4A–B), the neural arches are paired and unfused. However, in larger specimens of skull length more than 65 mm (Text-figs 1, 2A–B, 3B, 5A–B), neural arches only occur singly, suggesting that they have co-ossified by this stage of ontogeny.

Prezygapophyses and postzygapophyses are developed even in small specimens. A small diapophysis can be seen in two vertebrae (AMNH23363, Text-fig. 2A) beneath a weakly developed transverse process.

The intercentrum (Text-fig. 5A–B) is large, well-ossified and about half the total height of the whole neural arch. Its flattened outline suggests a crescentic three-dimensional shape. There is a transverse midline suture across its internal surface, where the two intercentrum halves have fused (MB Am81, Text-fig. 3D). Bipartite intercentra are also recorded for *Balanerpeton*.

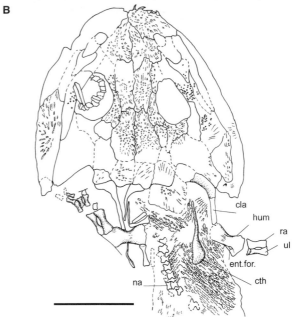

TEXT-FIG. 4. *Cochleosaurus bohemicus* Frič. A, MBAm50, skull and anterior postcranium in dorsal view, original bone, entepicondylar foramen visible on right humerus. B, interpretive drawing of A. Scale bar represents 10 mm.

Pleurocentra are ovoid and are approximately half the intercentrum height in the largest available specimen (NMPM631, Text-fig. 5A–B). As is common in temnospondyls, they were almost certainly paired bones, although no specimen shows this clearly.

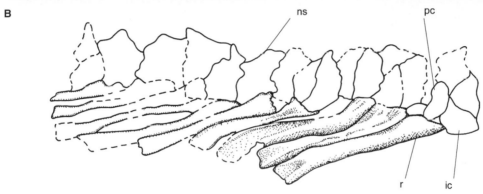

TEXT-FIG. 5. *Cochleosaurus bohemicus* Frič. A, NMPM631, plaster cast of posterior skull, partial pectoral girdle, vertebrae, ribs and ilium. B, interpretive drawing of A, anterior vertebrae and ribs. Scale bar represents 10 mm.

It is possible to estimate the presacral vertebral count from a series of associated vertebrae and ribs (Text-fig. 5A–B). A measurement taken along eight of these vertebrae and their associated ribs spans one-third of the distance from the back of the skull to the ilium. Given that vertebral dimensions remain constant along the spinal column, this strongly suggests that *C. bohemicus* had a total of 24 presacral vertebrae.

Caudal vertebrae. The caudal vertebrae are not preserved with any available specimens.

Ribs
The anterior 5–6 pairs of thoracic ribs are preserved in AMNH23363/BMNH R2823 and MM1 (Text-figs 2A–B, 3A).

These ribs are short, with an almost straight central shaft which differentiates into a broad proximal articular head and a slightly less expanded distal end. In the largest specimen (NMPM631, Text-fig. 5A–B), ribs near the pectoral girdle are very robust, with the distal portion of the ribs being as broad as their proximal articular heads. Ribs posterior to this series are narrower distally, but otherwise, their detailed outline is poorly preserved.

None of the ribs is differentiated into capitulum and tuberculum and there are no uncinate processes.

It has not been possible to definitely identify the sacral ribs. However, there may be a faint impression of a sacral rib on the cast of MBAm25 (Text-fig. 6A, C), near the disrupted hindlimb bones.

None of the caudal ribs has been preserved.

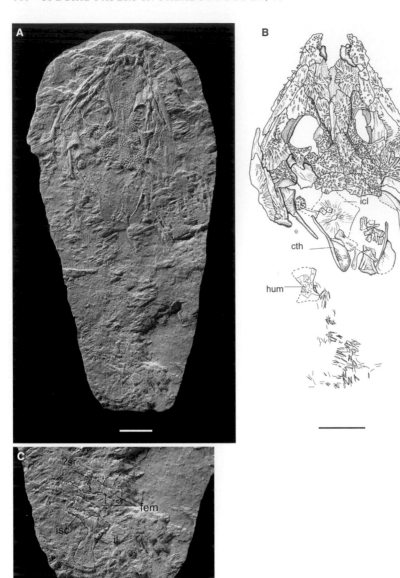

TEXT-FIG. 6. *Cochleosaurus bohemicus* Frič. A, MBAm25, skull in dorsal view, partial pectoral and pelvic girdles, including ilium and possible sacral rib. B, interpretive drawing of anterior region of A. C, MBAm25. Ilium, ischium, femora and possible sacral rib outlined. Scale bar represents 10 mm.

Pectoral girdle. The pectoral girdle comprises an unpaired, median, interclavicle and paired clavicles and cleithra. These bones are ossified even in the smallest individuals of skull length about 22 mm.

Interclavicle. The rhomboidal interclavicle is a relatively small bone, of length approximately one-third midline skull length. In the smaller specimens (Text-figs 1, 2A–B, 3A–B, 4A–B, 6A–B), the interclavicle is about 1.1–1.2 times longer than wide, but these proportions become equal in larger specimens (Text-figs 5A, 7A–C). The shape of the bone changes slightly with growth as the anterolateral edges fill out, giving an almost semicircular outline to its leading edge.

Dermal ornament of typical temnospondyl pattern is restricted to the ventral surface of the bone. Its centre of ossification is marked by a concentration of small dermal pits; these grade into narrow grooves which extend to the bone edges. This ornament pattern is modified in the anterior half of the interclavicle. Anterolateral to the ossification centre, triangular areas of smooth bone with some weakly developed striations mark the location of small articular facets for the clavicles. The smooth bone anteromedial to these facets is marked by narrow, fimbriate striae, which radiate to the anterior interclavicle margin (Text-figs 7A–C).

Clavicle. Each clavicle comprises an approximately triangular ventral blade, of maximum width about half the blade length and a very slender, posterodorsally orientated shaft (BMNH R2823, MBAm25, NMPM631, Text-figs 1, 2A–B, 3A, 4A–B, 5A, 6A–B, 7C). The narrow shaft is at least as long as the clavicle blade and angles posterodorsally upwards. The clavicle blades articulate with the interclavicle adjacent to its centre of ossifica-

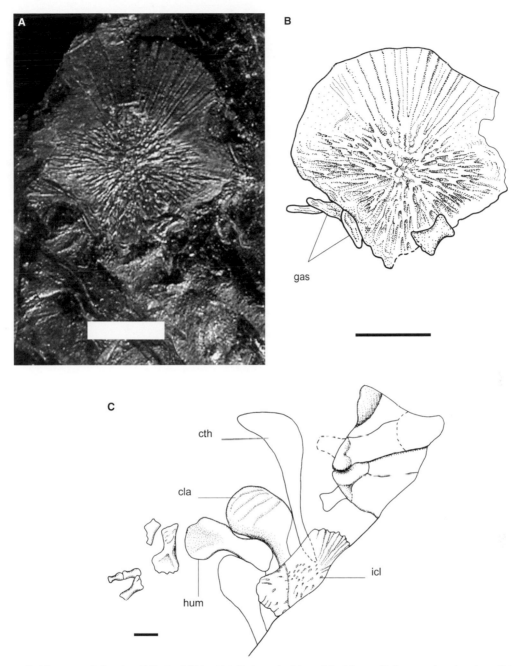

TEXT-FIG. 7. *Cochleosaurus bohemicus* Frič. A, MBAm83.1/2, interclavicle, original bone. B, interpretive drawing of A. C, NMPM631, interpretive drawing, partial pectoral girdle. Scale bar represents 10 mm.

tion, each blade occupying about one-third of the interclavicle width at this point (Text-figs 1, 2A–B, 7A–B). Dermal ornament is present only on the ventral surface of the clavicle blade.

Cleithrum. Cleithra differ only slightly in their shape and proportions from clavicles. The cleithral blade is narrower and there is no curvature of the shaft at its junction with the blade (Text-figs 1, 2A–B, 3A–B, 4A–B, 5A, 6A–B, 7C). No dermal ornament is present on the cleithra in any specimens.

Scapulocoracoid. The scapulocoracoid is only preserved in two specimens. In the smaller individual (MM1, Text-fig. 3A) of skull length *c.* 24 mm, the bone has a regular and broadly ovoid shape. There is no differentiation of the scapulocoracoid at this stage into the kidney-shaped outline characteristic of larger temnospondyls.

In the larger individual (AMNH23363/BMNH R2823, Text-figs 1, 2A–B), of skull length *c.* 67 mm, the bone has developed a more reniform shape, with a concave posteroventral edge continuing anterodorsally as a broadly convex blade. The thickened supraglenoid buttress grades dorsally into the more slender, semicircular blade. The incompletely ossified anterior margin of the blade suggests that it was comprised of cartilage in that region. No foramina can be seen in these specimens.

Pelvic girdle

Ilium. The shape of the ilium is typical for primitive temnospondyls; it comprises a broadly triangular ilium base grading into a simple dorsal blade with no dorsal process. The ilium is already well-ossified even in small specimens (Text-figs 2A, 6A, C, 8A–B). In these individuals, the dorsal blade is broad, being between 0.6 and 0.7 times the width of the ilium base, but in the largest specimen (NMPM631, Text-figs 5A, 8C) the blade is only 0.3 times

baseplate width. Dorsal blade height represents 0.5–0.6 total ilium height in all specimens.

The ventral margin of the baseplate is almost semicircular in MBAm25 (Text-figs 6A, C, 8A–B) but becomes more differentiated in larger specimens. During ontogeny, the baseplate edge becomes obliquely angled at the junction of its common suture with the pubis and ischium.

The acetabulum is visible on the lateral side of the ilia in the two larger specimens (Text-figs 5A, 8C–D). In NMPM631, the acetabulum has a narrow oval outline, which is strengthened by a raised bone ridge along its dorsal border. The fragmentary ilium of MBAm83.1/2 (Text-fig. 8D) has a broader and more circular acetabulum, but this may be a the result of crushing. The smallest ilium (MBAm25, Text-figs 6A, C, 8A–B) is also difficult to interpret, but it seems from its position and

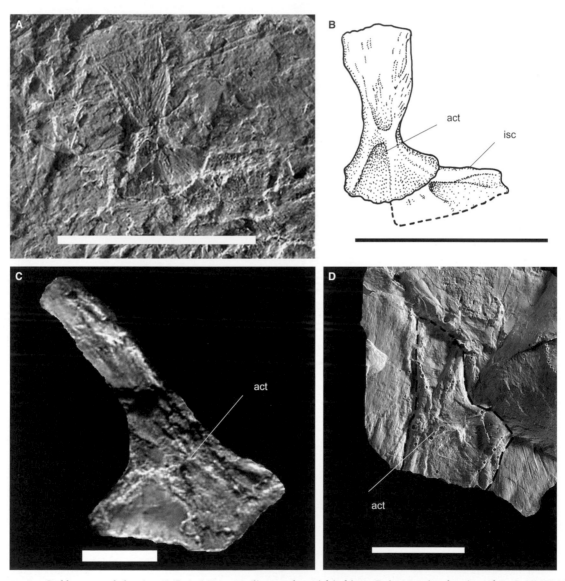

TEXT-FIG. 8. *Cochleosaurus bohemicus* Frič. A, MBAm25, ilium and partial ischium. B, interpretive drawing of A. C, NMPM631, right ilium, lateral view, showing acetabulum. D, MBAm83.1/2, fragmentary left ilium in lateral view, showing acetabulum. Scale bar represents 10 mm.

shape to be a left ilium seen in lateral view. The collapsed area at base of the dorsal blade has a thickened, arched, outline which may be the dorsal margin of the acetabulum.

Ischium. A faint impression of an ossified ischium is discernible on the cast of MBAm25, lying posterior to the baseplate of the ilium (Text-figs 6A, C, 8A–B). It is surrounded by a series of impressions which seem to be those of the associated hindlimb bones, but these are too indistinct to be certainly identifiable. One plausible interpretation shows a possible femur, tibia, fibula and sacral rib outlined.

Pubis. The pubis is not preserved in any specimen of *C. bohemicus* seen in this study.

Klembara (1985) described and figured temnospondyl postcranial remains comprising a pelvic girdle and rear limbs from Florence, Nova Scotia. As he noted, it seems likely that these are attributable to *Cochleosaurus florensis* (Rieppel 1980), since they derive from the same locality as that taxon. In the Florence specimen, the ilium is about 5 cm high and the pubis is large and fully ossified. This contrasts with a large *C. bohemicus* specimen with a 4 cm ilium (Text-figs 5A, 8C), from which an ossified pubis has not been recovered. This suggests that the pubes in *C. bohemicus* were unossified, even in larger specimens. It seems probable that in *C. florensis* the ossified pubes were an adaptation for its more terrestrial mode of life.

Limbs

Forelimb

The forelimbs are incomplete in all available specimens. The most informative specimen is the smallest individual (MBAm50, Text-figs 4A–B, 9A–B, 10A), which only lacks carpals.

Humerus. The 5 mm long humerus is already well-ossified in a small individual of skull length *c.* 22 mm (Text-figs 4A–B, 9A–B, 10A), in which the original bone remains preserved on the coal matrix. It is a stout bone, with a short, relatively wide shaft. Both its proximal and distal ends are equally expanded as broad, spatulate condyles in small specimens (Text-figs 3A, 4A–B, 6A–B, 9A–B). The ends of the condyles are incompletely ossified and would have been capped by cartilage in these small individuals. Condyle shape is modified in larger individuals, with the distal condyle becoming relatively wider (Text-figs 5A, 7C). The proximal and distal bone ends are only slightly rotated about the axis of the shaft and lie at an angle of about 30 degrees relative to each other.

An entepicondylar foramen is present on the right humerus in MBAm50 (Text-figs 4A–B, 9A–B, 10A). The bone has been slightly crushed and the distal condyle displaced, so that the foramen is visible on the posterior bone surface. It is probable that the foramen pierced the bone, to exit on its anterolateral surface, but this region is not exposed in either of the humeri. There is thus also no information about the supinator process and deltopectoral crest. There is apparently no entepicondylar foramen in the 3 cm humerus from a larger specimen (Text-figs 5A, 7C); however, the poor quality of that plaster cast limits certain interpretation.

Radius and ulna. The radius and ulna are of just over half the humerus length in small specimens and the ulna is marginally longer than the radius. There is already some obvious differentiation between the radius and ulna in MBAm50 (Text-figs 4A–B, 9A–B). The ulna shaft is about half the width of that of the radius, and its medial edge is more concave than the corresponding radius edge. There is a very small olecranon process on the proximal end of the ulna; the distal bone end is slightly broader. The radius is more symmetrical and has a broader shaft and articular condyles.

Manus. The manus is incompletely preserved in all available specimens, with most detail deriving from MBAm50 (Text-figs 4A–B, 9A–B, 10A, 12) and a larger individual, MBAm81 (Text-figs 3C–D, 10A, 12). In MBAm50, the retained left manus comprises four digits, which is the common temnospondyl condition. There are no ossified carpals in this or any other specimen examined. All four metacarpals are present, although that of digit 1 is damaged and has been displaced backwards towards the radius. The metacarpals are expanded at both their proximal and distal ends and have slightly waisted central shafts. An incomplete series of slender phalanges remains associated on digits 2, 3 and 4, but no terminal phalanges are preserved in this specimen. In MBAm81 (Text-fig. 3C–D), a disrupted but complete digit retains 3 phalanges, including the small terminal claw. This is thus either digit 3 or 4; it is probably digit 3 because of the relatively large sizes of its phalanges.

Combining the information above, the phalangeal formula for *C. bohemicus* is at least 1+, 1+, 3, 2+, and was probably 2, 2, 3, 3 when complete. This is the most common formula for the temnospondyl manus.

Among the fossils recovered from Nýřany is an isolated, almost complete, temnospondyl manus, and associated radius and ulna (NMPM1270, Text-fig. 10B–C). All the recovered bones in this left manus, including the carpals, are well-ossified, with only carpal 1 and distal carpals 1 (or 2) and 4 missing. This partial forelimb is about 38 mm long and would probably have a total length of 50 mm when complete with a humerus. By comparison to MBAm50, these ratios of limb length: skull length indicate that the manus came from a specimen with a msl of 70 mm or more.

Because of its relatively large size and degree of ossification, there are four realistic attributions within the Nýřany fauna for this manus. Two large temnospondyl genera, *C. bohemicus* and *Capetus palustris*, were both present at Nýřany, and are estimated to have grown to about 1.5 m total length, based on their maximum known skull lengths of 300–400 mm. They would have attained some degree of terrestriality at such sizes. Of these two taxa, *C. bohemicus* is by far the more common, with almost 100 specimens known from the locality. In contrast, *Capetus* is represented by only eight specimens. Two other tetrapods, *Baphetes* and *Mordex* must also be considered as possible sources for the manus. However, both are very rare components at Nýřany, with only two *Baphetes* and a few *Mordex* specimens recorded from the assemblage. The simplest explanation is that the manus probably derives from *C. bohemicus*.

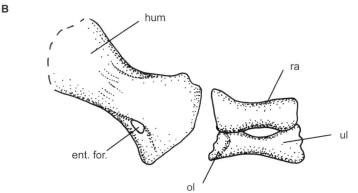

TEXT-FIG. 9. *Cochleosaurus bohemicus* Frič. A, MBAm50, right forelimb, humerus with entepicondylar foramen, radius and ulna. B, interpretive drawing of A. Scale bar represents 10 mm.

Hindlimb

Femur. The femur is described from three specimens (Text-figs 2A, 6A, C, 11A–C). It is the largest epipodial, being about twice the length of the humerus in AMNH23363/BMNH R2823 (Text-fig. 2A). In that individual, just the ends of the paired femora remain as firm impressions in the matrix and although their outline shape can be discerned, no detail remains. The femur of the smaller specimen MBAm25 (Text-fig. 6A, C) is slightly more informative. The bone is about the same length as the ilium and the femoral shaft is half the width of its distal condyle. No other detail can be discerned.

In MBAm83.1/2, a 38 mm long, more complete, femur is preserved in ventral view (Text-fig. 11A–C). There is a small internal trochanter just below its proximal articular head which is continuous with the narrow adductor crest extending down towards the distal condyle. Because the bone has been flattened, there is no indication of the popliteal fossa structure. Both the proximal and distal condyle margins are irregular in outline, implying that growth was still ongoing. In life they would have been capped with cartilage.

Cochleosaurus florensis. Klembara (1985, fig. 9) figured a femur associated with the probable *C. florensis* pelvis from Nova Scotia. He described the femur of *C. florensis* as having a distal condyle which is considerably wider than its proximal articulation; there is also a very broad and deep popliteal fossa. This contrasts with femur of *C. bohemicus*, in which the distal condyle is about two-thirds the width of the proximal condyle and the popliteal fossa was probably less robust, since it has apparently not withstood fossilization. These anatomical differences between the pelvis and femur support the concept of *C. florensis* as a more terrestrialized form than *C. bohemicus*.

Outlines of the tibia and fibula are visible in MBAm83.1 (Text-fig. 11A), where they lie adjacent to the incompletely preserved second femur. They are each of similar length and both are about three-quarters the femoral length. The proximal and distal condyles of the tibia are broad, being about twice the shaft width. The fibula is partly obscured and is too poorly preserved to yield more information.

Bones of the pes have not been preserved in the available specimens.

Gastralia

Several specimens, including the smallest individuals, retain a series of well-ossified gastralia (Text-figs 1, 2A–B, 3A, C–D, 4A–B, 5A, 6A–C, 7A–B, 9A). Individual scales are slender and ovoid, with distinct concentric growth rings on their ventral surface (Text-figs 3C–D, 7A–B, 9A). Their length ranges from

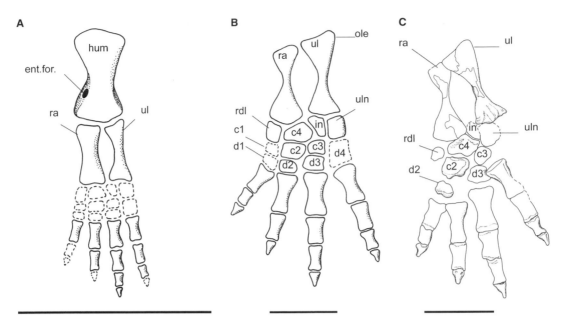

TEXT-FIG. 10. *Cochleosaurus bohemicus* Frič. A, reconstruction, small manus, based mainly on MBAm50 and MBAm81. B, NMPM1270, reconstruction, large manus. C, NMPM1270, redrawn from A. R. Milner (with permission), showing ossified carpals. Scale bar represents 10 mm.

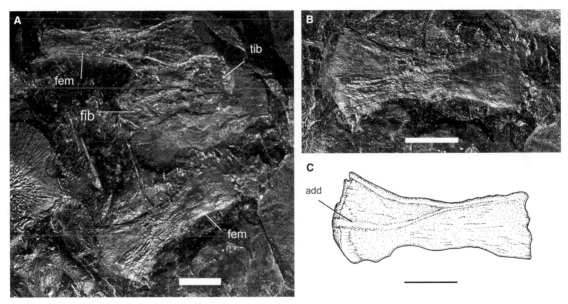

TEXT-FIG. 11. *Cochleosaurus bohemicus* Frič. A, MBAm83.1/2, femora, tibia and fibula, some original bone remaining. B, MBAm83.1/2, close-up of femur, showing internal trochanter and adductor fossa. C, interpretive drawing of B. Scale bar represents 10 mm.

2.5 mm to 7 mm. There are at least 48 rows of these scales, arranged in contralateral rows, which overlap at and radiate away from the ventral midline towards the dorsal surface. The most anterior 12–14 rows are aligned parallel to the posterior interclavicle edges, but there is an abrupt change in their angle of orientation at about the level of the forelimbs (Text-figs 1, 2A–B, 4A–B). Here, the scale rows have fanned out and lie at 90 degrees to the midline. Behind the level of the forelimbs,

the rows are posterodorsally angled and slope upwards at about 45 degrees to the midline. The first row of scales articulates directly with the posterior interclavicle margins; they are large and more strongly ossified than the succeeding gastralia. These scales seem to have been firmly attached to the interclavicle, and some remain *in situ* in two specimens (Text-figs 1, 2B, 7A–B).

No dorsal scutes have been identified.

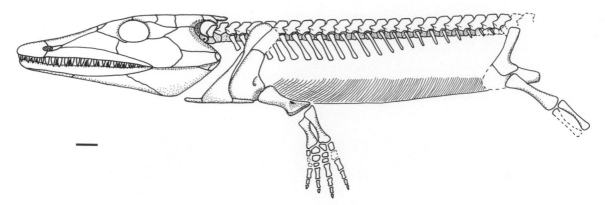

TEXT-FIG. 12. *Cochleosaurus bohemicus* Frič. Reconstruction. Axial skeleton and pelvis based mainly on NMPM631, manus based on NMPM1270, humerus based on small specimen MBAm50. Scale bar represents 10 mm.

ONTOGENY

The relatively large size-range (22–172 mm msl) of well-preserved specimens facilitates identification of some ontogenetic changes in *C. bohemicus*. Modifications affect either the degree of ossification or the shape of bones, specifically the anterior vertebrae, ilium and humerus. The rate of ossification varies slightly in the available specimens, but there is no evidence of a substantive metamorphosis in *C. bohemicus*. However, in the absence of very small specimens (msl less than 20 mm), their morphological changes remain unknown, so this is a speculative conclusion at present.

Anterior dorsal vertebrae in specimens of less than 36 mm msl are incompletely ossified and the bipartite neural arches are unfused. Ossification is more complete and neural arch fusion has occurred in larger individuals with skulls of 66 mm msl.

Primitively, an entepicondylar foramen is present in the smallest humerus, but this seems to be closed in the largest humerus from an individual of estimated skull length 172 mm. The great size disparity between the available humeri from basal temnospondyls (5–180 mm long) complicates their comparison, since it is not possible to identify size-dependent changes within each taxon with certainty. The 5 mm humerus of *C. bohemicus* retains an entepicondylar foramen which may be lost in larger individuals, but this interpretation requires further confirmation. In contrast, larger humeri in *Balanerpeton* and *Dendrerpeton* retain an entepicondylar foramen, but it is absent in *Edops* and *Trimerorhachis*. Absence of the entepicondylar foramen may be size-linked in the very large *Edops* humerus (180 mm), but without comparative smaller *Edops* specimens that remains speculative.

The ilium of *C. bohemicus* undergoes the greatest ontogenetic modification. The dorsal blade changes from its very wide, upright shape in the smallest specimen to become much more slender and posterodorsally angled in the largest individual.

DISCUSSION

Cochleosaurus bohemicus has no postcranial apomorphies and lacks specific terrestrial or aquatic adaptations present in several other basal taxa. The lifestyles of *Balanerpeton woodi* (Milner and Sequeira 1994) and *Dendrerpeton acadianum* (Carroll 1967; Holmes *et al.* 1998) are considered to be more or less terrestrial, whilst *Trimerorhachis insignis* (Williston 1916; Pawley 2007) is mainly aquatic. *C. bohemicus* has an estimated 24 presacral, rhachitomous, well-ossified vertebrae which, with the exception of the proatlas-atlas complex, show no regional size differences, contra *Balanerpeton*, *Dendrerpeton* and *Trimerorhachis*. Ribs, where known in these taxa, lack bicipital articular heads or uncinate processes. However, three large temnospondyl ribs found in Carboniferous deposits at East Kirkton, which may be from *Edops* (Milner and Sequeira 1994), have bicipital heads and uncinate processes, suggestive of a more highly developed degree of terrestriality. The simple axial skeleton of *C. bohemicus* suggests that its load-bearing and locomotor capability was limited, although sufficiently well-developed to allow terrestrial locomotion.

Pectoral girdle structure in *C. bohemicus* is most similar to *Balanerpeton*. Both taxa have relatively small interclavicles with wide, pectinate anterior plates and small clavicular facets. Significant modifications exist in *Dendrerpeton*, where the pectinate anterior interclavicle plate is narrower and more strongly arched and clavicles occupy about three-quarters of the central interclavicle area. *Trimerorhachis* has a highly modified interclavicle with a very broad posterior plate, a very tall clavicle blade and an incompletely ossified supraglenoid buttress. Again, the simple girdle in *C. bohemicus* suggests limited terrestriality, without great development of features associated with increased musculature.

The manus is known only for *C. bohemicus*, *Balanerpeton* and *Dendrerpeton*. The bones of the hand, including the carpals, are well-ossified in these taxa, but in the smallest *C. bohemicus* manus, the carpals were unossified. However, a larger, almost complete manus (probably *C. bohemicus*) has substantial carpals. The development of a fully ossified manus in larger *C. bohemicus* individuals would have allowed increased terrestrial locomotion.

The pubes are unossified in *C. bohemicus*, significantly reducing the terrestrial load-bearing capability of its pelvic girdle. Ossified pubes occur in *C. florensis* (Klembara 1985), *Balanerpeton* and *Dendrerpeton*. The simple dorsal iliac blade of *C. bohemicus* changes shape with growth, becoming more slender and more posterodorsally oriented. That final shape is shared with *Balanerpeton* and *Dendrerpeton*. However, *C. florensis* retains a relatively broad dorsal blade margin, and in *Trimerorhachis* the dorsal iliac process is very wide. Acetabulum structure is similar in all these taxa, except *Trimerorhachis*, where it is incompletely ossified ventrally, and *Edops craigi* (Romer and Witter 1942), where a large bone ridge passes diagonally above the dorsal acetabulum edge. The femora of *C. bohemicus* are strikingly unmodified and lack additional areas for muscle attachment developed in more derived taxa. In *Cochleosaurus florensis*, *Chenoprosopus milleri* (Hook 1993, as *Chenoprosopus lewisi*; Reisz *et al.* 2005), *Balanerpeton* and *Dendrerpeton*, condyles are significantly broader; in those taxa and in *Trimerorhachis* there is variable development of the intercondylar fossa and internal trochanter. The adductor crest expands as a broad sheet of bone in *Balanerpeton* and *Dendrerpeton*. Femora in *Edops* and *Trimerorhachis* lack flanges and condyles are modestly expanded. The isolated femur tentatively referred to *Nigerpeton ricqlesi* (Sidor *et al.* 2005; Steyer *et al.* 2006), is undescribed. The poorly preserved tibia and fibula of *C. bohemicus* resemble those of other primitive temnospondyls.

The simple postcranium of *C. bohemicus* contrasts with the range of adaptive morphologies found in the more derived sister-taxa. Greater terrestriality in *Balanerpeton* and *Dendrerpeton* is evidenced by increasing complexity of their vertebrae, interclavicles (in *Dendrerpeton*), pelves and femora, with consequent increases in musculoskeletal load-bearing capacity. The retention of specific aquatic adaptations at large size in *Trimerorhachis* (32 presacral vertebrae, massive interclavicle, incomplete ossification) differs from developmental patterns found in *C. bohemicus*, where gradual ossification occurs throughout growth. Habitat comparison also supports the morphological evidence for increasing specialization in taxa crownwards of *C. bohemicus*. *Dendrerpeton* mainly derives from fossilized stumps of lycopsid trees at Joggins, Nova Scotia, indicating an ability to move some distance on land; that locality has also yielded a predominantly terrestrial fauna (Carroll *et al.* 1972; Holmes *et al.* 1998). *Balanerpeton*, from Viséan deposits at East Kirkton, Scotland, occupied a humid, shallow-lake habitat influenced by hot springs and local volcanism. The habitat supported an aquatic, amphibious and terrestrial fauna; *Balanerpeton* is interpreted as a more terrestrial form (Milner and Sequeira 1994), based on its cranial characters and ossified carpals and tarsals. Most *Trimerorhachis* specimens were found associated with fish and aquatically adapted tetrapods at Early Permian, North American localities in which the palaeoenvironment comprised multiple waterbodies (Sander 1989; Pawley 2007), suggesting it was preferentially, if not an obligate, aquatic animal.

The palaeoenvironment at Nýřany was intermontane, lacustrine and swampy (Pešek 1974; Milner 1980) and somewhat similar to East Kirkton (Clarkson *et al.* 1994); there is also some evidence of possible volcanism at Nýřany, since tuff-like inclusions sometimes occur in the coal matrix near specimens. A sample of 72 *C. bohemicus* individuals from Nýřany includes about 50 small (msl less than 140 mm) specimens, none of which has specific aquatic adaptations. Larger specimens are also unmodified, but more completely ossified. The predominance of juveniles, with occasional larger animals, suggests that the lakes may have been breeding areas to which more terrestrially adapted adults returned seasonally. This evidence suggests that juveniles were obligately aquatic, whilst adults were more amphibious. Increasing longirostry seen in larger *Cochleosaurus* specimens is also suggestive of an amphibious lifestyle; an elongate snout and crocodiloid dentition adapted for piscivory is common amongst aquatic temnospondyls. The recent discovery of a lateral line system in an adult specimen of *Nigerpeton* (Sidor *et al.* 2005; Steyer *et al.* 2006) supports the concept that cochleosaurids were mainly amphibious.

PHYLOGENETIC IMPLICATIONS

Most recent phylogenetic analyses of temnospondyls have been compromised by two main problems: an almost total reliance on cranial character datasets, and difficulty in identifying stable characters unchanged by ontogeny. Outgroup choice for postcranial analyses is also problematic, since the primitive temnospondyl postcranium is so poorly known. As a result, the colosteids have frequently been chosen as outgroups on the basis of their completeness and morphological similarity to temnospondyls. However, it is widely accepted that colosteids form a separate clade, which is not necessarily closely related to temnospondyls (Milner 1990; Ruta *et al.* 2003; Witzmann and Schoch 2006). New information from the basal temnospondyl *C. bohemicus*, which lacks postcranial apomorphies, now provides an alternative outgroup against which to assess temnospondyl character polarity in future analyses. Using *C. bohemicus* as a reference taxon, it should be possible to resolve coding ambiguities resulting from current reliance on derived forms such as *Balanerpeton* and *Dendrerpeton*, in which increasing terrestriality is associated with significant character changes.

Growth changes which occur in the postcranium of *C. bohemicus* highlight the difficulty in identifying characters which are independent of either ontogeny or size-linkage. *C. bohemicus* shows changes in shape and orientation of the ilium blade and possible loss of the entepicondylar foramen in the humerus with increasing size. However, most recent analyses have considered these characters to be apparently stable and to have unambiguous polarity (Milner and Sequeira 1994; Ruta *et al.* 2003; Witzmann and Schoch 2006; Pawley 2007). In particular, retention of an entepicondylar foramen is considered to be plesiomorphic in basal temnospondyls, but its presence or absence may simply be ontogenetically dependent. This remains problematic for *Edops*, where absence of the foramen in a very large humerus may simply be a size-linked loss; there are no small *Edops* specimens for comparison. This complicates our understanding of basal temnospondyl relationships, including those of the Edopoidea.

The Edopoidea currently comprises the Cochleosauridae + *Edops* (Milner and Sequeira 1994; Godfrey and Holmes 1995; Holmes *et al.* 1998; Ruta *et al.* 2003; Sequeira 2004; Steyer *et al.* 2006). Whilst the cladistic integrity of the monophyletic Cochleosauridae is uncontroversial, the monotypic Edopoidea remains weakly defined on the basis of cranial synapomorphies which may be a consequence of large size in the case of the single specimen of *Edops*. Unfortunately, identification of size-linked characters remains problematic within the Edopoidea, since there are no comparably large specimens of *C. bohemicus* known to date. The Edopoidea thus remains poorly defined on the basis of cranial characters.

Debate has also focussed on the lifestyle of the earliest amphibians, particularly whether they were primarily terrestrial or aquatic (Vallin and Laurin 2004; Witzmann and Schoch 2006; Pawley 2007). The preponderance of small *C. bohemicus* individuals with incompletely ossified skeletons recovered from the lacustrine deposits at Nýřany indicates that *C. bohemicus* juveniles were probably primarily aquatic. Absence of ossified pubes in larger *C. bohemicus* specimens suggests that they remained poorly adapted for terrestrial locomotion and retention of enclosed lateral line sulci in *Nigerpeton* (Sidor *et al.* 2005; Steyer *et al.* 2006) implies that adult cochleosaurids were also predominantly aquatic.

Acknowledgements. Andrew Milner's enthusiasm for vertebrate palaeontology, together with his scientific rigour and ready wit combined to encourage, challenge and enlighten his many undergraduate students at Birkbeck, University of London. His stimulating series of lectures convinced me that I would enjoy palaeontological research, and the ideal opportunity to test my skills arrived with the East Kirkton project, when I joined Andrew as a research assistant, with joint responsibility for describing the new Palaeozoic temnospondyl amphibians. That immensely rewarding and successful experience was to be repeated in our subsequent collaborative research projects. I thank Andrew for his moral support and invaluable guidance, and for his confidence in employing me on those numerous grant-funded projects. I am delighted and honoured to join his friends and colleagues in contributing to his Festschrift volume.

I thank the following colleagues for generous advice and technical assistance with the preparation of this paper: Dr Angela Milner (Natural History Museum, London) for access to Museum facilities and permission to study specimens, Dr M. Ruta (Department of Earth Sciences, University of Bristol) and Dr P. Barrett (Natural History Museum, London), reviewers Dr R. Schoch (Staatliches Museum für Naturkunde, Stuttgart, Germany), Dr J. Klembara (Comenius University in Bratislava, Slovakia) and Dr J. S. Steyer (Muséum National d'Histoire Naturelle, Paris) for manuscript improvements. I am grateful to Dr R. Holmes (Redpath Museum, Montreal, Canada) for discussion and access to research specimens and I thank the following staff for access to specimen collections in their care: Dr B. Eckrt (Narodni Museum, Prague, Czech Republic), Ms C. Holton (American Museum of Natural History), Dr G. Hoeck (Naturhistorisches Museum, Vienna, Austria) and the late Dr H. Jaeger (Museum für Naturkunde, Humboldt Universität, Berlin, Germany). Photographs were provided by Mr P. Crabb (Photographic Unit, Natural History Museum, London) and Dr A. R. Milner. Overseas research trip funding was provided by the Natural Environment Research Council, University of London Central Research Fund and the Leverhulme Trust.

REFERENCES

BROILI, F. 1905. Beobachtungen an *Cochleosaurus bohemicus* Frič. *Palaeontographica*, **52**, 1–16.

CARROLL, R. L. 1967. Labyrinthodonts from the Joggins Formation. *Journal of Paleontology*, **41**, 111–142.

——BELT, D. S., DINELEY, D. and McGREGOR, C. 1972. *Vertebrate Paleontology of Eastern Canada.* Guide Book, Field Excursion A-59. International Geological Congress, Montreal, 113 pp.

CLARKSON, E. N. K., MILNER, A. R. and COATES, M. I. 1994. Palaeoecology of the Viséan of East Kirkton, West Lothian, Scotland. *Transactions of the Royal Society of Edinburgh, Earth Sciences*, **84** (for 1993), 417–425.

FRIČ, A. 1876. Ueber die Fauna der Gaskohle des Pilsner und Rakonitzer beckens. *Sitzungsberichte der Königlichen Böhmischen Gesellschaft der Wissenschaften*, **1875**, 70–79.

FRITSCH, A. (=Frič) 1879–1901. *Fauna der Gaskohle und der Kalksteine der Permformation Böhmens.* I–IV. Rivnać, Prague, 529 pp.

GODFREY, S. J. and HOLMES, R. 1995. The Pennsylvanian temnospondyl *Cochleosaurus florensis* Rieppel, from the lycopod stump fauna of Florence, Nova Scotia. *Breviora*, **500**, 1–25.

HOLMES, R. B., CARROLL, R. L. and REISZ, R. R. 1998. The first articulated skeleton of *Dendrerpeton acadianum* (Temnospondyli, Dendrerpetontidae) from the Lower Pennsylvanian locality of Joggins, Nova Scotia, and a review of its relationships. *Journal of Vertebrate Paleontology*, **18**, 64–79.

HOOK, R. W. 1993. *Chenoprosopus lewisi*, a new cochleosaurid amphibian (Amphibia: Temnospondyli) from the Permo-Carboniferous of North-Central Texas. *Annals of Carnegie Museum*, **62**, 273–291.

KLEMBARA, J. 1985. A new embolomerous amphibian (Anthracosauria) from the Upper Carboniferous of Florence, Nova Scotia. *Journal of Vertebrate Paleontology*, **5**, 293–302.

LANGSTON, W. Jr 1953. Permian amphibians from New Mexico. *University of California Publications in Geological Science*, **29**, 349–416.

LINNAEUS, C. 1758. *Systema naturae*, Tenth edition. vol **1**, Salvi, Stockholm, 824 pp.

LOHMANN, U. and SACHS, S. 2001. Observations on the postcranial morphology, ontogeny and palaeobiology of *Sclerocephalus haeuseri* (Amphibia: Actinodontidae) from the Lower Permian of Southwest Germany. *Memoirs of the Queensland Museum*, **46**, 771–781.

MILNER, A. R. 1980. The tetrapod assemblage from Nýřany, Czechoslovakia. 439–496. *In* PANCHEN, A. L. (ed.). *The Terrestrial Environment and the Origin of Land Vertebrates*. Systematics Association Special Volume, 15, Academic Press, London, 633 pp.

—— 1990. The radiations of temnospondyl amphibians. 321–349. *In* TAYLOR, P. D. and LARWOOD, G. P. (eds). *Major evolutionary radiations*. Clarendon Press, Oxford, 437 pp.

—— and SEQUEIRA, S. E. K. 1994. The temnospondyl amphibians from the Viséan of East Kirkton, West Lothian, Scotland. *Transactions of the Royal Society of Edinburgh, Earth Sciences*, **84** (for 1993), 331–361.

MOULTON, J. M. 1974. A description of the vertebral column of *Eryops* based on the notes and drawings of A. S. Romer. *Breviora*, **428**, 1–44.

PAWLEY, K. 2007. The postcranial skeleton of *Trimerorhachis insignis* Cope, 1878 (Temnospondyli: Trimerorhachidae): a plesiomorphic temnospondyl from the Lower Permian of North America. *Journal of Paleontology*, **81**, 873–894.

PEŠEK, J. 1974. Lateral passages between variegated and grey Carboniferous sediments. *Compte Rendu. 7 Congrès pour l'Avancement des Études de Stratigraphie Carbonifère*, **4**, 75–83.

REISZ, R. R., BERMAN, D. S. and HENRICI, A. C. 2005. A new skull of the cochleosaurid amphibian *Chenoprosopus* (Amphibia: Temnospondyli) from the Early Permian of New Mexico. 253–255. *In* LUCAS, S. G. and ZEIGLER, K. E. (eds). *The Nonmarine Permian*. New Mexico Museum of Natural History and Science Bulletin, 30, 362 pp.

RIEPPEL, O. 1980. The edopoid amphibian *Cochleosaurus* from the Middle Pennsylvanian of Nova Scotia. *Palaeontology*, **23**, 143–149.

ROMER, A. S. 1945. *Vertebrate Paleontology*, Second edition. Chicago University Press, Chicago, ix + 687 pp.

—— 1947. Review of the Labyrinthodontia. *Bulletin of the Museum of Comparative Zoology, Harvard College*, **99**, 1–368.

—— and WITTER, R. V. 1942. *Edops*, a primitive rhachitomous amphibian from the Texas red beds. *Journal of Geology*, **50**, 925–960.

RUTA, M., COATES, M. I. and QUICKE, D. L. J. 2003. Early tetrapod relationships revisited. *Biological Reviews*, **78**, 251–345.

SANDER, P. M. 1989. Early Permian depositional environments and pond bonebeds in central Archer County, Texas. *Palaeogeography, Palaeoclimatology, Palaeoecology*, **69**, 1–21.

SEQUEIRA, S. E. K. 2004. The skull of *Cochleosaurus bohemicus* Frič, a temnospondyl from the Czech Republic (Upper Carboniferous) and cochleosaurid interrelationships. *Transactions of the Royal Society of Edinburgh, Earth Sciences*, **94** (for 2003), 21–43.

—— and MILNER, A. R. 1993. The temnospondyl amphibian *Capetus* from the Upper Carboniferous of Nýřany, Czech Republic. *Palaeontology*, **36**, 657–680.

SIDOR, C. A., O'KEEFE, F. R., DAMIANI, R., STEYER, J. S., SMITH, R. M. H., LARSON, H. C. E., SERENO, P. C., IDE, O. and MAGA, A. 2005. Permian tetrapods from the Sahara show climate-controlled endemism in Pangaea. *Nature*, **434**, 886–889.

SKOČEK, V. 1968. Upper Carboniferous varvites in coal basins of central Bohemia. *Věstník Ústředního Ustavu Geologického*, **43**, 113–121. [In Czech].

STEEN, M. C. 1938. On the fossil Amphibia from the Gas Coal of Nýřany and other deposits in Czechoslovakia. *Proceedings of the Zoological Society of London*, **B108**, 205–283.

STEYER, J. S., DAMIANI, R., SIDOR, C. A., O'KEEFE, F. R., LARSON, H. C. E., MAGA, A. and IDE, O. 2006. The vertebrate fauna of the Upper Permian of Niger. IV. *Nigerpeton ricqlesi* (Temnospondyli:Cochleosauridae) and the Edopoid colonization of Gondwana. *Journal of Vertebrate Paleontology*, **26**, 18–28.

VALLIN, G. and LAURIN, M. 2004. Cranial morphology and affinities of *Microbrachis*, and a reappraisal of the phylogeny and lifestyle of the first amphibians. *Journal of Vertebrate Paleontology*, **24**, 56–72.

WILLISTON, S. W. 1916. The skeleton of *Trimerorhachis*. *Journal of Geology*, **24**, 291–297.

WITZMANN, F. and SCHOCH, R. R. 2006. The postcranium of *Archegosaurus decheni*, and a phylogenetic analysis of temnospondyl postcrania. *Palaeontology*, **49**, 1211–1235.

VON ZITTEL, K. A. R. 1888. *Handbuch der Paläontologie. Abteilung 1. Paläozoologie Vol. III. Vertebrata (Pisces, Amphibia, Reptilia, Aves)*. R. Oldenbourg, Munich and Leipzig, 900 pp.

—— and BROILI, F. 1923. *Grundzüge der Paläontologie (Paläozoologie) von Karl A. von Zittel.* 163–383 pp. 4th edn. Neubearbeitet von F. Broili und M. Schlosser. II. Abteilung: Vertebrata, R. Oldenbourg, Munich and Berlin, v + 689 pp.

[Special Papers in Palaeontology 81, 2009, pp. 155–160]

FIRST EVIDENCE OF A TEMNOSPONDYL IN THE LATE PERMIAN OF THE ARGANA BASIN, MOROCCO

by J. SÉBASTIEN STEYER* *and* NOUR-EDDINE JALIL†

*UMR 5143 CNRS Paléobiodiverstié et Paléoenvironnements, Département Histoire de la Terre, CP38, Muséum national d'Histoire naturelle, 8 rue Buffon, 75005 Paris, France; e-mail: steyer@mnhn.fr

†Université Cadi Ayyad, Faculté des Sciences Semlalia, "Vertebrate Evolution and Paleo-environments", B.P. 2390, Marrakech 40000, Kingdom of Morocco; e-mail: njalil@ucam.ac.ma

Typescript recieved 5 December 2008; accepted in revised form 15 March 2009

Abstract: The posterodorsal portion of a temnospondyl skull roof from the Late Permian Ikakern Formation, Argana Basin, High Atlas mountains of Morocco, is described. The specimen is fragmentary; it consists of either articulated left postparietal and tabular (with surrounding bone fragments) or articulated right supratemporal and intertemporal (with surrounding bone fragments). The sculpture of the dermal bones combined with either the straight posterior margin of the skull table or the presence of an intertemporal (respectively) suggest a non-stereospondyl and a non-euskelian temnospondyl. This specimen is the first record of a Palaeozoic temnospondyl from North Africa. Its co-occurrence with diplocaulid lepospondyl remains and its differences from Late Permian temnospondyls from Niger lend support to a palaeobiogeographical scenario positing amphibian migrations from Euramerica to Africa.

Key words: Temnospondyli, Palaeozoic, Africa, migration, Pangaea, palaeobiogeography.

TEMNOSPONDYLS from the Permian of Africa are relatively rare compared with those from the Triassic. Late Permian African temnospondyls are known from South Africa (e.g. *Uranocentrodon*; Latimer *et al.* 2002), Madagascar (*Rhinesuchus*; Piveteau 1926), Tanzania (*Peltobatrachus*; Panchen 1959) and, more recently, Niger (*Nigerpeton* and *Saharastega*; Sidor *et al.* 2005; Steyer *et al.* 2006; Damiani *et al.* 2006). In Morocco, the Permian to Early Jurassic Argana Basin (Text-fig. 1) has been the subject of extensive fieldworks conducted by joint French-Moroccan geological and palaeontological expeditions since the 1960s (e.g. Duffaud *et al.* 1966; Dutuit 1976; Jalil 1999). The Late Permian Ikakern Formation of the Argana Basin has yielded a few plants [e.g. *Voltzia heterophylla* (De Koning 1957)], ichnites (e.g. *Scoyenia* and *Synaptichnium*; Jones 1975; Hmich *et al.* 2006), and a remarkable tetrapod fauna consisting of a diplocaulid lepospondyl ('*Diplocaulus*' *minimus* Dutuit 1988; see also Dutuit 1976; under revision by D. Germain and the authors), at least one pareiasaur parareptile (*Arganaceras vacanti* Jalil and Janvier 2005), and several capthorinid reptiles (*Acrodonta irerhi* Dutuit 1976 and a moradisaurine; Jalil and Dutuit 1996). Here, we describe an incomplete skull recently found by us, representing the first record of a temnospondyl from the Ikakern Formation of the Argana Basin and the first Palaeozoic temnospondyl known from North Africa. This discovery adds to the Permian Moroccan tetrapod fauna and presents novel palaeobiogeographical implications for temnospondyl dispersal at the end of the Palaeozoic.

Institutional abbreviations. MHNM, Muséum d'Histoire naturelle de Marrakech, Morocco; MNN, Musée national du Niger, Niamey.

SYSTEMATIC PALAEONTOLOGY

AMPHIBIA Linnaeus, 1758
TEMNOSPONDYLI von Zittel, 1887–1890

Temnospondyli indet.
Text-figure 2

Referred specimen. MHNM-ARG01, posterodorsal portion of temnospondyl skull roof preserving either articulated left postparietal and tabular and fragments of surrounding bones (hypothesis 1, see text), or articulated right supratemporal and intertemporal and fragments of surrounding bones (hypothesis 2, see text).

Horizon. The specimen comes from a fine reddish siltstone level at the top of the 'T2 lithostratigraphic Unit' of Duffaud *et al.* (1966) and Tixeront (1973, 1974), Tourbihine Member, Ikakern Formation, Argana Basin, Late Permian (Ambroggi 1963;

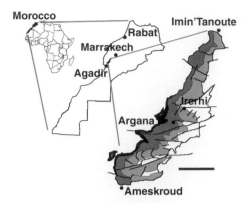

TEXT-FIG. 1. Location and simplified geology of the Moroccan Argana Basin (after Duffaud *et al.* 1966; Tixeront 1973, 1974; Tourani *et al.* 2000). Black colour corresponds to Early Jurassic, dark and light grey to Triassic, and white to Late Permian. Scale bar represents 20 km.

Tixeront 1973; Tourani *et al.* 2000; Olsen *et al.* 2000; Jalil and Janvier 2005).

Locality. MHNM-ARG01 was collected one hundred metres north of the diplocaulid site (Nr. XXII) of Dutuit (1976), about 1.5 km S-SW of the Irerhi village, Timezgadiwine District, Tikida region, Argana Basin, South-Western Atlas Mountain Chain, Kingdom of Morocco.

Description

The specimen is relatively flat (11 mm thick), and presents two major surfaces. One surface is slightly weathered and shows a dermal sculpture consisting of a honeycomb-like pattern at the centre of the bones (partly filled with micropebbles here) and striations at the periphery (radiating out from ossification centres), as well as finely interdigitated cranial sutures. This surface is therefore interpreted as belonging to the dorsal side of a skull roof. The dermal sculpture is typical for temnospondyls, and its strong development suggests that the specimen was an adult (Steyer 2000). The other surface shows muscle scars and foramina, as well as smoothly sinuous cranial sutures. It is interpreted as the ventral surface of the skull roof. In life, and by analogy with other temnospondyls, this surface would have been in contact with the otic or braincase region.

Six bones are preserved, two being almost complete:

1. The large almost complete bone (30 mm × 20 mm in dorsal view) is subrectangular. In ventral and lateral views, it shows a large, subcircular and deep natural foramen (3 mm in maximum diameter as well as depth) (?sf, Text-fig. 2B). As this foramen does not pierce the dorsal side of the bone and it is not associated with any cranial suture, we conclude that it cannot represent the pineal foramen; therefore the bone carrying the foramen is not a parietal. We interpret this foramen as being sensory (P. Janvier, pers. comm. 2008). A number of ventral depressions surround the foramen, and are likely to correspond to muscular attachment areas or

nutrient foramina, such as occur on the ventral side of the skull table in other tetrapods. By analogy with other temnospondyls, the large subrectangular bone represents either the postparietal (hypothesis 1, see below) or the supratemporal (hypothesis 2, see below).

2. The smaller almost complete bone (13 mm × 14 mm in dorsal view) is rhomboid in outline. It is slightly eroded dorsally and shows the same type of dermal sculpture and suture morphology as the large bone described above. These two bones share a long and finely interdigitated dorsal suture. They also form a long and straight natural margin which allows to position the specimen in the skull table: it consists of either articulated left postparietal and tabular (with surrounding bone fragments) (hypothesis 1) or articulated right supratemporal and intertemporal (with surrounding bone fragments) (hypothesis 2). These two hypotheses are discussed below.

DISCUSSION AND COMPARISONS

The specimen is a portion of an adult temnospondyl skull table. Because of its preservation, two hypotheses are proposed concerning its exact position in the skull table and the nature of its constituent bones.

Hypothesis 1

The large almost complete bone is the left postparietal and the adjacent smaller bone is the left tabular. Its width:length ratio is 1.5, indicating that the conjoined postparietals have a width:length ratio of about 3, as in non-euskelian temnospondyls, e.g. *Dendrerpeton* or *Neldasaurus* (Yates and Warren 2000, fig. 2, p. 86). If the above interpretation (Text-fig. 2A) is correct, then the large sensory foramen described above, visible in ventral and occipital views, would be oriented posteroventrally toward the otic region. The rhomboid tabular is smoothly pointed posteriorly but does not show a posterodorsal projection or 'horn' such as is common in stereospondyls (e.g. Schoch and Milner 2000). The tabular contacts the squamosal portion laterally (an ambiguous synapomorphy of the Trimerorhachidae according to Yates and Warren 2000, p. 90) and the supratemporal portion anteriorly. As these surrounding bones are very fragmentary, it is difficult to state whether the otic notch is absent or very shallow, again as in trimerorhachids (e.g. Milner and Sequeira 2004). Because of the poor preservation of the parietal, the sutural pattern between this bone and the tabular is not clear and could not be used for a more precise identification. In occipital view, the maximum thickness of the tabular (11 mm) is reached in its mid-part, at the level of a bumpy area which could be interpreted as a muscular or a tendon attachment area. Interestingly, MHNM-ARG01 also differs from the

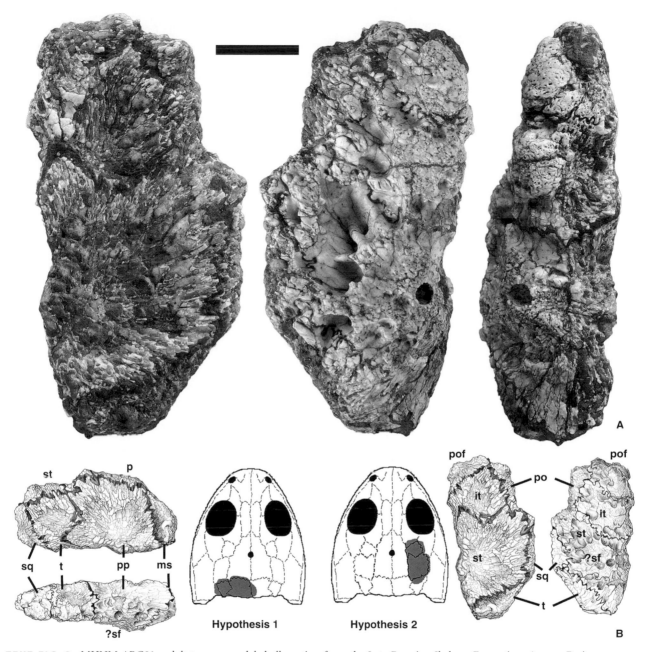

TEXT-FIG. 2. MHNM-ARG01, adult temnospondyl skull portion from the Late Permian Ikakern Formation, Argana Basin, Morocco. A, photos (dorsal, ventral and lateral views) of the specimen. B, interpretative drawings of the specimen with its possible locations in the skull table (hypotheses 1 and 2, see text). it, intertemporal; ms, midline suture; p, parietal; po, postorbital; pof, postfrontal; pp, postparietal; sq, squamosal; ?sf, possible sensory foramen; st, supratemporal; t, tabular. Scale bars represent 1 cm.

geographically and stratigraphically proximal *Nigerpeton* and *Saharastega* from the Permian of Moradi, Niger (Sidor *et al.* 2005). Thus, although *Nigerpeton* (MNN MOR70) resembles MHNM-ARG01 in having a subrectangular post-parietal, it differs from MHNM-ARG01 in having deeper otic notches and in lacking a squamosal-tabular contact (Steyer *et al.* 2006; JSS, personal observation, 2007). *Saharastega* (MNN MOR73) also has a shallow otic notch, but differs from MHNM-ARG01 in that its tabular is

considerably expanded laterally and does not contact the squamosal (Damiani *et al.* 2006; JSS, personal observation, 2007).

Hypothesis 2

The large almost complete bone is the right supratemporal and the smaller bone is the right intertemporal. In dorsal view, therefore, the natural straight border of the specimen would correspond to the sutural contact between these two bones and a relatively long parietal (not preserved) (Text-fig. 2B). The rhomboid intertemporal contacts the postfrontal portion anteriorly and the postorbital portion mesially. The subrectangular supratemporal contacts the tabular portion posteriorly and the squamosal portion mesially. In ventral view, the postorbital appears elongated, its posterior portion being relatively pointed and reaching the supratemporal. Given this alternative interpretation, the large sensory foramen described above, visible on the ventral side of the supratemporal, would thus be associated with the braincase region. The intertemporal is lost in stereospondylomorphs (ambiguous synapomorphy of this clade according to Yates and Warren 2000, p. 92), euskelians (ambiguous synapomorphy of this clade according to Yates and Warren 2000, p. 85) and dvinosaurids (unambiguous synapomorphy of clade -or node- nr. 10 of Yates and Warren 2000, p. 91). The intertemporal is present in a number of unrelated Palaeozoic temnospondyls, such as *Capetus* (Sequeira and Milner 1993), *Dendrerpeton* (Holmes *et al.* 1998), *Balanerpeton* (Milner and Sequeira 1994), *Neldasaurus* (Chase 1965), and edopoids (Damiani *et al.* 2006). Further occurrences have been documented by Gubin *et al.* (2000). In *Nigerpeton*, the intertemporal and the supratemporal are very reduced and extremely wide (conditions linked to the extreme shortening of the skull table), and the intertemporal is larger than the supratemporal. This is not the case here. In MHNM-ARG01, the shape and proportions of the intertemporal and supratemporal are closer to those of *Dendrerpeton* and *Saharastega*, although in the latter, the supratemporal is comparatively wider and the intertemporal more lozenge-shaped.

CONCLUSION

In both hypotheses, the dermal sculpture and sutural pattern of MHNM-ARG01 conform to those of adult temnospondyls which are neither stereospondyls nor euskelians. The dimensions of the specimen, as preserved, suggest a relatively small or middle-sized skull.

PALAEOBIOGEOGRAPHICAL IMPLICATIONS

MHNM-ARG01 is the first record of a Palaeozoic temnospondyl from North Africa. In Western Africa (Niger), the Moradi temnospondyl fauna resembles Euramerican faunas (Sidor *et al.* 2005; Steyer *et al.* 2006). The Moroccan Permian fauna also includes a diplocaulid nectridean,

the only occurrence of a lepospondyl outside Euramerica (Dutuit 1976 1988). Concerning the temnospondyls, there is now evidence for two distinct Late Permian African faunas: a North-Western fauna (*Nigerpeton*, *Saharastega* and the Moroccan temnospondyl described here), endemic and with Euramerican features; and a South-Eastern fauna (from South Africa, Tanzania, and Madagascar), dominated by rhinesuchids and *Peltobatrachus* (e.g. Milner 1993). The discovery of MHNM-ARG01 in the Late Permian of Morocco corroborates the hypothesis of a Late Permian separation between tetrapod faunas. In addition, it strengthens ecological similarities between the Moradi and the Argana tetrapod faunas, both of which include herbivorous amniotes (moradisaurine capthorinids and pareiasaurs) and carnivorous temnospondyl amphibians.

The Moroccan and Nigerien localities occupied Central Pangaea during the Late Permian (e.g. Scotese 2001). In order to explain the occurrence of 'Euramerican-type' temnospondyls in Niger, Steyer *et al.* (2006) hypothesized that several taxa from Laurussia colonized Gondwana, using one or two possible migration routes, namely a pericontinental (coastal) route and/or an intracontinental (fluvial) route. Dispersal events are likely to have occurred over a long time period, because there is a temporal gap of at least 40 Ma between the Euramerican and Nigerien faunas (Sidor *et al.* 2005). Together with the occurrence of a diplocaulid from the Ikakern Formation, the presence of a temnospondyl in the Late Permian of Morocco tends to support the hypothesis that amphibians took an intracontinental migration route throughout the Hercynian mountain chain, presumably via river systems of tropical or sub-tropical areas (e.g. Fluteau *et al.* 2001; Rees *et al.* 2002). Ongoing fieldworks in the Argana basin will provide more material that will shed light on continental vertebrates distribution and evolution during the Late Permian.

Acknowledgements. The authors are delighted to contribute to this Festschrift volume in honour of Andrew Milner. Andrew was a Reviewer and Jury Member for JSS' PhD dissertation in 2001. He taught JSS tetrapod anatomy and osteology when the senior author visited collections at Birkbeck College (University of London) and the Natural History Museum, London, and has provided him with constant and valued help in his studies of 'these road-killed temnospondyls'.

The authors thank Marcello Ruta (University of Bristol) for his help in the identification of the specimen and for his remarks which highly increased the quality of the manuscript. We also thank Jennifer Clack (University Museum of Cambridge), Anne Warren (La Trobe University) and an anonymous reviewer for their constructive comments. Many thanks also to Renaud Vacant for his help in the field and for the preparation of the specimen, Philippe Loubry for the photographs, and Charlène Letenneur for the interpretative drawings (all from the

UMR 5143 CNRS/MNHN, Paris). This research was supported by the UMR 5143 CNRS, Paris, France, and by the University Cadi Ayyad, Faculté des Sciences Semlalia, Marrakech, Kingdom of Morocco. We thank Philippe Janvier (UMR 5143 CNRS/MNHN) for fruitful discussions and for his help in the interpretation of a possible ventral sensory foramen, and the Ministry of Mines of Morocco for the field authorizations.

REFERENCES

AMBROGGI, R. 1963. Étude géologique du versant méridional du Haut-Atlas occidental et de la plaine du Souss. *Notes et Mémoires du Service géologique du Maroc*, **157**, 1–32.

CHASE, J. N. 1965. *Neldasaurus wrightae*, a new rhachitomous labyrinthodont from the Texas Lower Permian. *Bulletin of the Museum of Comparative Zoology at Harvard University*, **133**, 153–225.

DAMIANI, R., SIDOR, C. A., STEYER, J. S., SMITH, R. M. H., LARSSON, H. C. E., MAGA, A. and IDE, O. 2006. The Vertebrate fauna of the Upper Permian of Niger. IV. The primitive temnsopondyl *Saharastega moradiensis*. *Journal of Vertebrate Paleontology*, **26**, 559–572.

DE KONING, G. 1957. Géologie des Ida ou Zal (Maroc). Stratigraphie, pétrographie et tectonique de la partie Sud-Ouest du bloc occidental du Massif ancien du Haut-Atlas (Maroc). *Leidse Geologische Mededelingen*, **23**, 215.

DUFFAUD, M. F., BRUN, L. and PLANCHUT, B. 1966. Le bassin du sud-ouest Marocain. *In* REYRE, D. (ed.). Bassins sédimentaires du littoral africain. Symposium New Delhi 1964, 1ère Partie: Littoral Atlantique. Vol 1966. Publications de l'Association des Services géologiques d'Afrique, *Paris*, 5–25.

DUTUIT, J. M. 1976. Introduction à l'étude paléontologique du Trias continental marocain. Description des premiers Stégocéphales recueillis dans le couloir d'Argana (Atlas occidental). *Mémoires du Muséum national d'Histoire naturelle*, **36**, 1–253.

—— 1988. *Diplocaulus minimus* n. sp. (Amphibia : Nectridea), lépospondyle de la formation d'Argana dans l'Atlas occidental marocain. *Comptes Rendus de l'Académie des Sciences, Paris*, **307**, 851–854.

FLUTEAU, F., BESSE, J., BROUTIN, J. and RAMSTEIN, G. 2001. The Late Permian climate. What can be inferred from climate modelling concerning Pangea scenarios and Hercynian range altitude? *Palaeogeography, Palaeoclimatology, Palaeoecology*, **167**, 39–71.

GUBIN, Y. M., NOVIKOV, I. V. and MORALES, M. 2000. A review of anomalies in the structure of the skull roof of temnospondylous labyrinthodonts. *Paleontological Journal*, **34** (Suppl. 2), 154–164.

HMICH, D., SCHNEIDER, J. W., SABER, H., VOIGT, S. and EL WARTITI, M. 2006. New continental carboniferous and permian faunas of Morocco: implications for biostratigraphy, palaeobiogeography and palaeoclimate. *In* LUCAS, S. G., CASSINIS, G. and SCHNEIDER, J. W. (eds). *Non-marine permian biostratigraphy and biochronology. Geological Society, London, Special Publications*, **265**, 297–324.

——CARROLL, R. L. and REISZ, R. R. 1998. The first articulated skeleton of *Dendrerpeton acadianum* (Temnospondyli, Dendrerpetontidae) from the Lower Pennsylvanian locality of Joggins, Nova Scotia, and a review of its relationships. *Journal of Vertebrate Paleontology*, **18**, 64–79.

JALIL, N. E. 1999. Continental Permian and Triassic vertebrate localities from Algeria and Morocco and their stratigraphical correlations. *Journal of African Earth Sciences*, **29**, 219–226.

—— and DUTUIT, J. M. 1996. Permian captorhinid reptiles from the Argana Formation, Morocco. *Palaeontology*, **39**, 907–918.

—— and JANVIER, P. 2005. Les pareiasaures (Amniota, Parareptilia) du Permien supérieur du Bassin d'Argana, Maroc. *Geodiversitas*, **27**, 35–132.

JONES, D. F. 1975. *Stratigraphy, environments of deposition, petrology, age, and provenance, basal red beds of the Argana valley, Western High Atlas Mountains, Morocco*. M.S. thesis, New Mexico Institute of Mining and Technology, Socorro.

LATIMER, E. M., HANCOX, P. J., RUBIDGE, B. S., SHISHKIN, M. A. and KITCHING, J. W. 2002. The temnospondyl amphibian *Uranocentrodon*, another victim of the end-Permian extinction event. *South African Journal of Science*, **98**, 191–193.

LINNAEUS, C. 1758. *Systema naturae*, 10 edn. Vol 1, Salvi, Stockholm, 824 pp.

MILNER, A. R. 1993. Biogeography of palaeozoic tetrapods. 324–353. *In* LONG, J. A. (ed.). *Palaeozoic vertebrate biostratigraphy and biogeography*. Belhaven Press, London, 384 pp.

—— and SEQUEIRA, S. E. K. 1994. The temnospondyl amphibians from the Visean of East Kirkton, West Lothian, Scotland. *Transactions of the Royal Society of Edinburgh, Earth Sciences*, **84** (for 1993), 331–361.

—— —— 2004. Slaugenhopia texensis (Amphibia: Temnospondyli) from the Permian of Texas is a primitive Tupilakosaurid. *Journal of Vertebrate Palentology*, **24**, 320–325.

OLSEN, P. E., KENT, D. V., FOWELL, S. J., SCHLISCHE, R. W., WITHJACK, M. O. and LE TOURNEAU, P. M. 2000. Implications of a comparison of the stratigraphy and depositional environments of the Argana (Morocco) and Fundy (Nova Scotia, Canada) Permian-Jurassic basins. 165–183. *In* OUJIDI, M. and ET-TOUHAMI, M. (eds). *Le Permien et le Trias du Maroc: Actes de la Première Réunion du Groupe marocain du Permien et du Trias*. Hilal Impression, Oujda, 183 pp.

PANCHEN, A. L. 1959. A new armoured amphibian from the Upper Permian of East Africa. *Philosophical Transactions of the Royal Society of London. Series B, Biological Sciences*, **242**, 207–281.

PIVETEAU, J. 1926. Paléontologie de Madagascar, XIII. Amphibiens et reptiles permiens. *Annales de Paléontologie*, **15**, 1–128.

REES, P. M., ZIEGLER, A. M., GIBBS, M. T., KUTZBACH, J. E., BEHLING, P. J. and ROWLEY, D. B. 2002. Permian phytogeographic patterns and climate data/model comparisons. *Journal of Geology*, **110**, 1–31.

SCHOCH, R. R. and MILNER, A. R. 2000. *Handbuch der Paläoherpetologie. Teil 3B. Stereospondyli*. Verlag Friedrich Pfeil, Munich, 220 pp.

SCOTESE, C. R. 2001. *Atlas of Earth History, Volume 1, Paleo-geography*. Paleomap Project, Arlington, Texas, 52 pp.

SEQUEIRA, S. E. K. and MILNER, A. R. 1993. The temnospondyl amphibian *Capetus* from the Upper Carboniferous of the Czech Republic. *Palaeontology*, **36**, 657–680.

SIDOR, C. A., O'KEEFE, F. R., DAMIANI, R., STEYER, J. S., SMITH, R. M. H., LARSSON, H. C. E., SERENO, P. C., IDE, O. and MAGA, A. 2005. Permian tetrapods from the Sahara show climate-controlled endemism in Pangaea. *Nature*, **434**, 886–889.

STEYER, J. S. 2000. Ontogeny and phylogeny in temnospondyls amphibians : a new method of analysis. *Zoological Journal of the Linnean Society*, **130**, 449–467.

—— DAMIANI, R., SIDOR, C. A., O'KEEFE, F. R., LARSSON, H. C. E., MAGA, A. and IDE, O. 2006. The Vertebrate fauna of the Upper Permian of Niger. V. *Nigerpeton ricqlesi* (Temnospondyli: Cochleosauridae), and the edopoid colonization of Gondwana. *Journal of Vertebrate Paleontology*, **26**, 18–28.

TIXERONT, M. 1973. Lithostratigraphie et minéralisations cuprifères et uranifères stratiformes, syngénétiques et familières des formations détritiques permo-triasiques du couloir d'Argana, Haut-Atlas occidental (Maroc). *Notes et Mémoires du Service géologique du Maroc*, **33**, 147–177.

—— 1974. Carte géologique et minéralisations du Couloir d'Argana. *Notes et Mémoires du Service géologique du Maroc*, **205**, 1–1.

TOURANI, A., LUND, J. J., BENAOUISS, N. and GAUPP, R. 2000. Stratigraphy of Triassic syn-rift deposition in Western Morocco. *Zentralblatt für Geologie und Paläontologie*, **9**/10, 1193–1215.

YATES, A. M. and WARREN, A. A. 2000. The phylogeny of the 'higher' temnospondyls (Vertebrata: Choanata) and its implications for the monophyly and origins of the Stereospondyli. *Zoological Journal of the Linnean Society*, **128**, 77–121.

VON ZITTEL, K. A. R. 1887–1890. *Handbuch der Paläontologie. Abteilung 1. Paläozoologie Vol. III. Vertebrata (pisces, amphibia, reptilia, aves)*. Oldenbourg, Munich and Leipzig, 900 pp.

[Special Papers in Palaeontology 81, 2009, pp. 161–173]

UNIQUE STEREOSPONDYL MANDIBLES FROM THE EARLY TRIASSIC PANCHET FORMATION OF INDIA AND THE ARCADIA FORMATION OF AUSTRALIA

by ANNE WARREN*, ROSS DAMIANI* *and* DHURJATI P. SENGUPTA†

*Department of Zoology, La Trobe University, Melbourne, 3086 Vic., Australia; e-mail: a.warren@latrobe.edu.au
†Geological Studies Unit, Indian Statistical Institute, 203 Barrackpore Trunk Road, 700108 Kolkata, India; e-mail: dhurjati@isical.ac.in

Typescript received 26 June 2008; accepted in revised form 27 August 2008

Abstract: A new genus of stereospondyl amphibian, *Capulomala* gen. nov., is described on the basis of postglenoid areas of the mandible, which possess a uniquely hypertrophied postglenoid process. Separate species are recognised, *C. panchetensis* gen. nov., et sp. nov. found in the Panchet Formation of India and *C. arcadiaensis* gen. nov. et sp. nov. from the Arcadia Formation of Australia. *Capulomala* cannot be reliably associated with any known cranial material, but together with other tetrapods, can be used to correlate the Panchet and Arcadia formations, which are placed at the base of the Triassic.

Key words: Temnospondyli, Stereospondyli, Panchet, Arcadia, *Capulomala*, mandible.

T H E Panchet Formation of the Damodar Basin, India and the Arcadia Formation of the Bowen Basin, Australia, preserve a suite of fossil vertebrates from the earliest Triassic, especially stereospondyls. Those from the Arcadia Formation have been described in a series of publications between 1972 and 2007; knowledge of the Panchet fauna stems from the 19th century and little has been added to the recent literature with the exception of a review of Panchet temnospondyls (Tripathi 1969) and the description of a new taxon (Yates and Sengupta 2002). The described stereospondyls from the Panchet and Arcadia formations and their current taxonomic status are listed in Table 1. Some of the Panchet stereospondyls are fragmentary and not determinable to family level, having been re-designated several times. The Panchet and Arcadia faunas overlap at the higher taxonomic level but to date no genera common to both faunas have been identified.

This paper describes fragmentary mandibular remains from the Panchet and Arcadia formations (Text-figs 1–2), and emphasises the potential of scrap collections not only for the identification of new taxa but also for correlation between strata. The history of fossil collection in the Arcadia Formation may be unique in that all bone, however small and seemingly insignificant, was collected over a period exceeding 30 years. Analysis of this scrap collection revealed components of the fauna that might otherwise have remained undetected. Especially important in this context was the recognition of very small

members of the stereospondyl *Watsonisuchus aliceae*, resulting in a growth series for that taxon (Warren and Hutchinson 1988; Warren and Schroeder 1995), the description of the thumbnail sized *Lapillopsis* (Yates 1999) and *Nanolania* (Yates 2000) and the discovery of new elements of the fauna preserved within coprolites (Northwood 2005). Additionally, Northwood (1997) recognised among the collection further specimens of a taxon described by Warren (1985*b*) as *Plagiobatrachus australis*. Material attributed to *Plagiobatrachus* by Warren (1985*b*) consisted of a series of vertebral centra similar to the gastrocentrous centra found in members of the Plagiosauridae, as well as fragments of the posterior ends (postglenoid areas or PGAs, Jupp and Warren 1986) of dorsoventrally flattened mandibles, usually with tubercular (pustular) ornament. Although the mandibles and centra were not associated Warren (1985*b*) placed both in the Plagiosauroidea. Included in the new material from the Arcadia Formation is one specimen in which the surangular of the mandible is ornamented both ventrolaterally and dorsally (Text-fig. 2), an apparently unique feature for temnospondyls, and subsequently several additional fragments ornamented on two surfaces were found in the collection.

During an examination of scrap collected from the Panchet Formation by DS and associates, AW recognised an isolated surangular bone with an elongate postglenoid process that was ornamented on both sides. The surangular is comparable with three other postglenoid areas from

TABLE 1. Stereospondyl fauna from the Panchet Formation of India and the Arcadia Formation of Australia. Authorities cited are to the original description and the current taxonomic position.

Higher taxon	Panchet formation	Arcadia formation
Lapillopsidae	*Manubrantlia khaki* Yates and Sengupta, 2002	*Lapillopsis nana* Yates, 1999
Mastodonsauridae	Present (this paper)	*Watsonisuchus gunganj* Warren, 1980 (Damiani 2001)
		Watsonisuchus rewanensis Warren, 1980 (Damiani 2001)
		Watsonisuchus aliceae Warren and Hutchinson, 1988 (Damiani 2001)
Rhytidosteidae	*Indobrachyops panchetensis* Huene and Sahni, 1958 (Warren and Black, 1985)	*Rewana quadricuneata* Howie, 1972a
		Arcadia myriadens Warren and Black, 1985
		Acerastea wadeae Warren and Hutchinson, 1987
		Nanolania anatopretia Yates, 2000
Trematosauridae	*Gonioglyptus longirostris* Huxley, 1865	Trematosauridae (Warren, 1985a)
	Gonioglyptus huxleyi Lydekker, 1882	*Tirraturhinus smisseni* Nield *et al.*, 2006
Trematosauridae *incertae sedis*	*Glyptognathus fragilis* Lydekker, 1882 (Schoch and Milner, 2000)	
	Indolyrocephalus panchetensis Tripathi, 1969 (Schoch and Milner, 2000)	
	Panchetosaurus panchetensis Tripathi, 1969 (Schoch and Milner, 2000)	
Brachyopidae		*Xenobrachyops allos* (Howie, 1972b; Warren and Hutchinson 1983)
Chigutisauridae	Present (this paper)	*Keratobrachyops australis* Warren, 1981
?Tupilakosauridae	Tupilakosauridae (Huxley, 1865; Ochev and Shishkin, 1989)	?Tupilakosauridae (Damiani and Warren, 1996)
?Plagiosauridae		*Plagiobatrachus australis* Warren, 1985b
Indet	*Pachygonia incurvata* Huxley, 1865 (Damiani, 2001)	*Capulomala arcadiaensis* n. gen. n. sp.
	Indobenthosuchus panchetensis (Tripathi, 1969; Damiani, 2001)	
	Lydekkerina panchetensis (Tripathi, 1969; Hewison 2007)	
	Capulomala panchetensis n. gen. n. sp.	

the same formation, and is similar to the postglenoid areas of *Plagiobatrachus australis*. This new material, to date found only in the Panchet and Arcadia formations, is described below (Text-figs 1–3) and a reconstruction is provided (Text-fig. 4).

Institutional abbreviations. GSI, Geological Survey of India, Kolkata, India; ISI, Indian Statistical Institute, Kolkata, India; MCZ, Museum of Comparative Zoology, Harvard University; QM F, Queensland Museum Fossil; QM L, Queensland Museum Locality, Queensland Museum, Brisbane, Queensland, Australia.

Anatomical abbreviations. a, angular; ar, articular; bpgp, base of postglenoid process; f, foramen; gf, glenoid fossa; ms, mandibular sulcus; os, oral sulcus; pgp, postglenoid process of the surangular; pra, prearticular; r, ridge of surangular marking transition from postglenoid process to edge of adductor fossa; sa, surangular; safl, surangular flange; tr, triangular depression on the anterolabial face of the postglenoid process.

SYSTEMATIC PALAEONTOLOGY

TEMNOSPONDYLI Zittel, 1887–1890
STEREOSPONDYLI Zittel, 1887–1890 *incertae sedis*

CAPULOMALA gen. nov.

1841   *Labyrinthodon* Owen, p. 504.
1969   *Pachygonia* (Huxley); Tripathi, p. 31 (*pars*).
1985b *Plagiobatrachus* Warren, p. 236 (*pars*).

Derivation of name. From the Latin *Capulus*, sword or handle, *mala* jaw.

Type species. Capulomala panchetensis Tripathi, 1969.

Diagnosis. Stereospondyl mandible with the following features of the PGA. Autapomorphies of the genus: a

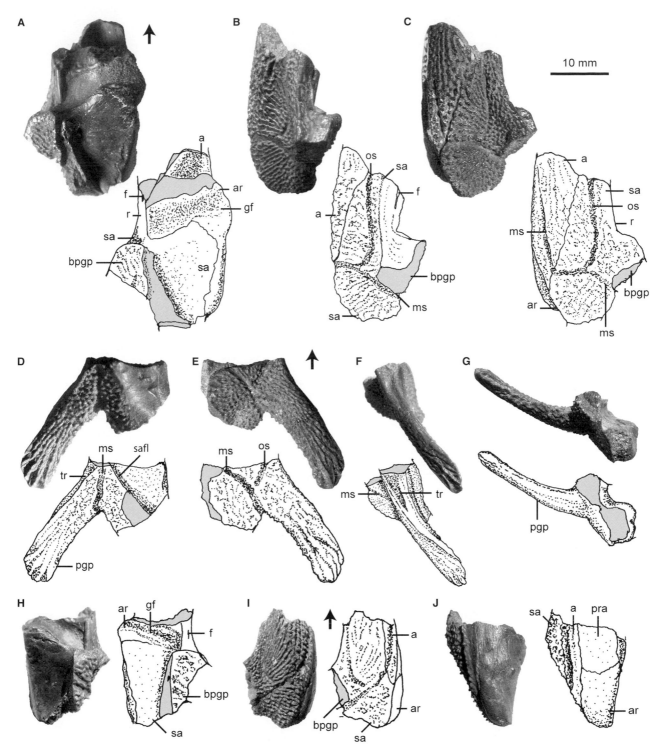

TEXT-FIG. 1. Photographs and drawings of *Capulomala panchetensis* gen. nov et sp. nov. A–C, the holotype (GSI 17886), a left PGA. A, dorsal. B, lateral. C, ventrolateral. D–G the almost complete left surangular with hypertrophied postglenoid process (ISI A179). D, dorsolingual. E, ventrolateral. F, lateral. G, posterior. H–J, the right PGA complete posteriorly (ISI A 178). H, dorsolingual. I, ventrolateral. J, ventral. Arrows indicate anterior, broken surfaces filled with grey.

hypertrophied postglenoid process of the surangular directed posterolaterally; ornament of the postglenoid process extensively exposed on the dorsal surface of the PGA; sensory sulci well defined; and the dorsal part of the mandibular sensory sulcus curves around the base of the postglenoid process onto the dorsal ornamented

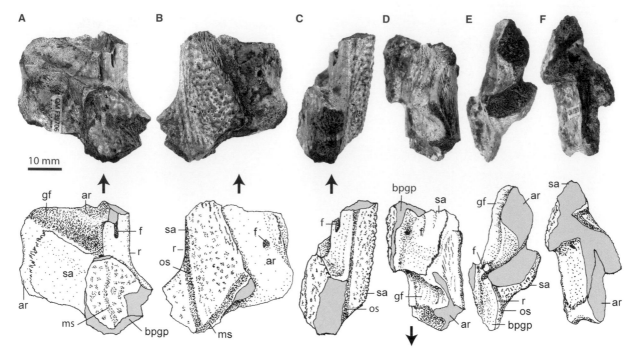

TEXT-FIG. 2. Photographs and drawings of the holotype of *Capulomala arcadiaensis*, a right PGA, (QM F39706), QM L1111, Arcadia Formation. A, dorsal. B, ventral. C, labial. D, lingual. E, anterior. F, posterior. Note that the angular and prearticular bones are missing from this specimen and that the ornamented base of the postglenoid process is preserved on the dorsal surface. Arrows indicate anterior, broken surfaces filled with grey.

surface of the PGA. *Capulomala* species have tubercular ornament borne on reticulate ridges, a feature shared with some Plagiosauridae and Rhytidosteidae.

Remarks. Among the previously described scrap material from the Panchet Formation are a number of PGAs that have been named but that have been considered indeterminable by subsequent authors. Among these is the holotype of '*Labyrinthodon*' *panchetensis*, GSI 17886, a dorsoventrally flattened PGA figured in dorsal, labial and lingual views (Tripathi 1969, pl. 1, figs 11–13; Text-fig. 1A–C). The dorsal view clearly shows the ornamented base of a postglenoid process of the surangular; in no temnospondyl other than the ones described as *Capulomala* (in the present paper) is an extensive ornamented area of the surangular visible in dorsal view. Further inspection of GSI 17886 by DS confirmed that this is indeed the base of a postglenoid process of the surangular, from a PGA that is not distinguishable from the material described below. The base of the postglenoid process is visible also in GSI 17887, originally referred to '*Labyrinthodon*' *panchetensis* (Tripathi 1969). The ventral part of the mandibular sulcus curves posteriorly initially in GSI 17886 and GSI 17887, and also in GSI 17885, a fragmentary PGA lacking the base of the postglenoid process, placed in *Pachygonia incurvata* by Tripathi (1969).

In his original description Tripathi (1969) placed *Labyrinthodon panchetensis* in the Capitosauridae. Subse-

quently, Schoch and Milner (2000) declared it a *nomen dubium*, listing it as *incertae sedis* within Stereospondyli, and Damiani (2001) considered it Stereospondyli indet. Since the specimen is clearly from the same taxon as the new material from the Panchet Formation, and is close to that from the Arcadia formation, it is not a *nomen dubium*. *Labyrinthodon*, a genus originally erected by Owen (1841) for mandibular material from the English Triassic, has had a number of species associated with it, but none of these are now considered valid (see review *in* Moser and Schoch 2007). The original material of *Labyrinthodon* is an objective junior synonym of *Mastodonsaurus* and hence the name *Labyrinthodon* is not available. GSI 17885, described as *Pachygonia incurvata*, is clearly distinct from the holotype of that species, GSI 2257, in which the lower part of the mandibular sulcus curves anteriorly.

Capulomala panchetensis comb. nov.
Text-figure 1

1969 *Labyrinthodon panchetensis* Tripathi, p.32, pl. 1, figs 11–13.
1969 *Pachygonia incurvata* (Lydekker); Tripathi, p.29 (*pars*), pl. IV, fig.3.

Holotype. GSI 17886 (Text-fig. 1A–C), the postglenoid area of a left mandibular ramus (Tripathi 1969).

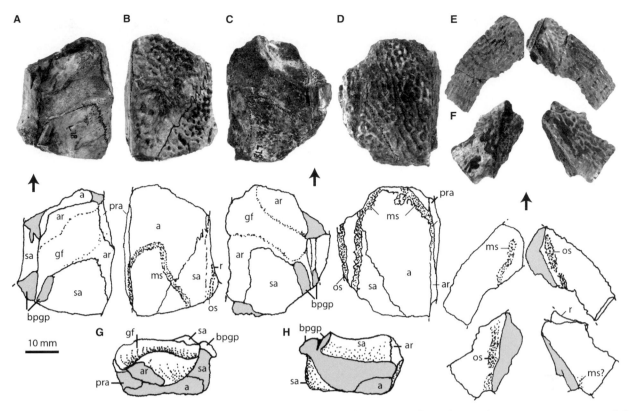

TEXT-FIG. 3. Photographs and interpretive drawings of *Capulomala arcadiaensis* PGAs from the Arcadia Formation. A–B, G–H, left PGA QM F12270. C–D, right PGA QM F39708. A, C, dorsal. B, D, ventral. G, anterior. H, posterior. E, left postglenoid process QM F39707, dorsal (left), ventral (right). F, right postglenoid process QM F39709, dorsal (right), ventral (left). In B, D, note the course of the mandibular sulcus passing anteriorly from the surangular to the angular, then turning posteriorly to exit the ornamented area onto the lingual surface. Arrows indicate anterior, broken surfaces filled with grey.

Referred material. *Labyrinthodon panchetensis* GSI 17887, a right PGA; *Pachygonia incurvata* GSI 17885, a left PGA; ISI A 178, a right PGA complete posteriorly; ISI A 179, a partial left PGA including a complete postglenoid process; ISI A 180, a left PGA.

Occurrence. Dumdumi (23°38′N, 86°53′E), Panchet Formation, Damodar Basin, India; Induan.

Diagnosis. As for the genus.

Description of new material

Postglenoid areas. The ornamentation in all specimens consists of tubercles, sometimes arranged along the summit of ridges. In many places the ridges are without tubercles, perhaps because of abrasion. The sensory sulci are distinct from those of most other stereospondyls in that they are well incised but extremely narrow. On the surangular an oral sulcus traverses the bone antero-posteriorly from a few millimetres below the rim of the adductor fossa to the posterior margin of the postglenoid process where it curves ventrally to join the mandibular sulcus. From this junction the mandibular sulcus curves dorsally, posterior to the postglenoid process, to extend anteriorly across the dorsal ornamented surface of the process (seen clearly on ISI A

178 and 179 below). In no specimen of *Capulomala* is a chorda tympanic foramen present.

ISI A 178 (Text-fig. 1H–J) is the posterior end of a right mandibular ramus, broken transversely part way through the glenoid fossa. It is the only *Capulomala* PGA that is complete posteriorly. Overall the ramus is dorsoventrally flattened. The PGA is surprisingly short with the retroarticular and arcadian processes (Warren and Black 1985) not discernable separately, but with the broken base of the postglenoid process of the surangular (Jupp and Warren 1986) present.

In a dorsal view of ISI A 178, significantly, an area of ornamented bone can be seen on the labiodorsal surface just posterior to the glenoid fossa. This is the broken proximal end of the postglenoid process of the surangular, hereafter referred to as the postglenoid process. The postglenoid process is ornamented dorsally and ventrally, an autapomorphy of *Capulomala*; its full extent is described below in ISI A 179. The anterolabial face of the postglenoid process carries dorsal and ventral ridges which demarcate an elongate triangular surface that is fully preserved only in ISI A 179 (below). The ventral ridge is a clearly-defined smooth curve, linking it with the labial margin of the adductor fossa; the dorsal ridge marks the edge of the ornamented area. The medial margin of the postglenoid process is incomplete in ISI A 178 and is separated from the unornamented dorsal surface of the surangular by a deep recess, pierced by a foramen.

A suture is present between the surangular and articular, passing transversely along the posterior rim of the glenoid and then turning posteriorly near the lingual margin of the PGA. Antero-labially the surangular forms an almost horizontal shelf marking the margin of the adductor fossa. The shelf is pierced by the posterior border of an elongate foramen.

The ventral surface of ISI A 178 is composed of the ornamented surangular and angular, which terminates just posterior to the level of the glenoid fossa, and, on the lingual margin, a small exposure of articular posterior to a strip of prearticular.

In labial view the broken base of the postglenoid process is visible dorsally, with below it the oral sensory canal. The suture between the surangular and angular crosses the anterior half of the labial surface, angled towards the posteroventral surface.

In lingual view the articular expands posteriorly to occupy the whole side of the PGA below the surangular; more anteriorly the articular sutures with the prearticular. There is no prearticular-surangular suture, the two bones being separated by a sliver of articular.

ISI A 179 (Text-fig. 1D–G.) is broken immediately posterior to the glenoid and is incomplete posteriorly. It is composed wholly of surangular. This is the only specimen in which the postglenoid process is complete. This spectacular process curves laterally and a little posteriorly from the PGA. It is ornamented on the dorsal and ventral surfaces, with the tubercles of the ornament arranged along the apex of ridges that radiate towards the distal half. The distal end of the process is rounded in out-line; in section the process is an elongate oval distally but slightly more rounded proximally. The proximal anterolabial surface of the process is deep and unornamented with an elongate, triangular depression, marked by dorsal and ventral ridges. The dorsal ridge (marking the edge of the ornament) describes a convex curve proximally (seen especially in the holotype) to continue in a posterior direction as the ornamented surangular crest (new term) – possibly homologous to the crista muscularis of mastodonsauroids (Novikov and Shishkin 1992) – that is complete in this specimen only. The surangular flange clearly projects lingually so that it overhangs the unornamented part of the surangular. The ventral ridge below the anterior margin of the postglenoid process curves convexly to continue anteriorly as the margin of the adductor fossa. A portion of the mandibular sensory sulcus passes immediately posterior to the postglenoid process from the ventral surface and curves anteriorly onto the dorsal surface to terminate near the anteromedial margin of the process. Ventrally the mandibular sulcus traverses the surangular in an anteroventral line from the posterior margin of the process, with the oral sulcus arising from it to pass anteriorly towards the adductor fossa.

Capulomala arcadiaensis sp. nov.
Text-figures 2–3

1985 *Plagiobatrachus australis* Warren, p. 237 (*pars*), figs 1–4.

Derivation of name. From the Arcadia Formation, from which all the material derives.

Holotype. QM F39706 (Text-fig. 2), the postglenoid area of a right mandibular ramus with the base of the postglenoid process preserved, Tank Locality (QM L1111).

Referred material. Postglenoid areas as follows: The Crater (QM L78): QM F12269 (Warren 1985*b*), QM F12270 (Warren 1985*b*), QM F39708, QM F39710, QM F39711, QM F39712; Duckworth Creek (QM L215): QM F39713. Postglenoid processes as follows: The Crater (L78): QM F39707, QM F39709.

Occurrence. All three localities: QM L1111 (Tank Locality), QM L78 (The Crater), and QM L215 (Duckworth Creek), are from the Arcadia Valley, Queensland; Arcadia Formation, Rewan Group; Bowen Basin, Griesbachian.

Diagnosis. Distinguished from *Capulomala panchetensis* by the following: dorsoventral flattening in the PGAs (in section) is extreme so that the PGA is rectangular in section; PGA barely tapers posterior to the glenoid; a more rounded (in cross section) postglenoid process; a different arrangement of the sensory canals in that the oral sulcus originates from the mandibular sulcus immediately below the postglenoid process, rather than from a more ventral position on the surangular, and the mandibular sulcus extends anteriorly across the surangular and angular before curving backwards to leave the ornamented surface on the lingual margin.

Remarks. The species *Plagiobatrachus australis* Warren 1985*b* was erected for a series of vertebral centra, one of which (QM F12267) is the holotype. These centra still appear to be plagiosaurid in affinity, but the PGAs referred to *P. australis* clearly are not plagiosaurid. For this reason we here separate the PGAs from the centra, placing the PGAs in *C. arcadiaensis* but leaving the centra and a newly discovered neural arch (QM F39714) in *P. australis*.

Description

Postglenoid areas. In total eight posterior mandibles that can be assigned to *Capulomala arcadiaensis* have been recovered from the Arcadia Formation, from three localities. *C. arcadiaensis* PGAs are larger than those of *C. panchetensis* except for QM F39713 which is of a similar size. Of these the most significant is QM F39706 because it retains the ornamented base of the postglenoid process (Text-fig. 2). This ornamented area on the dorsal surface of the PGA was not preserved in the Arcadia Formation specimens described previously, being broken off in an almost horizontal plane immediately behind the glenoid fossa. Additionally, two isolated partial postglenoid processes were found among the scrap. Although neither of these can be fitted onto any of the eight PGAs, it is clear that they are similar processes to that described above from *C. panchetensis*. *Capulomala arcadiaensis* PGAs are distinctive in that they are even more dorsoventrally flattened than those of *C. panchetensis*. Perhaps as a result of this flattening the PGA is almost as wide as the glenoid fossa, only tapering slightly towards the posterior. While some of the external surface of the dermal bones is covered by tubercular ornament, usually arranged along ridges, the ornament of other fragments appears to be more reticular. Careful inspection of specimens with reticular ornamentation always reveals some tubercular areas and it is most probable that the tubercles have been abraded off the ridges. The sulci of the sensory canal system are deeply incised and narrow as in *C. panchetensis*. Perhaps because of the extreme flattening of these PGAs the arrangement of the canals differs from *C. panchetensis*. The oral sulcus lies immediately below the rim of the adductor fossa and joins the mandibular sulcus more dorsally. Above this junction, the dorsal part of the mandibular sulcus passes behind the postglenoid process onto the dorsal surface of the postglenoid process, as in *C. panchetensis*. Below the junction it passes anteromedially across the suture between the surangular and angular. Continuation of the mandibular sulcus on the angular can be seen in QM F12270 and QM F39708 (Text-fig. 3), where it continues anteriorly before turning posteriorly through the ornament of the angular and on to the unornamented lingual surface. Here a groove immediately below the ventral margin of the mandible may have been the continuation of the canal towards the posterior of the PGA. This posterior alignment of the mandibular sulcus and posterior continuation of the course of the sulcus on unornamented bone has not been described in other temnospondyls.

QM F39706 (Text-fig. 2), which is missing the angular and prearticular, is broken part way through the glenoid fossa and posterior to the base of the postglenoid process. In dorsal view a large ornamented area elevated above the unornamented part of the surangular is the base of the postglenoid process. Three foramina penetrate the surangular: an elongate foramen on the shelf-like margin of the adductor fossa, a smaller foramen at the extreme posterolabial edge of the glenoid, and another on the lingual-facing wall of the surangular beneath the broken margin of the surangular flange. The ventral ridge marking the anterior margin of the base of the postglenoid process has the characteristic smooth concave curve extending anteriorly beside the shelf of

surangular bordering the adductor fossa, as seen in *C. panchetensis*. The articular is present as a partially preserved glenoid fossa, a narrow lingual margin to the PGA, and a marked anterolingual flange on the floor of the adductor fossa. In all but QM F39706, where the postglenoid process broke off, the break in the remaining PGAs show a posterolabially directed sliver of articular that may have contributed to a weakness of the PGA at this point.

In ventral view, because the angular and prearticular are missing, the surface of a well-ossified articular is visible on the lingual side, passing above the surangular. A foramen penetrates the articular near the centre of the exposed surface.

In a labial view of QM F39706 only the surangular is visible. In lingual view the thick edge of the articular that formed the suture with the (missing) angular and prearticular is present below a narrow exposure of surangular, with the broken margin of the surangular flange visible dorsally.

Postglenoid processes. Both partial postglenoid processes of the surangular (Text-fig. 3) are from the proximal end, with each displaying part of the oral sulcus on their ventral surface. The more complete (QM F39707) is sub-rounded where it is broken distally but the processes are more oval in proximal sections. QM F39709 preserves a little more of the proximal end of the process. This shows a slight projection in the angle between the margin of the adductor fossa and the more distal part of the process, which results in a slight forward curvature. This angle marks the distal extent of the unornamented anterior face of the postglenoid process, suggesting that this surface is much less extended distally than it is in *C. panchetensis*. The slight forward curvature differs from QM F39707 and *C. panchetensis* which both curve slightly posteriorly.

Reconstruction

In the reconstruction (Text-fig. 4) a temnospondyl mandible from the Rhytidosteidae, that of *Arcadia myriadens*, was combined with the PGAs and isolated postglenoid

TEXT-FIG. 4. Composite reconstructions of left mandibles of *Capulomala*. A, *C. arcadiaensis*, based on two PGAs (QM F39706, QM F12270), and a postglenoid process (QM F39707). B, *C. panchetensis*, based on the complete postglenoid process (ISI A 179) and the PGA with an unbroken posterior end (ISI A 178). Dorsal view to the left, ventral to the right. Grey shading represents the approximate preserved area of the PGAs; dotted line represents the course of the lateral line sulci.

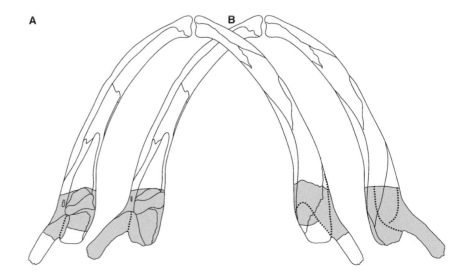

processes of *Capulomala arcadiaensis*. This reconstruction incorporates the unique arrangement of the sensory sulci and the extreme flattening of *C. arcadiaensis*. A reconstructed PGA from *C. panchetensis* used the complete postglenoid process and posteriorly complete specimens, and shows the differing arrangement of the sensory sulci in this less flattened species.

DISCUSSION

Systematic affinity of Capulomala

In the original description Warren (1985b) associated the *C. arcadiensis* mandibles with centra (the holotype of *Plagiobatrachus australis*) from the Arcadia Formation that are similar morphologically to those of members of the Plagiosauridae. The mandibles are dorsoventrally compressed and flattened ventrally; this is true also in some plagiosaurs, but in them the compression is less marked. Also similar is a thickened extension of the articular anteriorly on the floor of the adductor fossa, and the nature of the ornamentation, which is tubercular. The many PGAs from the Arcadia Formation show that the anterior extension of the articular on the floor of the adductor fossa is present in other PGAs, especially Type II (below); it may be more marked in *C. arcadiensis* PGAs because of their dorsoventral flattening.

In an overview of Triassic temnospondyl mandibles Jupp and Warren (1986) divided these into two types, based largely on PGA morphology. The primitive Type I PGA, present in, for example, mastodonsauroids and rhytidosteids, is characterised as follows: the prearticular does not extend onto the PGA posterior to the glenoid so that the lingual surface of the PGA is largely formed by an exposure of the articular, with the labial surface covered by surangular; the angular does not extend far onto the PGA; the posterior extremity of the PGA usually forms an arcadian process (Warren and Black 1985) separated from the retroarticular process (formed by articular only) by an arcadian groove; a chorda tympanic foramen is always present; and the most prominent sensory canal is the mandibular that often separates a postglenoid process from an arcadian process. The more derived Type II PGA, found in Brachyopoidea, Metoposauridae and Plagiosauridae is characterised as follows: the prearticular usually extends onto the PGA (leaving a tongue of articular exposed on the lingual side of the PGA); the angular participates in the PGA; the articular never forms a distinct process; the surangular always forms the posteriormost extremity of the PGA; an arcadian groove is never present; the chorda tympanic foramen is often absent; the oral sensory canal is dominant; and a postglenoid process is absent.

The *Capulomala* PGAs from the Panchet and Arcadia formations are Type I in that both oral and mandibular sensory sulci are present, with the mandibular sulcus passing between the postglenoid and retroarticular processes. An arcadian process is absent. Conversely, they exhibit some more derived features found in the Type II PGA in that the angular extends ventrally beneath the PGA posterior to the glenoid, the prearticular extends posteriorly to the glenoid, and a chorda tympanic foramen is absent. Also similar to the Type II PGA is the tongue of articular exposed on the lingual margin of the PGA so that the surangular forms most of the dorsal surface. The Indian specimen in which the posterior of the PGA is complete shows that the *Capulomala* PGAs are shorter than most stereospondyl PGAs and especially shorter than the typical Type II PGAs of the Brachyopidae, Chigutisauridae and Plagiosauridae.

Other stereospondyls with marked processes on the PGA include two members of the Australian Rhytidosteidae, *Arcadia myriadens* and *Rewana quadricuneata* (Warren and Black 1985). In these taxa all three processes of the PGA are present; in *R. quadricuneata* the processes are sub-equal in size, whereas in *A. myriadens* the arcadian process is longer than the retroarticular or postglenoid processes. The Lydekkerinidae (e.g. Jeannot *et al.* 2006; Hewison 2007) and Lapillopsidae (Yates 1999; Yates and Sengupta 2002) have a long posteriorly-directed process but this is the arcadian process rather than the postglenoid process, which is small in lapillopsids. The origin of the surangular processes (the postglenoid and arcadian processes) described above can be found in the less marked processes seen in *Benthosuchus sushkini* where the mandibular sulcus clearly separates a postglenoid process (= tuberculum postglenoidale of Bystrow and Efremov 1940) from a smaller arcadian process; both surangular processes are present in Type I PGAs, but absent in Type II PGAs, because in the Type II PGAs the mandibular sensory sulcus that defines the processes by passing between them, is missing.

The combination of characters described in the *Capulomala* PGAs is found in no other stereospondyl taxon. The PGAs of *Capulomala panchetensis* and *C. arcadiaensis* are unique in their combination of features from Type I and Type II PGAs, but most obviously in the extreme development of the postglenoid process. In no other stereospondyl does such a hypertrophied process exist. These mandibles are unique also in the presence of an ornamented area of the surangular on the dorsal surface of the PGA, in the morphology of the mandibular sensory sulcus that passes onto the dorsal surface of the postglenoid process, and more ventrally passes posteriorly across the ornamented parts of the surangular and angular, in the extremely lingual suture of the surangular with the articular and in the marked dorsoventral flattening of the PGA.

It seems unlikely that the scrap collection from both the Arcadia and Panchet formations should contain mandibles, now totalling eight (Arcadia) and seven (Panchet) for which no cranial material has been collected or identified. Yet all of the known taxa from both formations have readily identifiable mandibles, apart from the enigmatic *Indobrachyops panchetensis*, a single skull (without a mandible) from the Panchet Formation, which has variously been aligned with the Brachyopidae (Huene and Sahni 1958), the Indobrachyopidae (Cosgriff and Zawiskie 1979), and the Rhytidosteidae (Warren and Black 1985).

Tubercular (pustular) ornament of the dermal bones is a characteristic of most members of the Plagiosauridae, some members of the Rhytidosteidae and the Permian *Peltobatrachus* (Panchen 1959). It is logical to align the *Capulomala* PGAs with one of these taxa, but the *Capulomala* PGAs clearly differ from all of the known mandibles. Two South African taxa, *Pneumatostega potamia* (Cosgriff and Zawiskie 1979), a rhytidosteid and *Laidleria gracilis* (Kitching 1957; Warren 1998), a rhytidosteid or plagiosaurid relative, have mandibles that are a little dorsoventrally flattened and in which the PGAs are not preserved. We can not rule out the possibility that the *Capulomala* PGAs described here belong to one of these taxa or to *Indobrachyops*. The ornamentation of the *Indobrachyops* skull is similar to that of *Capulomala* and some rhytidosteids, and some Australian members of the Rhytidosteidae have enlarged processes on the PGAs, suggesting that *Capulomala* could be related to them. A skull with associated PGA is needed to confirm this.

Function of the hypertrophied postglenoid process

The hypertrophied postglenoid process of *Capulomala* is ornamented on both dorsal and ventral surfaces indicating that both surfaces were skin-covered and hence projected from the mandible. Two other instances of projections from the mandible are known: the zatrachydid temnospondyls *Acanthostomatops* (Witzmann and Schoch 2006) and *Zatrachys* (Langston 1953; Schoch 1997) have a fan of small ornamental spines projecting ventrally from the angular of the mandible, and some fossil suines such as *Entelodon* have one or more ventrolateral projections. On the other hand, projections from the skull are relatively common, especially in mammals. Among temnospondyls projections of dermal bone that are skin-covered on both sides also occur on the skull, on the quadratojugal of some zatrachydids (Schoch 1997) and some plagiosaurids (Hellrung 2003), and involving a number of posterolateral bones in nectridians (e.g. Beerbower 1963). The projections in *Capulomala* are unique in emanating

from the surangular, in their extreme length and in the posterolateral angle of projection. They are relatively fragile, so unlikely to have been used for defence and would better have served as hydroplanes were they longer anteroposteriorly. Possibly they had the effect of visually enlarging the head to discourage predators, although were this effective it could have been expected to be expressed more often among temnospondyls. The unornamented anterolateral face on the postglenoid process of the Panchet specimens may indicate the base of a skin flap that could be extended if required, but this was not present in the Arcadia specimens.

Without an associated skull it is difficult to speculate on the function of the postglenoid process. Nevertheless, the projections from the mandible found in *Capulomala* are of interest not only because they show that a temnospondyl genus was common to the Arcadia and Panchet formations, suggesting a close age relationship, but because their structure is unique among known temnospondyls and perhaps among vertebrates.

It is of interest that the taxa with the most pronounced processes on the PGA, namely the postglenoid process of *Capulomala* and the arcadian processes of the Lydekkerinidae and Lapillopsidae are found in both the Panchet and Arcadia formations, strata in which the temnospondyl taxa tend to be of small body size, compared with Permian and later Triassic faunas.

Systematic affinity of Plagiobatrachus australis

The gastrocentrous vertebral centra of *P. australis* originally associated with the mandibles can not be allocated to any described taxon from the Arcadia Formation. Such centra have not been found in the Panchet Formation. Ten additional centra have been collected and all appear to have articulated with intercentral neural arches and possibly intercentral ribs, as is typical of plagiosaurs, but also found in *Gobiops desertus*, a brachyopid from the Jurassic (Shishkin 1991) and some metoposaurs. Intercentral neural arches have been described in part of the vertebral column of the cynodont *Thrinaxodon* (Jenkins 1971), but although possible cynodont material has been found in the Arcadia formation (Thulborn 1990), it is unlikely that, if a cynodont was present, all the centra would have been associated with intercentral neural arches.

Among the scrap collection from the Arcadia Formation a single neural arch could be plagiosaurid (Text-fig. 5). This arch (QM F39714) is unlike any other from the Arcadia Formation. Although the neural spine is incomplete posteriorly and anteriorly in that the area that would have supported the zygapophyses is missing, it must have been anteroposteriorly short and squat. It is

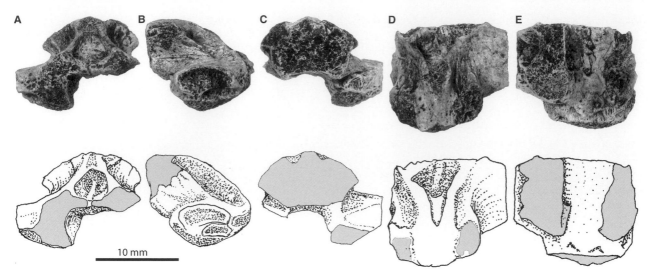

TEXT-FIG. 5. Photographs and drawings of a neural arch from the Arcadia Formation (QM F39714) referred to *Plagiobatrachus australis*. A, posterior. B, left lateral. C, anterior. D, dorsal. E, ventral. Note the short, laterally-directed transverse process, and expanded lateral buttresses on the neural spine, both characteristic of plagiosaur neural arches. Broken surfaces filled with grey.

well buttressed laterally, as are most plagiosaurid neural spines, for support of dermal scutes. The transverse process, preserved on the right, is exceptionally short and positioned above the paired facets on the ventral surface of the arch beside the neural canal. It is double recessed with a larger more dorsal recess and a smaller ventral recess. Such an arrangement could have accommodated a triple faceted rib head that articulated with the transverse process and two adjacent centra, as in, for example, some regions of the vertebral column of the plagiosaurid *Gerrothorax*. QM F39714 resembles neural arches of *Laidleria* (Warren 1998) in the short, buttressed neural spine, but this could be because *Laidleria* was heavily armoured. The neural arch bears some resemblance to those of rhytidosteids from the Arcadia Formation; however, unlike them, it is well ossified and shows no indication of having been ossified in right and left halves. On the anterior surface QM F39714 has a marked supraneural pit above the neural canal. A supraneural canal is present in a neural arch of *Gerrothorax* from the Middle Triassic of Germany (MCZ 9524), but is rare elsewhere among stereospondyls. Hence, it is still possible that a plagiosaur was present in the Arcadia Formation, although the Griesbachian is early for that taxon, the next known plagiosaur being *Melanopelta* from the Olenekian of Russia (Shishkin 1987).

Stratigraphic position of the Arcadia and Panchet formations

A typical *Lystrosaurus* Assemblage Zone fauna occurs in the Panchet Formation, with *Proterosuchus*, *Lystrosaurus* and *Thrinaxodon*, as well as a variety of stereospondyls

(Table 1); procolophonids have been reported (Lucas 1998; Yates and Sengupta 2002) but the material can not now be verified as procolophonid by the third author. Palynological studies indicate that the contact between the Panchet Formation and the underlying Raniganj Formation marks the Permo/Triassic boundary (Veevers and Tewari 1995; Ghosh *et al.* 1996), hence the Panchet Formation has been placed at the base of the Triassic (Bandyopadhyay *et al.* 2002).

A fauna similar to that of the Panchet Formation occurs in the Arcadia Formation of Queensland, Australia, although with impoverished *Lystrosaurus* and a more diverse stereospondyl fauna. Correlation of the Arcadia Formation, and part of the Rewan Formation of the adjacent Galilee Basin, with the early part of the *Lystrosaurus* Assemblage Zone of the Karoo Basin was made by Warren *et al.* (2006) on the basis of their vertebrate assemblages. Palynological evidence from the Galilee Basin suggested a Griesbachian age (APT1) for the Rewan Formation fauna and (by faunal correlation) for the Arcadia Formation. This stratigraphic placement of the Arcadia Formation is supported by the presence of an undescribed basal procolophonid, close to *Sauropareion anoplus* from the lowermost Triassic of South Africa (Modesto and Damiani 2007). The Arcadia Formation has been placed in a later position in the Early Triassic by some authors (e.g. Anderson and Cruickshank 1978; Ochev and Shishkin 1989; Lucas 1998), but there is little justification for this given the plesiomorphic nature of the brachyopid *Xenobrachyops*, the stem stereospondyl *Lapillopsis*, the three species of the mastodonsaurid *Watsonisuchus*, the presence of *Lydekkerina huxleyi*, and the basal procolophonids.

Relationship between the Panchet and Arcadia formations

A comparison of components of the stereospondyl fauna of the Panchet and Arcadia formations (Table 1) shows two faunas that are similar, but different in the apparent absence in the Panchet Formation of the commonest stereospondyls found in the Arcadia Formation, early members of the Mastodonsauridae such as *Watsonisuchus*. Also notable for their absence are the rarer Chigutisauridae and Brachyopidae. This disparity may be because there has not been an intensive study based on the identification of scrap from the Panchet Formation comparable with the study by Northwood (1997) of Arcadia Formation scrap. Scrap bone from the Panchet Formation is rarer and has been documented but not identified (Patra 2006). Hence, identification of components of the Panchet scrap should add further evidence for the close affinity of the Panchet and Arcadia formations in the earliest Triassic. For instance, contemporary knowledge of the morphology of PGAs, (e.g. Warren and Black 1985; Damiani 1999; Novikov and Shishkin 1992) enabled AW and DPS to identify two mastodonsaurid PGAs and a plesiomorphic brachyopoid PGA among the Panchet scrap, thus filling two obvious faunal gaps in that record.

The faunal components common to the Panchet and Arcadia formations that stand out are those that are extremely rare elsewhere (Table 1): the Tupilakosauridae are also known from isolated occurrences in the late Permian of the USA (Milner and Sequeira 2004) and France (Werneburg *et al.* 2007), and the earliest Triassic of Russia (e.g. Shishkin 1973), Greenland (e.g. Nielsen 1954) and South Africa (Warren 1999); while the Lapillopsidae are found only in the Arcadia and Panchet formations, as is *Capulomala*. The common presence of these rare stereospondyls clearly suggests not only a close biogeographic but also a close biostratigraphic relationship between the Arcadia and Panchet formations. Additionally, it emphasises the importance of scrap collections for increasing our knowledge of diversity within faunas, and for biostratigraphic correlation.

Acknowledgements. All three authors would like to acknowledge the debt that we owe to Andrew Milner for his constant advice, especially on temnospondyl taxonomy and the intricacies of systematic zoology. Among his many diverse publications Andrew was the first to attempt a comprehensive cladogram of temnospondyls (Milner 1990), a study that is still cited and which largely has been corroborated by more recent publications. We hope that he may be intrigued by *Capulomala* and await his opinion as to its identity.

We thank the Geological Survey of India for allowing DPS access to the holotype of *Capulomala panchetensis* and other material, the Indian Statistical Institute for hospitality to AW and for technical and financial support to DPS, Thomas Sulej (Institute of Paleobiology, Polish Academy of Science) for initial photography, Farish Jenkins (Museum of Comparative Zoology, Harvard University) for comparative material of *Gerrothorax* neural arches, and Caroline Northwood (La Trobe University) for discussions. Field work in the Arcadia Formation was supported by an Australian Research Council Large Grant to AW.

REFERENCES

ANDERSON, J. M. and CRUICKSHANK, A. R. I. 1978. The biostratigraphy of the Permian and the Triassic. Part 5. A review of the classification and distribution of Permo-Triassic tetrapods. *Palaeontologia Africana*, **12**, 15–44.

BANDYOPADHYAY, S., ROY-CHOWDHURY, T. K. and SENGUPTA, D. P. 2002. Taphonomy of some Gondwana vertebrate assemblages of India. *Sedimentary Geology*, **147**, 219–245.

BEERBOWER, J. R. 1963. Morphology, paleoecology, and phylogeny of the Permo-Pennsylvanian amphibian *Diploceraspis*. *Bulletin of the Museum of Comparative Zoology, Harvard*, **130**, 33–108.

BYSTROW, A. P. and EFREMOV, J. A. 1940. *Benthosuchus sushkini* Efr. A labyrinthodont from the Eotriassic of Sharjenga River. *Trudy Paleozoologicheskogo Instituta*, **10**, 1–152.

COSGRIFF, J. W. and ZAWISKIE, J. M. 1979. A new species of the Rhytidosteidae from the *Lystrosaurus* Zone and a review of the Rhytidosteoidea. *Palaeontologia Africana*, **22**, 1–27.

DAMIANI, R. J. 1999. Giant temnospondyl amphibians from the Early to Middle Triassic Narrabeen Group of the Sydney Basin, New South Wales, Australia. *Alcheringa*, **23**, 87–109.

—— 2001. A systematic revision and phylogenetic analysis of Triassic mastodonsauroids (Temnospondyli: Stereospondyli). *Zoological Journal of the Linnean Society*, **133**, 379–482.

—— and WARREN, A. A. 1996. A new look at members of the Superfamily Brachyopoidea (Amphibia, Temnospondyli) from the Early Triassic of Queensland and a preliminary analysis of brachyopoid relationships. *Alcheringa*, **20**, 277–300.

GHOSH, S. C., NANDI, A., AHMED, G. and ROY, D. K. 1996. Study of Permo-Triassic Boundary in Gondwana Sequence of Raniganj Basin, India. 179–193. *In* MITRA, N. D., ACHARYA, S. K., CHANDRA, P. R., GOSH, A., GOSH, S. and GUHA, P. K. S. (eds). *Gondwana Nine, Ninth International Gondwana Symposium*. Volume 1. A. A. Balkema, Rotterdam/Brookfield, xxi + 585 pp.

HELLRUNG, H. 2003. *Gerrothorax pustuloglomeratus*, ein Temnospondyle (Amphibia) mit knöcherner Branchialkammer aus dem Unteren Keuper von Kupferzell (Süddeutschland). *Stuttgarter Beiträge zur Naturkunde Serie B (Geologie und Paläontologie)*, **39**, 1–130.

HEWISON, R. H. 2007. The skull and mandible of the stereospondyl *Lydekkerina huxleyi*, (Tetrapoda: Temnospondyli) from the Lower Triassic of South Africa, and a reappraisal of the family Lydekkerinidae, its origin, taxonomic relationships and phylogenetic importance. *Journal of Temnospondyl Palaeontology*, **1**, 1–80.

HOWIE, A. A. 1972a. On a Queensland labyrinthodont. 50–64. In JOYSEY, K. A. and KEMP, T. S. (eds). *Studies in vertebrate evolution*. Oliver Boyd, Edinburgh, 284 pp.

—— 1972b. A brachyopid labyrinthodont from the Lower Trias of Queensland. *Proceedings of the Linnean Society of New South Wales*, **96**, 268–277.

HUENE, F. von and SAHNI, M. R. 1958. On *Indobrachyops panchetensis* gen. et sp. nov. from the Upper Panchets (Lower Trias) of the Raniganj coalfield. *Monograph of the Palaeontological Society of India*, **2**, 1–17.

HUXLEY, T. H. 1865. On vertebrate fossils from the Panchet rocks, near Ranigunj, Bengal. *Palaeontologia Indica (4)*, **1**, 3–24.

JEANNOT, A. M., DAMIANI, R. and RUBIDGE, B. S. 2006. Cranial anatomy of the Early Triassic stereospondyl *Lydekkerina huxleyi* (Tetrapoda: Temnospondyli) and the taxonomy of South African lydekkerinids. *Journal of Vertebrate Paleontology*, **26**, 822–838.

JENKINS, F. A. Jr 1971. The postcranial skeleton of African cynodonts. *Bulletin of the Peabody Museum of Natural History, Yale University*, **36**, x + 216.

JUPP, R. and WARREN, A. A. 1986. The mandibles of the Triassic temnospondyl amphibians. *Alcheringa*, **10**, 99–124.

KITCHING, J. W. 1957. A new small stereospondylous labyrinthodont from the Triassic beds of South Africa. *Palaeontologia Africana*, **5**, 67–82.

LANGSTON, W. Jr 1953. Permian amphibians from New Mexico. *University of California Publications in Geological Science*, **29**, 349–416.

LUCAS, S. G. 1998. Global Triassic tetrapod biostratigraphy and biochronology. *Palaeogeography, Palaeoclimatology, Palaeoecology*, **143**, 347–384.

LYDEKKER, R. 1882. On some Gondwana labyrinthodonts. *Records of the Geological Society of India*, **15**, 24–28.

MILNER, A. R. 1990. The radiations of temnospondyl amphibians. 321–349. *In* TAYLOR, P. D. and LARWOOD, G. P. (eds). *Major evolutionary radiations*. Systematics Association Special Volume 42. Clarendon Press, Oxford, xi + 437 pp.

—— and SEQUEIRA, S. E. K. 2004. *Slaugenhopia texensis* (Amphibia: Temnospondyli) from the Permian of Texas is a primitive tupilakosaurid. *Journal of Vertebrate Paleontology*, **24**, 320–325.

MODESTO, S. and DAMIANI, R. 2007. The procolophonid reptile *Sauropareion anoplus* from the lowermost Triassic of South Africa. *Journal of Vertebrate Paleontology*, **27**, 337–349.

MOSER, M. and SCHOCH, R. R. 2007. Revision of the type material and nomenclature of *Mastodonsaurus giganteus* (Jaeger) (Temnospondyli) from the Middle Triassic of Germany. *Palaeontology*, **50**, 1245–1266.

NIELSEN, E. 1954. *Tupilakosaurus heilmani* n. g. et n. sp. an interesting batrachomorph from the Triassic of East Greenland. *Meddelelser om Grønland*, **72**, 4–33.

NORTHWOOD, C. 1997. *Palaeontological interpretations of the early Triassic Arcadia Formation, Queensland*. Unpublished PhD thesis, La Trobe University, Melbourne, xxii + 479 pp.

—— 2005. Early Triassic coprolites from Australia and their palaeobiological significance. *Palaeontology*, **48**, 49–68.

NOVIKOV, I. V. and SHISHKIN, M. A. 1992. New middle Triassic labyrinthodonts from Pechorian Cisuralia. *Paleontological Journal*, **26**, 92–102.

OCHEV, V. G. and SHISHKIN, M. A. 1989. On the principles of global correlation of the continental Triassic on the tetrapods. *Acta Palaeontologica Polonica*, **34**, 149–173.

OWEN, R. 1841. On the teeth of species of the genus *Labyrinthodon* (*Mastodonsaurus* of Jaeger), common to the German Keuper formation and the Lower Sandstone of Warwickshire and Leamington. *Transactions of the Geological Society of London, Series 2*, **6**, 503–514.

PATRA, A. R. 2006. *Significance of the bone fragments collected from the Early Triassic beds of Panchet Formation, Damodar basin, south of Asansol*. Unpublished M.Sc. thesis. University of Calcutta, Kolkata, India.

PANCHEN, A. L. 1959. A new armoured amphibian from the Upper Permian of East Africa. *Philosophical Transactions of the Royal Society of London, Series B*, **691**, 207–281.

SCHOCH, R. R. 1997. Cranial anatomy of the Permian temnospondyl amphibian *Zatrachys serratus* Cope 1878, and the phylogenetic position of the Zatrachydidae. *Neues Jahrbuch für Geologie und Paläontologie, Abhandlungen*, **206**, 223–248.

—— and MILNER, A. R. 2000. *Stereospondyli*. Handbuch der Paläoherpetologie, Teil 3B. Verlag Dr Friedrich Pfeil, Munich, xii + 203 pp.

SHISHKIN, M. A. 1973. The morphology of the early Amphibia and some problems of the lower tetrapod evolution. *Trudy Paleontologicheskogo Instituta, Doklady Akademii Nauk SSSR*, **137**, 1–257. [In Russian].

—— 1987. Evolution of ancient amphibians (Plagiosauroidea). Transactions of the Paleontological Institute, Academy of Sciences of the Union of Soviet Socialist Republics, **225**, 1–144. [In Russian].

—— 1991. A Late Jurassic labyrinthodont from Mongolia. *Paleontological Journal*, **1991**, 81–95.

THULBORN, R. A. 1990. Mammal-like reptiles of Australia. *Memoirs of the Queensland Museum*, **28**, 169.

TRIPATHI, C. 1969. Fossil labyrinthodonts from the Panchet series of the Indian Gondwanas. *Memoirs of the Geological Survey of India*, **38**, 1–45.

VEEVERS, J. J. and TEWARI, R. C. 1995. Gondwana master basin of peninsular India, between Tethys and the interior of the Gondwanaland Province of Pangea. *Geological Society of America Memoir*, **187**, v + 72.

WARREN, A. A. 1980. *Parotosuchus* from the Early Triassic of Queensland and Western Australia. *Alcheringa*, **4**, 25–36.

—— 1981. A horned member of the labyrinthodont superfamily Brachyopoidea from the Early Triassic of Queensland. *Alcheringa*, **5**, 273–288.

—— 1985a. Two long-snouted temnospondyls (Amphibia, Labyrinthodontia) from the Triassic of Queensland. *Alcheringa*, **9**, 293–295.

—— 1985b. Triassic Australian plagiosauroid. *Journal of Paleontology*, **59**, 236–241.

—— 1998. *Laidleria* uncovered: a redescription of *Laidleria gracilis* Kitching (1957), a temnospondyl from the *Cynognathus* Zone of South Africa. *Zoological Journal of the Linnean Society*, **122**, 167–185.

—— 1999. Karroo tupilakosaurid: a relict from Gondwana. *Transactions of the Royal Society of Edinburgh, Earth Sciences*, **89**, 145–160.

—— and BLACK, T. 1985. A new rhytidosteid (Amphibia, Labyrinthodontia) from the Early Triassic Arcadia Formation of Queensland, Australia, and a consideration of the relationships of Triassic temnospondyls. *Journal of Vertebrate Paleontology*, **5**, 303–327.

—— and HUTCHINSON, M. N. 1983. The last labyrinthodont? A new brachyopoid (Amphibia, Temnospondyli) from the Early Jurassic Evergreen Formation of Queensland, Australia. *Philosophical Transactions of the Royal Society of London, Series B*, **303**, 1–62.

—— —— 1987. The skeleton of a new hornless rhytidosteid (Amphibia, Temnospondyli). *Alcheringa*, **11**, 291–302.

—— —— 1988. A new capitosaurid amphibian from the Early Triassic of Queensland, and the ontogeny of the capitosaur skull. *Palaeontology*, **31**, 857–876.

—— and SCHROEDER, N. 1995. Changes in the capitosaur skull with growth: an extension of the growth series of *Parotosuchus aliciae* (Amphibia, Temnospondyli) with comments on the otic area of capitosaurs. *Alcheringa*, **19**, 41–46.

—— DAMIANI, R. and YATES, A. M. 2006. The South African stereospondyl *Lydekkerina huxleyi* (Tetrapoda, Temnospondyli) from the Lower Triassic of Australia. *Geological Magazine*, **143**, 877–886.

WERNEBURG, R., STEYER, J. S., SOMMER, G., GAND, G., SCHNEIDER, J. W. and VIANEY-LIAUD, M. 2007. The earliest tupilakosaurid amphibian with diplospondylous vertebrae from the Late Permian of southern France. *Journal of Vertebrate Paleontology*, **27**, 26–30.

WITZMANN, F. and SCHOCH, R. R. 2006. Skeletal development of the temnospondyl *Acanthostomatops vorax* from the Lower Permian Döhlen Basin of Saxony. *Transactions of the Royal Society of Edinburgh, Earth Sciences*, **96**, 365–385.

YATES, A. M. 1999. The Lapillopsidae: a new family of small temnospondyls from the Early Triassic of Australia. *Journal of Vertebrate Paleontology*, **19**, 302–320.

—— 2000. A new tiny rhytidosteid (Temnospondyli: Stereospondyli) from the Early Triassic of Australia and the possibility of hidden temnospondyl diversity. *Journal of Vertebrate Paleontology*, **20**, 484–489.

—— and SENGUPTA, D. P. 2002. A lapillopsid temnospondyl from the Early Triassic of India. *Alcheringa*, **26**, 201–208.

ZITTEL, K. A. von 1887–1890. *Handbuch der Paläontologie. Abteilung 1. Paläozoologie Band 3. Vertebrata (Pisces, Amphibia, Reptilia, Aves)*. Oldenbourg, Munich and Leipzig, xii + 900 pp.